Manual Materials Handling

by

M.M. Ayoub

Texas Tech University, USA

and

Anil Mital

University of Cincinnati, USA

Taylor & Francis
London · New York · Philadelphia
1989

UK	Taylor & Francis Ltd, 4 John St., London WC1N 2ET
USA	Taylor and Francis Inc., 1900 Frost Road, Suite 101, Bristol, PA 19007

Copyright © M. M. Ayoub and A. Mital 1989

All rights reserved. No part of this publication may be reproduced, stored in a retrieval system, or transmitted, in any form or by any means, electronic, electrostatic, magnetic tape, mechanical, photocopying, recording or otherwise, without the prior permission of the copyright owner.

The authors and publishers have made all reasonable efforts to secure permission for the reproduction of figures and tables within this work.

British Library Cataloguing in Publication Data

Ayoub, M.M.
   Manual materials handling.
   1. Materials handling. Manual lifting. Safety measures
   I. Title   II. Mital Anil
   612'.76

   ISBN 0-85066-383-0

Library of Congress Cataloging in Publication Data

is available

Typeset in 11/12pt Bembo by Chapterhouse, The Cloisters, Formby
Printed in Great Britain by Taylor & Francis (Printers) Ltd.

Contents

Preface	xi
Chapter 1. The Materials Handling Problem	1
1.1 Introduction	1
1.2 Manual Materials Handling Injury Frenquency and Cost	1
1.3 Affected Industrial Populations	4
1.4 Scope of the Book	7
Chapter 2. Variables in Manual Materials Handling	9
2.1 Worker–Task–Environment 'System' Concept	9
2.2 The 'Worker' Component	12
2.2.1 Age	13
2.2.2 Sex	15
2.2.3 Body-Weight	16
2.2.4 Anthropometry or Body Build, and Body Composition	17
2.2.5 Static and Dynamic Endurance	18
2.2.6 Strength	19
2.2.7 Training/Experience	21
2.2.8 Physical Fitness	23
2.2.9 Psychological Factors	23
2.3 The 'Task' Component	26
2.3.1 Frequency	31
2.3.2 Task Duration	35
2.3.3 Object Size	37
2.3.4 Object Shape	40
2.3.5 Couplings	41
2.3.6 Object Weight/Force Application (Exertion)	46
2.3.7 Load Distribution and Stability	46
2.3.8 Vertical Lift Height	48
2.3.9 Workplace Geometry	49
2.3.10 Distance Travelled/Slope and Traction	50
2.3.11 Posture/Technique	50
2.3.12 Asymmetrical Lifting/Carrying	54
2.4 The 'Environment' Component	54
2.4.1 Temperature and Humidity	54
2.4.2 Noise, Illumination and Vibration	56
2.4.3 Altitude	56
2.5 Interactive Effects of System Components	57
2.6 System Response Measures	59

Chapter 3. Design Criteria		60
3.1 Introduction		60
3.1.1 Stress/Strain Concept		60
3.1.2 MMH Stresses		60
3.2 The Biomechanical Approach		60
3.2.1 Biomechanical Analysis for MMH		60
3.2.1.1 Definitions and Applications of Biomechanics		60
3.2.1.2 The Body as a System of Levers		62
3.2.1.3 Stress on the Musculoskeletal System		63
3.2.1.3.1 Static Analysis		64
3.2.1.3.2 Analysis of One-Segment Link		64
3.2.1.3.3 Analysis of Two Links		66
3.2.1.3.4 Analysis of Multiple Links		68
3.2.1.3.5 Analysis of Internal Forces		69
3.2.1.4 Dynamic Analysis		72
3.2.1.5 Three-Dimensional Modelling		78
3.2.1.6 Stress on the Lumbosacral Spine		80
3.2.1.6.1 Symmetric Lumbar Loading		81
3.2.1.6.2 Asymmetric Lumbar Loading		84
3.2.1.6.3 The Role of Intra-Abdominal Pressure in Spinal Loading		85
3.2.1.6.4 Use of EMG in Spinal Loading		88
3.2.1.7 Factors Affecting Biomechanical Stress on MMH		88
3.2.1.7.1 Tissue Characteristics		88
3.2.1.8 Task Characteristics		90
3.2.1.8.1 Load Size and Shape		90
3.2.1.8.2 Weight of Load		92
3.2.1.8.3 Position of Load Relative to the Spine		92
3.2.1.9 Biomechanical Design Criterion		92
3.2.1.9.1 Design Criterion		92
3.2.1.9.2 Models Using the Biomechanical Criteria		95
3.2.1.9.3 Lifting Capacity Determinations		95
3.3 The Physiological Approach		97
3.3.1 Classification of Activity		97
3.3.1.1 Dynamic Activity		97
3.3.1.2 Isometric Activity		99
3.3.1.3 Semi-dynamic Activity		99
3.3.2 Physiological Response to Lifting		99
3.3.2.1 Energy Cost		100
3.3.2.2 Heart Rate		102
3.3.2.3 Blood Pressure		102
3.3.2.4 Blood Lactate Level		102

	3.3.3	Physiological Fatigue		103
		3.3.3.1 Endurance Time		104
			3.3.3.1.1 Bink's Logarithmic Formula	104
			3.3.3.1.2 Graphical Interpretation, With Example	104
	3.3.4	Physical Fitness for Lifting		107
		3.3.4.1 Aerobic Capacity		107
			3.3.4.1.1 Aerobic Lifting Capacity	108
		3.3.4.2 Anaerobic Threshold		109
	3.3.5	Design Criterion		111
		3.3.5.1 Criterion Limit		111
		3.3.5.2 Factors Affecting the Design Criterion		111
		3.3.5.3 Models Using the Physiologic Criterion		111
			3.3.5.3.1 Frederick's Model	111
			3.3.5.3.2 Garg's Model	112
			3.3.5.3.3 Asfour's Model	113
			3.3.5.3.4 Intaranont's Model	113
	3.3.6	Comparison of Models with Limitations		114
3.4	The Psychophysical Approach			116
	3.4.1	Rationale for Use of the Psychophysical Design Criterion		116
	3.4.2	Principles of the Psychophysical Design Criterion		118
	3.4.3	Factors Affecting the Psychophysical Design Criteria		119

Chapter 4. Lifting and Lowering Activity Data Bases — 121
4.1 Strength versus Capacity — 121
4.2 Strength Data for Two-Handed Lifting and Lowering Activities — 122
4.3 Capacity Data for Two-Handed Lifting and Lowering Activities — 130
4.4 Strength Data for One-Handed Lifting and Lowering Activities — 148
4.5 Capacity Data for One-Handed Lifting and Lowering Activities — 152
4.6 Work-Rate Recommendations for Two-Handed Lifting Tasks — 159
    4.6.1 Box Size Effects — 159
    4.6.2 Frequency Effects — 163
    4.6.3 Effects of Height Level — 163
    4.6.4 Comprehensive Maximum Acceptable Weight of Lift Data Base — 163
4.7 Performance Ceilings for One- and Two-Handed Lifting Tasks — 163
    4.7.1 One-Handed Lifting Tasks — 166
    4.7.2 Two-Handed Lifting Tasks — 166

Chapter 5. Pushing, Pulling, Carrying and Holding Data Bases — 167
5.1 Strength Data for Two-Handed Pushing and Pulling Activities — 167
5.2 Capacity Data for Two-Handed Pushing and Pulling Activities — 174
5.3 Strength Data for One-Handed Pushing and Pulling Activities — 179
5.4 Capacity Data for One-Handed Pushing and Pulling Activities — 180
5.5 Strength Data for Two-Handed Carrying Activities — 180

5.6	Capacity Data for Two-Handed Carrying Activities	183
5.7	Strength Data for One-Handed Carrying Activities	184
5.8	Capacity Data for One-Handed Carrying Activities	185
5.9	Holding Data Base	185

Chapter 6. Job Design/Redesign and Screening Procedure — 189

- 6.1 Principles of Job Design/Redesign — 189
  - 6.1.1 Eliminating MMH — 190
  - 6.1.2 Decreasing Job Demands — 190
  - 6.1.3 Minimizing Body Movements — 191
- 6.2 Principles of Screening Procedures — 193
  - 6.2.1 Back X-Ray Films — 193
  - 6.2.2 Strength Testing — 194
  - 6.2.3 Medical Examinations — 194
  - 6.2.4 Psychological Tests — 196
  - 6.2.5 Job Simulators — 196
  - 6.2.6 Rating Methods — 196
- 6.3 Criteria for Screening of Personnel — 197
- 6.4 Tools for Job Design/Redesign and Screening with Examples — 198
  - 6.4.1 Job Severity Index — 198
    - 6.4.1.1 Using JSI as a Job Design/Redesign Device — 198
    - 6.4.1.2 Using JSI as a Job Screening/Placement Device — 206
    - 6.4.1.3 JSI Validation — 208
  - 6.4.2 Job Design/Redesign in Conjunction with JSI — 208
  - 6.4.3 NIOSH Guidelines for Job Design/Redesign — 209
  - 6.4.4 Comparing JSI Method and NIOSH Guidelines for Job Design/Redesign — 210
  - 6.4.5 Lift Strength Ratio — 211
    - 6.4.5.1 Using the Lift Strength Ratio as a Job Design/Redesign Tool — 212
  - 6.4.6 Mital's Model for Evaluating Manual Handling Jobs — 214
    - 6.4.6.1 A Numeric Example — 216
  - 6.4.7 Lifting Optimization Model for Job Design/Redesign — 224

Chapter 7. Pre-employment Strength Testing — 226

- 7.1 Why Measure Human Strengths? — 226
- 7.2 Classification and Definition of Strengths — 226
- 7.3 Measurement of Strengths — 228
  - 7.3.1 Assessment of Isometric (Static) Strengths — 229
  - 7.3.2 Assessment of Isotonic Strengths — 233
  - 7.3.3 Assessment of Isokinetic Strengths — 234
  - 7.3.4 Assessment of Isoinertial Strengths — 241
- 7.4 Prediction of Strengths — 242
  - 7.4.1 Prediction of Isometric Strengths — 243
  - 7.4.2 Prediction of Isokinetic Strengths — 251
  - 7.4.3 Prediction of Isoinertial Strengths — 255

7.5	Additivity of Strengths	256
7.6	Relationship Between Strengths and Acceptable Weights of Lift	260
7.7	Repetitive Dynamic Strengths and Manual Lifting Capabilities	262

Chapter 8. Training and Manual Handling — 265

- 8.1 Introduction — 265
- 8.2 The Concept of Safe Lifting — 265
- 8.3 Training Programmes: Methods and Approaches — 268
  - 8.3.1 Training Programmes Within Industrial Organizations — 268
  - 8.3.2 Training/Rehabilitation Programmes Outside the Industrial Organization — 270
- 8.4 Effectiveness of Training Programmes — 272
- 8.5 Summary — 273

Chapter 9. Determination of Rest Allowances — 274

- 9.1 The Need for Rest Allowances — 274
- 9.2 Methods of Determining Rest Allowances — 274
  - 9.2.1 Metabolic Energy Expenditure Rate Models for Determining Rest Allowances — 276
- 9.3 Limitations of the Metabolic Energy Expenditure Rate Method — 277
- 9.4 A Comprehensive Metabolic Energy Model for Determining Rest Allowances — 278
  - 9.4.1 Data Collection and Results — 278
- 9.5 Model Development — 281
  - 9.5.1 Concept — 281
  - 9.5.2 Assumption — 281
  - 9.5.3 Model Structure — 283
  - 9.5.4 Model Validation — 284
  - 9.5.5 Example — 285
- 9.6 Computer Program — 286

Appendix — 290
Software Program: Calculating Rest Period Duration — 290

References — 297
Index — 321

Preface

For nearly four decades, manual handling of materials has been a major topic of interest to professionals from a number of disciplines, including engineering, ergonomics, physical therapy and rehabilitation, orthopaedic surgery, work physiology and biomechanics. The primary reason for this is the devastating cost and human suffering caused by the severity of material handling related injuries. Prevention and control of these inuries is a global concern, shared by many researchers and organizations.

Over the years, the application of ergonomics and ergonomic principles has resulted in a vast scientific literature dealing with various aspects of manual materials handling problems, published in numerous scientific journals and conference proceedings worldwide. With the added contribution of much data from the authors' published and unpublished work, this book is the first significant effort to review an extensive amount of published literature and condense relevant findings for use by researchers and practitioners. The book, as a whole, highlights the problems and hazards of manual materials handling and provides ergonomic and engineering solutions for alleviating them.

The nine chapters of the book provide a comprehensive overview of manual materials handling. The first chapter deals with the problem and its importance. Research literature dealing with the vast number of variables that influence materials handling capacities of individual workers are reviewed in the second chapter. The third, and the primary focus of the book, discusses design criteria based upon biomechanics, physiology and psychophysics. The next two chapters are devoted to design data for activities such as lifting, pushing and carrying. In the sixth chapter the focus is on job design, redesign and screening procedures that are presently available. The relevance of pre-employment strength testing and training are dealt with in the next two chapters, and in the last chapter the need for providing rest allowances is discussed along with a method for determining them.

We hope that this book fills a critical void and proves useful to both researchers and practitioners who are committed to solving the multifaceted manual materials handling problem. The book should also be useful to individuals who need to have background information in this area.

Finally, we realize that even though, over the two-year writing period, every effort was made to include all available information, it is possible that some relevant literature was inadvertently overlooked. We hope that such omissions, if any, are not serious and in no way undermine what we believe to be the importance of this book.

Acknowledgements

A number of individuals provided invaluable assistance during the writing of this book. In particular, the authors wish to thank Dr Joseph Selan, who lent his support by assembling a considerable amount of material for the text, Dr H. C. Chen for his considerable efforts in preparing the artwork and collecting material, and Mr Hong-Ki Kim and Ms Sarah Farnsworth who worked diligently to collect and assemble support materials for certain chapters.

The authors are also indebted to all their graduate students, especially Ms Marta Miller who patiently traced and checked all the references for their correctness for publication, and last, but not least, to Ms Beverly Strickland for typing the manuscript and making numerous changes.

M. M. Ayoub and A. Mital
Lubbock and Cincinnati
July 1989

Chapter 1
The Materials Handling Problem

1.1 Introduction

Research in the area of manual materials handling (MMH) has been conducted for over a quarter of a century and continues to be performed today. The research, basically, has entailed the establishment of acceptable handling limits using several different approaches, and the application of ergonomic principles to job design, employee placement and employee training. The reason for this immense quantity of research is that MMH, in particular manual lifting, represents a major cause of injury to industrial workers and cost to industry. In this chapter, the scope of the MMH problem confronting industry is discussed.

The question of sex and age differences has been addressed with regard to back injury statistics. According to Laughery and Schmidt (1984), age and sex variables do not affect back injury rates significantly. Possible explanations given by the authors include: (1) job assignments may be based on age and/or sex, and (2) older workers are generally more experienced at their job. Figure 1.1 presents age and sex-specific ratios of compensation claims according to the Bureau of Labor Statistics' Supplementary Data System (SDS). Contradicting the Laughery and Schmidt (1984) conclusions, Figure 1.1 indicates a significant decrease in injury claims as worker age increases. Less work experience associated with younger workers may be a partial explanation. Also, a process of self-selection may occur on many physically demanding jobs, with the individual lacking the necessary capacities to perform the job thus dropping out (either by choice or because of injury), which leaves the job for workers with capacities better suited to the job demands.

1.2 Manual Materials Handling Injury Frequency and Cost

Back injuries, particularly injuries to the lower back, occur with alarming frequency. Caillet (1981) estimates that 70 million Americans have suffered back injuries, and that this number will increase by 7 million people annually. Approximately 5 million people are partly disabled because of back injuries, and 2 million people are not able to work at all because of back injuries. Other estimates indicate that 6·5 million Americans lie in bed on any given day because of back pain. This estimate increases by 1·5 million monthly (Keim, 1981). The cost of such a prevalent pain problem is exorbitant for industries. Klein *et al.* (1984) estimate that 19·0–25·5% of all workers' compensation claims are due to back pain. Table 1.1 details types of events or exposures

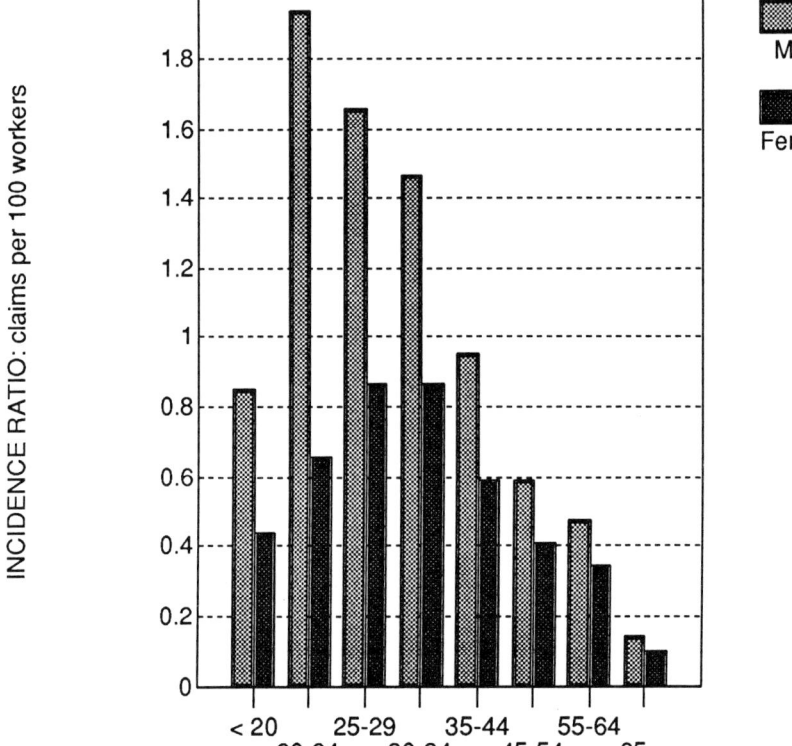

Figure 1.1. Age- and sex-specific ratios of compensation claims in 26 SDS States due to strains/sprains of the back (Klein et al., 1984) with permission of American Society of Occupational Medicine

Table 1.1. 1979 workers' compensation claims in the 26 SDS States for back strains/sprains according to type of event or exposure (after Klein et al., 1984)

Type	No. of claims	% of all back strains/sprains
Lifting objects	137 327	48·1
Over-exertion, not elsewhere classified	25 823	9·0
Pulling or pushing objects	25 557	9·0
Voluntary bodily motions	18 909	6·6
Holding, wielding, throwing or carrying objects	16 181	5·7
Involuntary bodily motions	13 353	4·7
Falls on working surfaces	12 150	4·3
Over-exertions, not elsewhere classified	6 433	2·3
Others, individually less than 2%	29 735	10·4
Total	285 468	100·1

leading to back strains/sprains in addition to the number of claims and percentages of all back sprain/strains according to SDS. As Table 1.1 indicates, nearly 50% of all back strains/sprains were precipitated by the manual lifting of objects.

According to Khalil et al. (1984) low-back pain is the second largest pain problem, headaches being the first. White (1983) reported that more than 70 million people, in the USA alone, see a physician annually complaining of low-back pain, while 17 million return to their physicians due to chronic low-back pain. Loesser (1979) estimated that, because of low-back pain, approximately 170 million working days are lost each year in the USA. This is indeed an expensive industrial problem considering that not only do industries pay workmen's compensation, but also spend billions of dollars on tests, treatments, claims, lawsuit awards, settlement and surgeries. Table 1.2 includes the type of injury, the number of cases and the mean cost per case according to SDS. Additionally, Figure 1.2 summarizes the National Safety Council's work injury and cost statistics from 1972–84 (National Safety Council, 1972–84).

Figure 1.2 reveals two intriguing facts regarding MMH-related injuries. First, despite improved medical care, increased automation in industry, and more extensive use of pre-employment examinations, only a marginal decline in worker injuries is observed. Secondly, the cost of these injuries has increased at an alarming rate, although the figure suggests that these costs may be levelling off. Although outside the realm of this text, the possibility that the lack of decline in worker injuries is related to the increased potential monetary gains associated with a work-related injury cannot be dismissed.

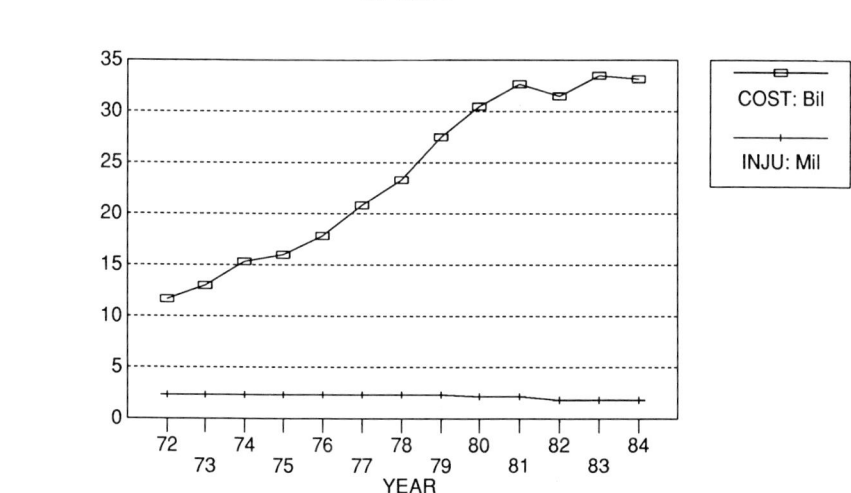

Figure 1.2. *Work injury and cost statistics*

Table 1.2. Mean medical and indemnity compensation costs for 1979 workers' compensation claims due to back injuries in five "Closed" SDS States by type of injury (after Klein et al., 1984)

Type	No. of cases reporting cost data	Mean cost per case ($)
	Medical payment	
Inflamed joints	200	4 689
Dislocation	762	3 533
Fracture	344	1 888
Strains/sprains	11 740	470
Laceration	24	425
Contusion	578	303
	Indemnity compensation	
Dislocation	2 906	19 536
Inflamed joints	235	7 120
Fracture	848	6 710
Nerve involvement	132	5 045
Strains/sprains	33 794	3 036
Laceration	40	2 712
Contusion	1 101	1 439
Burn/scald	23	891

1.3 Affected Industrial Populations

Back pain has been categorized into occupational and non-occupational origins. The context of this text dictates the exploration of occupational origins of back pain. The origin is categorized as occupational if the relationship between the work tasks and back injuries can be established.

Nurses experience more back injuries than most occupational groups (Jensen, 1985, 1986; Klein *et al.*, 1984), and are particularly vulnerable to low-back pain by virtue of their occupation (Harber *et al.*, 1985). Harber *et al.* (1985) studied a varied American population of nurses which included a large spectrum of back pain, not just severe cases of back pain to which researchers often restrict their investigations. It was found that the prevalence, incidence and lost work time were high, and interference with effective work was great. According to Harber *et al.* (1985), these parameters had been underestimated by the employee health service and need to be addressed.

In the Harber *et al.* (1985) study, questionnaires were sent to 1000 nursing staff members which included nurses, vocational nurses and administrative staff. Five hundred and fifty persons responded by the closing date. Overall, 52% of nurses reported the development of lower back pain in the previous six months, while only 20% of the unit service co-ordinators, who conduct clerical activities in the nursing units, reported the development of lower back pain. Twenty-nine percent of the nurses reported taking medicine for back pain, while 9% reported missing work due to back pain. While these staggering statistics delineate a widespread problem, the reader should be aware that the prevalence of back injuries for nurses may not be reflected in traditional injury statistics. Only 4% of the respondents had been hospitalized for back pain and 2% reported a history of surgery for these disorders. Furthermore, only 125 nurses had reported having seen a physician because of occupational back pain.

Nurses are not the only occupational group characterized by back injuries. Construction and mine workers also are plagued by back injuries. Descriptive data from the 1979 workers' compensation claims from the Bureau of Labor Statistics' SDS reveals that construction workers have the highest incidence ratio; 1·6 claims per 100 workers over a year, while they are ranked fourth for number of claims according to Klein et al. (1984). Mine workers are rated second, having an incidence ratio of 1·5 claims per 100 workers. Table 1.3 describes the distribution of compensation claims for back sprains/strains by industry employment. Estimated employment in each industry is given as are the claims per 100 employees.

Table 1.3. *1979 ratios (claims per 100 workers) of compensation claims for back strains/sprains in 26 SDS States, by industry (after Klein et al., 1984)*

Industry	No. of claims	Estimated employment in (1000s)	Claims per 100 employees
Construction	31 028	1964·2	1·6
Mining	4946	319·5	1·5
Transportation	25 338	2166·7	1·2
Manufacturing	92 556	8871·4	1·0
Agriculture	6077	677·0	0·9
Services	53 313	7371·5	0·7
Wholesale/retail trade	56 745	8956·3	0·6
Government (state & local)	14 463	5859·9	0·2
Finance	4032	2063·6	0·2
Total	285 468	38 250·1	0·7

Additional research reveals that back injuries account for 21% of all injuries in underground coal mines, 15% of all injuries in underground metal mines and 18% of all injuries in underground non-metal mines. When these injuries are categorized into various injury types, the strain/sprain injury type is dominated by back injuries (Stobbe et al., 1986).

On average, 3–4 weeks of work are lost per case because of strain/sprain injuries. According to Stobbe et al. (1986), back strains/sprains are the most frequent kind of strain/sprain, although not necessarily the most severe in terms of days lost from work.

Back strains/sprains in the underground coal mining industry accounted for an average of 21·1 and 19·5 days lost in 1983 and 1984, respectively. The number of days lost accounted for 52% of all days lost associated with strains/sprains, while back injuries accounted for 55% of all strain/sprain injuries. Back strains/sprains in underground metal/non-metal mining accounted for an average of 7·9 and 5·8 work days lost in 1983 and 1984, respectively. The number of days lost accounted for 61% of all days lost associated with strains/sprains, while back injuries accounted for 52% of the strain/sprain injuries.

Establishing causal relationships for back injuries and pain is often a difficult chore due to the intertwining causal variables. However, researchers have been able to isolate certain broad areas of concern.

The main causes of nurses' high frequency of back injuries are believed to be lifting and manoeuvring patients. This belief is based on personal experience of individual authors, unpublished analyses of hospital injury records and surveys (Jensen, 1985). In the survey by Harber *et al.* (1985), respondents associated lifting a patient in bed (48%), helping patients out of bed (30%), moving beds (27%) and lifting a patient to a gurney [trolley] (22%) with back pain. This study does suggest that much of the back pain is work-related since there is a significant difference between nurses and administrative employees, and that the results are consistent among different time frames.

The study by Snook *et al.* (1978) supports the contention that the main source of back pain is work related. In this study, a correlation of 0.88 was found between a specific act or movement at work and compensable back injuries. However, it may be that the nurses have been trained such that they believe that back pain is associated with patient handling and thus attribute pain to patient handling by virtue of their training; a self-fulfilling prophecy. According to Jensen (1985), with the available information, it can be estimated that 38–46% of all back episodes, and 73–81% of all compensable back pain are attributed to patient handling. However, Jensen (1985) warns the reader that these statistics are based on the few available studies, and suggests that more rigorous studies be conducted to verify such statistics.

Unlike accident data from the nursing occupation, no causal inferences have been made with regard to the construction or mining occupations. However, Stobbe *et al.* (1986) offer accident classification information of strains and sprains (Table 1.4), and the sources of strains and sprains (Table 1.5) and activities at the time of back strain or sprain (Table 1.6) for underground coal, metal and non-metal miners. The tables indicate that manual handling is pre-eminently associated with worker injuries.

Table 1.4. *Accident classification strain and sprain injuries (after Stobbe* et al., *1986)*

	Coal		Metal/non-metal	
	1983	1984	1983	1984
Roof fall	1·9*	2·1	1·0	3·2
Material handling	55·1	51·8	41·2	45·4
Hand tools	4·0	4·7	12·1	8·5
Powered haulage	9·0	9·8	3·3	5·0
Machinery	5·0	5·2	7·5	9·6
Slips/falls	16·2	18·5	25·8	20·9
Step/kneel on object	4·6	4·1	7·5	3·5
Strike/bump	1·8	2·1	0·3	0·4
Other	2·3	1·8	1·3	3·5
Baseline	2475	2776	306	282

* Figures are percentage of baseline at bottom of column.

Table 1.5. Sources of strain and sprain injuries in mining (after Stobbe et al., 1986)

	Coal		Metal/non-metal	
	1983	1984	1983	1984
General supplies	10·6*	9·7	7·5	6·0
Belt/conveyor	4·0	4·0	0·0	0·0
Cable handling	8·8	9·8	2·3	1·4
Tool use	3·8	4·4	10·5	12·1
Mining machinery	4·2	4·0	10·8	8·9
Metal NEC,-doors, pipes, etc.	6·0	6·4	6·9	6·7
Rock, coal, ore, etc.	13·1	11·3	13·7	11·7
Mine transport	5·1	5·8	2·0	3·5
Post, crib, timber, etc.	7·4	8·9	4·9	6·7
Bottom condition	13·7	15·7	21·2	15·6
Other	23·3	20·1	20·3	27·3
Baseline	2475	2776	306	282

* Figures are percentage of baseline at bottom of column.

Table 1.6. Activity at time of injury strain and sprain back injuries (after Stobbe et al., 1986)

	Coal		Metal/non-metal	
	1983	1984	1983	1984
Material handling	38·0*	35·0	29·9	38·2
Timbering	9·4	9·6	10·4	6·9
Cable handling	10·4	12·1	3·0	2·8
Shovel/move rock, coal, etc.	9·1	8·1	5·5	4·9
Machine repair	6·0	5·0	5·5	2·8
Equipment RLTD	10·6	13·1	12·8	13·2
Walking/running	4·2	3·8	6·1	6·3
Other	12·2	13·3	26·8	25·0
Baseline	1367	1507	164	144

* Figures are percentage of baseline at bottom of column.

1.4 Scope of the Book

The theme of this book is to expose the reader to the available information directly concerning manual materials handling. A wide range of topics are covered from variables influencing the MMH activity, to designing optimal MMH tasks, to determining rest durations. More specifically, Chapter 2 lists and describes variables that affect MMH activities. Chapter 3 addresses criteria necessary to design MMH activities which originate from three different approaches; the biomechanical, physiological and psychophysical approaches. Chapter 4 describes different data bases from which lifting and lowering activities can be designed which include strength and capacity data for one- or two-handed MMH activities. Similarly, Chapter 5 describes different data bases from which pushing, pulling, carrying and holding activities can be designed. Chapter 6 lends insight into screening procedures and/or design/redesigning MMH tasks. Chapter 7 is devoted to strength testing procedures which include assessment of isokinetic, isotonic and isoinertial strengths. Chapter 8 continues with a discussion about training and known effects of training. And finally, Chapter 9 describes the determination of rest allowances.

Chapter 2
Variables in Manual Materials Handling

2.1 Worker–Task–Enivronment 'System' Concept

A 'system' means a grouping of parts, or entities, that operate together for a common purpose. The system may consist of people or physical parts or both. Thus, a group of people responsible for resource allocation is a management system, as is an automobile which consists of various components which work together to provide ground transportation between two specified points. Similarly, a telephone, transmission and switching equipment and operator form a system for communicating with other people. On a much larger scale, we have weapons guidance systems, transportation systems, economic systems and political systems. A system could be as small and simple as a warehouse and a loading platform or as large and complex as a space shuttle.

Humans live and work within systems, but do not always fully understand them. The need to understand a system was not always compelling. For instance, in the days of primitive society, humans simply adapted to systems without feeling the need to understand their work or its rationale for existence. Even today, the use of the expression 'when in Rome, behave like Romans' is very pervasive.

The emergence of industrial societies in the last three centuries, or so, has resulted in systems which are complex and orderly, and require complete understanding of cause and effect relationships for efficient and total human adaptation. However, this is not to say that total adaptation is always possible. An ill-conceived and improperly organized system leads to inefficient system behaviour which must be tolerated by its human components, often at great cost, suffering and pain, if the system is to remain operational. Manual materials handling system is such a system.

In order to design an efficient MMH system, it is critical that the input of all its components on the capability of humans to adapt, or adjust, be clearly understood. Only when all cause and effect relationships become known, is it possible to design an orderly MMH system structure.

The MMH system consists of three main components:

(i) the worker,
(ii) the task, and
(iii) the environment.

It is a system which is known as a 'closed' or 'feedback' system. Such a system has a closed loop structure and its outputs (responses) influence inputs in such a way that the goal sought is achieved (negative feedback). In the context of materials handling, the goal might be to select a workload that does not lead to excessive fatigue or injury. If workload (input) exceeds a certain level, it will trigger responses (output) that will be unsafe and will lead to excessive fatigue or injury. These response levels are compared with the acceptable response levels (criteria) and the workload is adjusted by either the worker or the designer to avoid excessive fatigue or injury. Thus, the output influences the input. Figure 2.1 shows a simple negative-feedback loop for an elementary manual lifting task. For a given frequency and height of lift, as the weight of a specified size object increases, the corresponding work-rate increases as well. This work-rate evokes certain human body responses. The level of these human responses determines the acceptable, or safe, level of weight that can, or should, be lifted. As the weight lifted changes, the work-rate also changes. This, in turn, influences human responses. The weight, and consequently the work-rate, goes up or down depending upon the response levels. Thus, outputs (e.g. heart rate, oxygen uptake) influence the input (weight) until the desired level of weight (goal), a weight which does not cause excessive fatigue or injury, is reached.

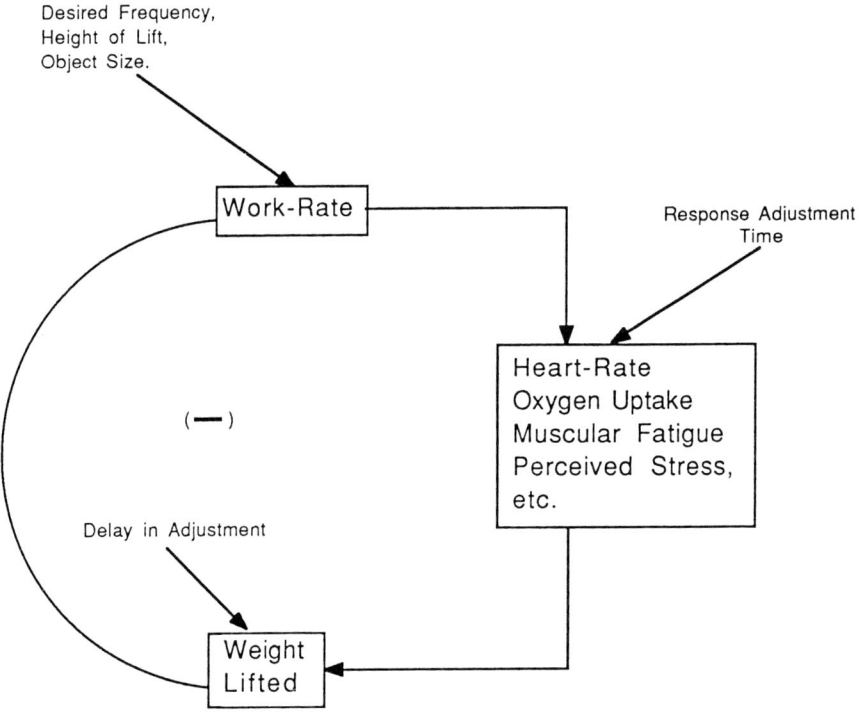

Figure 2.1. *Simple negative-feedback loop for an elementary manual lifting task*

In a realistic MMH system, the levels of various responses are not only influenced by the task itself, but also by the environment in which the task is being performed and the physical characteristics of the person performing the task. Thus, environmental characteristics, such as temperature and humidity, and physical characteristics of the worker, such as age, sex and physical strength, also influence the responses (i.e., output; see Figure 2.1). The goal (safe or acceptable weight in the example) is reached by adjusting the input for the task component, as well as environmental and worker factors (Figure 2.2).

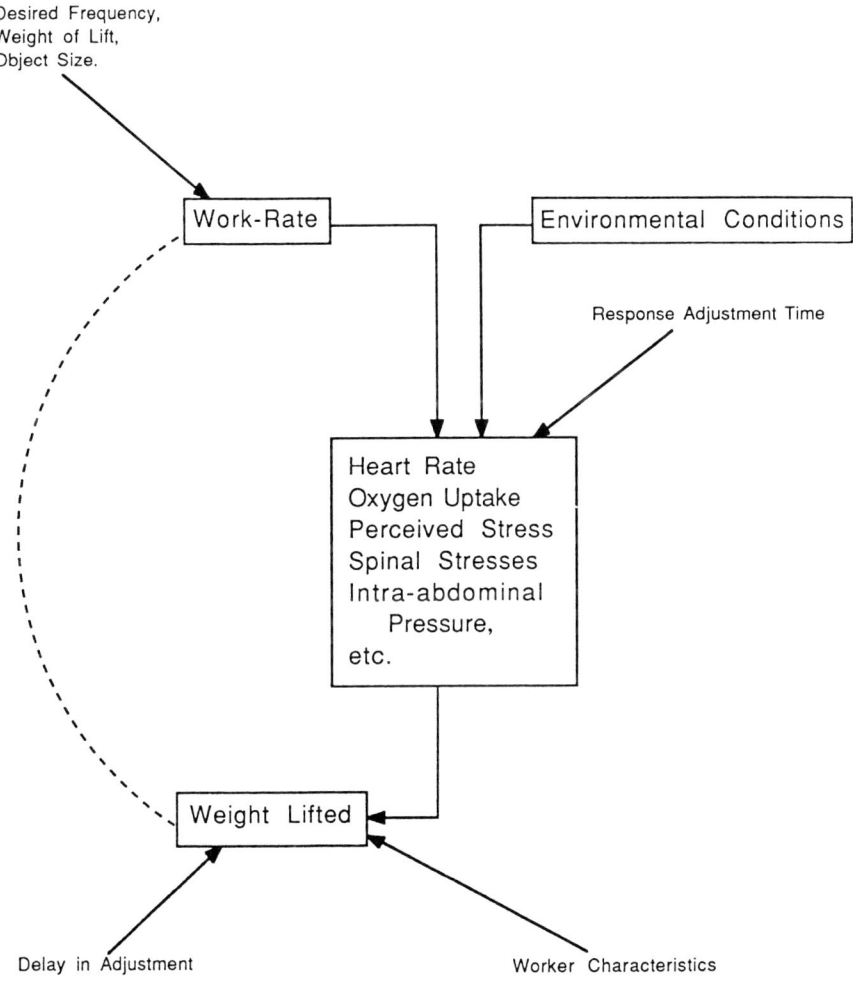

Figure 2.2. Inter-relationships of the components of a simple MMH system

Thus, in a MMH system, all three components (the task, the worker and the environment) influence outputs, the system goal, and system performance and/or accuracy. System accuracy is determined by the accuracy of the interrelationships between the various system component characteristics and the resulting responses.

Each of the three MMH system components has several elements. For example, the elements of the task component are all those factors which describe the job, such as the frequency of lift, size of the object being handled, origin and destination of the move. Similarly, factors describing the working environment or the physical capability of the worker are the elements of the environmental component and the worker component, respectively. Figure 2.2 is a simplified representation of the MMH system and the relationships of its components.

In order to design efficient MMH systems, it is necessary to understand the relationships between the elements of the three system components and the extent of human adaptation (or tolerance). The rest of this chapter is devoted to reviewing our knowledge about cause and effect relationships of various elements and MMH system capabilities. The system capabilities are described by the goals (safe or acceptable limits) and acceptable levels of responses (outputs such as heart rate, spinal stresses and metabolic energy expenditure rate).

2.2 The 'Worker' Component

The worker component of the MMH system comprises all elements (characteristics or factors) that define or describe a person's working capability. In terms of MMH activities, safe working capabilities are expressed in terms of responses to weight, or force, that can be lifted, or exerted, such that physiological costs can be sustained for the working duration without excessive fatigue, and physical stresses on the spine and other body joints that can be sustained without injury.

To date, researchers have recognized several different worker characteristics that limit a worker's material handling capability. While the effects of some of these characteristics are well known and have been found to be consistent, there exists both conflicting and contradictory information on other characteristics. This section provides an overview of what we know presently about the effects of worker characteristics on a person's capability to handle materials manually.

Table 2.1 shows a listing of the various worker characteristics that have been recognized by researchers and practitioners to be hazardous or limit a person's material handling capability (Herrin et al., 1974; Chaffin and Ayoub, 1975). Table 2.2 lists worker characteristics and their relative importance by percentage as cited by various researchers (Herrin et al., 1974). While the research base supports the notion that all these characteristics, singularly and collectively, determine a person's material handling capability, the research base is insufficient from which to derive specific quantitative information in many cases. For instance, the research base is insufficient to describe the effect of worker personality on worker's physical capability to handle materials.

In the following review, only those worker characteristics about which we have specific information are included.

Table 2.1. Worker characteristics comprising the worker component of the MMH system (Chaffin and Ayoub, 1975). Reprinted from Industrial Engineering. © Institute of Industrial Engineers

Physical	General worker measures such as age, sex, posture, anthropometry
Sensory	Measures of worker sensory processing capabilities such as visual, auditory, tactual, kinesthetic, vestibular and proprioceptive
Motor	Measures of worker motor capabilities such as strength, endurance, range of movement, kinematic characteristics and muscle training state
Psychomotor	Measures of worker capabilities interfacing mental and motor processes such as information processing, reaction/response time and co-ordination
Personality	Measures of worker values and job satisfaction by attitude profiles, attribution, risk acceptance and perceived economic need
Training/experience	Measures of worker education level in terms of formal training or instruction in MMH skills, informal training and work experience
Health status	Measures from worker general health appraisal such as previous medical complaints, diagnosed medical status, emotional status, regular drug usage, pregnancy, diurnal variations and deconditioning
Leisure time activities	Measures of the persons choosing to be involved in physical activities such as holding a second job or regular participation in sports

Table 2.2. Worker characteristics cited important by various researchers in determining personal risk of injury in MMH (Herrin et al., 1974)

Worker characteristic	Percentage of citations
Physical	38
Sensory	2
Motor	13
Psychomotor	3
Personality	6
Training/experience	8
Health status	30

2.2.1 Age

Generally, it is assumed that a person's working capability decreases after 20 years of age (Aberg, 1961). Frequently, older workers are restricted from load handling jobs and, in these situations, the advanced age of the worker is considered as a valid criterion for placing such restrictions (Chaffin et al., 1977a). In fact, the practice of awarding the softest job to the oldest worker is extremely pervasive in industries throughout the world. This practice may very well be the cause of fewer incidences of low-back pain among workers over 50–60 years of age (Hirsch, 1966; Brown, 1971). However, the reason is not clear. It is possible that the high rate of low-back pain incidence among 30–50-year-old workers (Herndon, 1927; Hult, 1954; Kosiak et al., 1968; Schein, 1968; Magora and Taustein, 1969; Rowe, 1969, 1983; Brown, 1974) is the combined result of screening older workers from material handling jobs and overloading the bodies of younger, inexperienced workers. Supporting this hypothesis, Blow and Jackson (1971) and Brown (1971) state that heavy physical work performed in early ages (twenties) accelerates the rate of injury and musculoskeletal injuries with ageing. A logical consequence is the high incidence of low-back pain among workers 30–50 years of age.

The effect of ageing on the material handling capability of workers is not clear. While a decline in operator's physical capabilities after an age of 50–60 years generally takes place (Astrand and Rodahl, 1977; Brown, 1977), it does not appear to lead to reduced manual lifting capabilities. In a study of 146 male and female industrial workers, ranging in age from 18–50 years, performed by Ayoub et al., (1978), no differences in the maximum acceptable weight of lift, due to age, were observed (Table 2.3). Similar observations were made by Mital (1984a, b) and Mital et al. (1984), on industrial workers ranging in age from 18–61 years (Table 2.3). The heart rates and metabolic energy expenditure rates at the maximum acceptable weight of lift were also unaffected by age (Mital et al., 1984).

Table 2.3. *Maximum weight of lift (kg) acceptable to industrial workers of different age groups*

Study		Age group (years)		
		Up to 29	30–39	40 and above
Ayoub et al. (1978)	Male	21·47	20·96	21·42
	Female	11·25	12·38	12·08
Mital (1984a, b) and	Male	10·22	12·10	12·02
Mital et al. (1984)	Female	9·18	9·38	8·96

The lack of influence age has on what weights of lift are acceptable is quite surprising since ageing tends to reduce overall physical capabilities. Reductions in physical work capacity (Bink, 1962; Malhotra et al., 1966; Astrand et al., 1973; Hartung, 1974; Ilmarinen and Rutenfraz, 1980), range of lumbar spinal motion (Bakke, 1931; Alvik, 1949; Tanz, 1953; Johck and Van Niekerk, 1961; Twomey, 1979; Taylor and Twomey, 1980), muscle strength (Asmussen and Heeboll-Nielsen, 1961, 1962; Astrand and Rodahl, 1977; Kamon and Goldfuss, 1978; Larsson and Karlsson, 1978; Larsson et al., 1979; also Figure 2.3), muscle contraction speed (Syrovy and Gutmann, 1970; Jennekens et al., 1971), shock absorbing characteristic of the lumbar disc (Perey, 1957; Happey, 1980; Hansson and Roos, 1981) and increases in disc degeneration (Ball, 1978; Hansson and Roos, 1981) and intra-abdominal pressure, for the same force application (Davis and Stubbs, 1980; Stubbs, 1985) are the general effects of ageing which are now well known. Ageing also contributes to the feeling of excessive fatigue (Reinberg et al., 1970).

While maximal oxygen uptake (physical work capacity) declines with age, submaximal work load oxygen consumption is not affected by age (Muller, 1962; Durnin and Passmore, 1967; Adams, 1967; Sheppard, 1974). The reduction in maximal strength with age is uncertain, however. Petrofsky and Lind (1975a) did not observe any changes in isometric strength with age. Mital and Ayoub (1980) and Nyland et al. (1978), on the other hand, did observe that isometric arm and leg strength increases with age. Furthermore, Montoye and Lamphiear (1977) did report some decrease of strength with age.

Although research reveals conflicting information, the fact that the stress supporting capacity of the spine declines with age, age should be treated as a potential

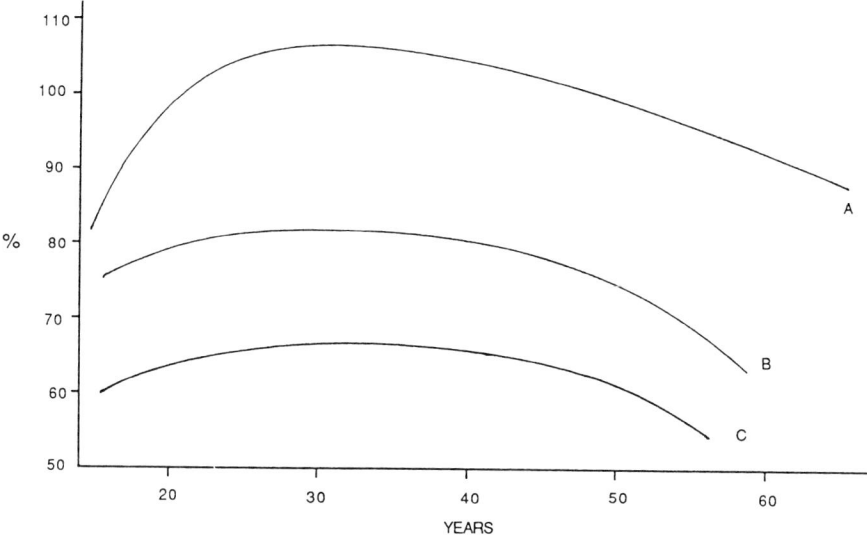

Figure 2.3. Isometric muscle strength in 25 different muscle groups in relation to age (expressed as percentage of strength of 21-year-old males: A-360 men; B-250 women; C-250 women; corrected stature 178 cm) (Asmussen and Heeboll-Nielson, 1961, 1962)

risk factor and extreme care should be taken in assigning older workers, especially those above 50 years of age, to demanding materials handling jobs.

2.2.2 Sex

Sex is the single most important worker characteristic that divides the working population because of differing anthropometric, biomechanical and physiological variables between males and females (Grasley *et al.*, 1978). The physical capability of females is approximately two-thirds the physical capability of males. However, this figure is not consistent. Isometric strengths, for instance, may be as low as 35% or, when corrected for body-weight, 88% of the male capability (Pheasant and Grieve, 1981). According to Doelen (1981), differences in physical capability between genders are due to differences in body-weight and body size.

In general, the body dimensions of both males and females are proportional to their stature. Additionally, there are only minor stature differences between genders, the average adult female stature being approximately 93% of the male's stature (Pheasant, 1982).

While stature differences are minimal between males and females, the differences in strengths are profound. On the average, a woman's lifting strength is 60–76% of a man's lifting strength (Asmussen and Heeboll-Nielsen, 1961, 1962; Troup and Chapman, 1969b; Chaffin, 1974; Snook and Ciriello, 1974a; Petrofsky and Lind, 1975b; Ayoub *et al.*, 1978; Mital *et al.*, 1978; Snook, 1978a, b; Pheasant and Grieve, 1981; Mital, 1984a, b). For specific strengths, however, the female strength may be as low as 33% or as high as 86% (Nordgren, 1972; Mortimer, 1974; Laubach, 1976;

Yates et al., 1980; Griffin et al., 1984). A similar relationship has been observed for muscle power (measured in kpm/s) by Komi and Karlsson (1979). According to them, female muscle power is 68% of male muscle power. Interestingly, when corrected for lean body mass, the difference of muscle power between genders almost disappears (a female–male muscle power ratio of 0·96 was obtained).

Biomechanical linkage mechanism differences between males and females contribute to differences between male and female lifting capacity (Chaffin and Moulis, 1969; Tichauer, 1973). This is probably the cause of different intra-abdominal pressure found between genders; females having 50% more intra-abdominal pressure than males when the same force is applied (David, 1985). Greater abdominal pressure found in females is reflected in the large number of complaints made by females when required to perform physically demanding tasks (Magora, 1970; Brown, 1971, 1974).

The pushing and pulling capabilities of females vary between 60–75% and 90–100% of the pushing and pulling capabilities of males, respectively (Snook, 1978a, b). The carrying capabilities of females, as determined from Snook's data, ranges from 67–73% of the carrying capabilities of males.

As with strength differences, there exists a significant difference in the aerobic capacity between men and women after puberty. Astrand and Rodahl (1977) have reported female aerobic capacity to be 70–75% that of male aerobic capacity. Vital capacity and stroke volume in females is also 75% that of males. A trained young female athlete may, however, have a maximal aerobic capacity equal to or greater than that of an untrained male (Doelen, 1981).

At submaximal workloads, several studies have shown that the metabolic cost of a certain task is significantly smaller in women than in men. However, when the metabolic cost is expressed as a function of the body-weight, contrasting results have been reported. Adams (1967) and Durnin and Namyslowski (1958) did not find any difference between men and women performing submaximal tasks when the metabolic cost was expressed as kcal/lb. Booynes and Keating (1957) and McDonald (1961), on the other hand, reported the existence of differences.

In hot and humid working conditions, females tend to sweat less than males according to Shapiro et al. (1980) and Frye and Kamon (1981). It has also been reported that women have higher heart rates for given lifting tasks (Snook and Ciriello, 1974a; Mital, 1984a, b). Higher metabolic rates for women performing lifting tasks have also been reported (Tichauer, 1970; Mital, 1984a, b). This indicates that women, in general, work closer to their aerobic capacities then men do and, therefore, may be at a greater risk performing the same task.

Female work capacity also tends to fall in pregnancy, due to increases as large as 30% in basic metabolic rate (Troup and Edwards, 1985), and increases in abdominal and pectoral girths. An increase in the ratings of perceived exertion, at a given level of work, has also been observed (Gamberale et al., 1975).

2.2.3 Body-Weight

The gross body-weight is an important determinant of a person's manual materials handling capability. As the body-weight increases so does the metabolic energy expenditure rate (Wyndham et al., 1963; Kamon and Belding, 1971; Garg, 1976). This means that for the same task, heavier people are physiologically more stressed. This could lead to quick exhaustion or other cardiovascular problems. On the other hand, heavier individuals are usually stronger and have enough mass to handle large objects (Snook and Irvine, 1967; Laubach and McConville, 1969; Troup and Chapman, 1969a; Konz et al., 1973; Larsson et al., 1979). These individuals, however, get tired easily and quickly (Petrofsky and Lind, 1975a). Everytime a person lifts an object, he may be also lifting up to 60% of his body-weight (Ayoub et al., 1978; Mital, 1984a). Thus, a heavier individual ends up lifting more weight and, therefore, gets fatigued easily and quickly.

Heavier individuals generally have larger abdominal and pelvic girth compared with their lighter counterparts. This causes postural problems and at times obese individuals have great difficulty in picking up objects from the floor. The horizontal distance between the load and the spine also increases, resulting in large spinal stresses. Perhaps for these reasons, the gross body-weight has been suggested as a limiting factor for work (Jorgensen, 1970).

The effect of body-weight is also significant in predicting the maximum acceptable weight of lift (Ayoub et al., 1978; Mital et al., 1978, 1984; Mital and Ayoub, 1980, 1986; Mital, 1983b). Additionally, body-weight is the major limiting factor for most pushing and pulling activities (Troup and Edwards, 1985).

2.2.4 Anthropometry or Body Build, and Body Composition

The anthropometric variables describe the physique of workers and therefore provide some indication of their work capacity. Switzer (1962) found that the lifting strength decreases steadily with increase in stature. Supporting this contention, many studies have reported that taller people are more prone to back pain (Tauber, 1970; Gyntelberg, 1974; Merriam et al., 1983). Watson (1977) determined the relationship between muscular strength (defined as the sum of left and right isometric hand grip and back strength) and 13 anthropometric measurements in 52 male students aged 16–18 years. Strength was significantly related to all the anthropometric measurements. Multiple regression analysis indicated that approximately 57% of variance in strength is accounted for by measures of body size and 26% by body shape. Laubach (1969) found that stature, body surface area, and mesomorphy were significantly correlated with muscle strength, with anthropometry and body composition accounting for 26–56% of strength performance variance. Fleishman et al. (1961) found body height was negatively correlated with dynamic strength (-0.39). Chaffin et al. (1977a) have outlined the biomechanical reasons for excessive stress in tall people; a taller person has to lean and reach further to pick up or set down a load resulting in a relatively larger moment arm for the load. On the other hand, it is stressful for shorter people to handle loads away from the body or above their normal reach.

Since large variations in strength and endurance can be observed within any body-size group, the use of body size to determine worker's working capacity is, at best, only a rule of thumb (Troup and Edwards, 1985). Several specific measures, such as shoulder height, chest width and depth, knee and knuckle heights and abdominal depth, have been found useful in predicting the maximum acceptable weight of lift (Snook and Irvine, 1967; Ayoub et al., 1978; Mital et al., 1978, 1984; Mital and Ayoub, 1980, 1984; Mital, 1983c). In general, these studies have found that a proportionately built person (big chest and narrow waist) has greater materials handling capability. Individuals with large torsos and abdominal depths have difficulty handling loads repetitively although they tend to be stronger than other body types.

2.2.5 Static and Dynamic Endurance

Almost all MMH activities require a combination of dynamic and static force exertions. The exact effect of static and dynamic endurance is only now being understood. It appears that static endurance is influenced by the posture (Lind et al., 1978; Warwick et al., 1980), and dynamic endurance (exertion) influences oxygen consumption and heart rates (Wald and Harrison, 1975). Duration of exertion and type of task also affect static endurance (Kroemer, 1970). Figure 2.4 describes the effects of time duration on strength requirements, where shorter time durations require greater strength requirements.

McGlynn (1969) studied the relationship between maximum voluntary wrist flexion and endurance using the Absolute Endurance Index. There was a significant relationship between endurance and maximum strength reported both before and after a 30-day training programme. McGlynn concluded that, due to occluded circulation during a maximum isometric contraction, most of the energy required for a sustained contraction depended upon the amount of aerobic energy reserves in the muscle.

Both static and dynamic components are integral to dynamic lifting tasks. Thus, a

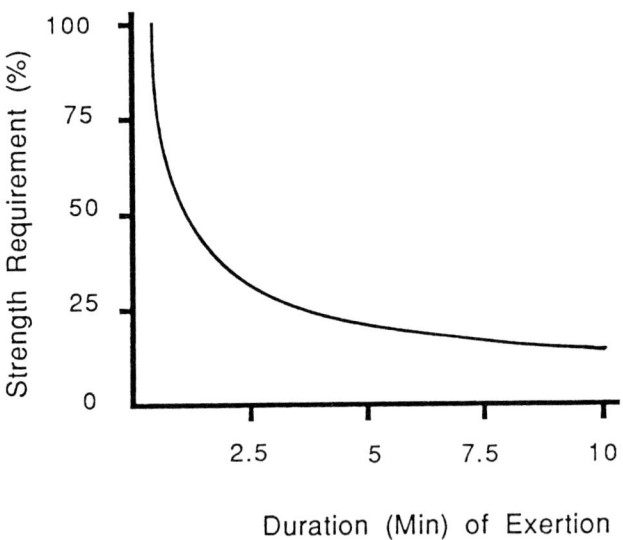

Figure 2.4. Endurance time as a function of strength requirement (Kroemer, 1970). From *Human Factors*, Vol. 12. © 1970 The Human Factor Society Inc., and reproduced by permission

person's working capacity can be determined by his static and dynamic endurance (Pytel and Kamon, 1981; Kamon et al., 1982; Mital, 1983c; Mital et al., 1986c). In fact, several studies have utilized dynamic endurance to predict isometric strengths and maximum acceptable weights of lift (Mital and Ayoub, 1980; Mital and Manivasagan, 1984). Because of their predictive capabilities, effects of static and dynamic endurance become increasingly important. Dynamic endurance also significantly affects dynamic strengths (A. Mital, 1981, unpubl. data). According to Petrofsky and Phillips (1981), static endurance decreases with body fat, however, it does not change with the length of sarcomere muscles of the hand for grip strength (Petrofsky et al., 1980).

2.2.6 Strength

Using a person's static (isometric) or dynamic (isokinetic, isotonic, or isoinertial) strength is now a widely accepted measurement for determining his/her material handling capability. Therefore, many believe that some kind of strength testing should be an integral part of pre-employment examination for applicants of MMH tasks (Kraus, 1967; Koyl and Hanson, 1969; Rowe, 1969, 1971, 1983; Chaffin et al., 1977a, 1978a, b; Pytel and Kamon, 1981; Kamon et al., 1982; Mital and Manivasagan, 1984; Mital et al., 1986c). More specifically, several researchers have shown that abdominal strength is a major factor in reducing compressive forces acting on the spine during manual lift (Bartelink, 1957; Morris et al., 1961; Alston et al., 1966; Davis, 1969). It has also been shown that individuals who do not have adequate isometric strength to perform lifting tasks are especially prone to low-back pain (Chaffin, 1974).

The protocol for reliable isometric strength testing has been well established (Kroemer, 1970; Caldwell et al., 1974; Chaffin, 1975a, b; Chaffin et al., 1977a). The testing itself is considered safe and it routinely produces a test–retest coefficient of variation of 10% or less (Chaffin et al., 1977a; Griffin et al., 1984).

As pointed out by Troup and Edwards (1985), two different methods of isometric strength testing have been used. The method more commonly used in Scandinavia involves measuring trunk flexor/extensor strength (Asmussen et al., 1959; Asmussen and Heeboll-Nielsen, 1961, 1962; Poulsen, 1970, 1971, 1981; Karvonen et al., 1980a, b; Yates et al., 1980), while the method most commonly used in North America involves isometric lifting strength (Chaffin and Park, 1973; Chaffin et al., 1977a, 1978; Ayoub et al., 1978; Keyserling et al., 1978, 1980a, b; Mital et al., 1978, 1984; Garg et al., 1980; Yates et al., 1980). The measurement of isometric strength has been successfully used in predicting materials handling capability of individuals in the case of infrequent tasks (Poulsen, 1970, 1971, 1981; Garg and Chaffin, 1975; Chaffin et al., 1977b; Keyserling et al., 1978, 1980a, b) as well as repetitive tasks (Ayoub et al., 1978, Mital et al., 1978, 1984, 1986c; Mital and Ayoub, 1980, 1986; Pytel and Kamon, 1981; Kamon et al., 1982; Mital, 1983b) (details of these procedures are given in Chapters 6 and 7). Isometric arm, back and shoulder strengths have been used to predict maximum acceptable weights of lift, while isometric shoulder, leg and back strengths have been used to estimate pushing and pulling capabilities (Ayoub and McDaniel, 1974). Arm and torso (composite) strengths have also been used to estimate job position strength (Chaffin et al., 1977a, 1978).

For manual lifting activities, isometric shoulder strength has been suggested as a limiting factor (Yates et al., 1980; Mital and Manivasagan, 1983). The result of other investigations, however, suggest that it is the back strength that limits capacity for these activities (Chaffin et al., 1977a, 1978; Poulsen, 1970, 1971, 1981; Keyserling et al., 1978, 1980a, b).

Isometric strengths are influenced by posture (Lind et al., 1978; Kroemer, 1969, 1974), visual feedback (Peacock et al., 1981) and the rate of development of muscle force (Kroemer and Marras, 1981). Visual feedback, for instance, results in a 10% increase in isometric strength. Conversely, the slow speed of isometric force development may produce only one-quarter of the peak force. The impact of these findings on working capability of individuals and job design is very clear.

In the last five years, the emphasis has shifted from the measurement of isometric strengths for estimating materials handling capabilities of individuals, to the measurement of dynamic strengths, simulated job dynamic strengths (SJDS) in particular. The rationale for this switch is that since most materials handling tasks are dynamic in nature, the dynamic strength testing can better estimate the contribution of the inertial component. The success of dynamic strength in determining material handling capabilities of workers in recent studies indeed justifies the switch (Pytel and Kamon, 1981; Kamon et al., 1982; Aghazadeh, 1982; Kroemer, 1983, 1985; Jiang, 1984; Mital et al., 1986c). Mital et al. (1986c) found that the correlation coefficients between simulated job dynamic strengths and maximum acceptable weight of lift in horizontal and vertical planes were substantially higher than those between isometric strengths and weight lifted (Table 2.4). This is supported by Aghazadeh (1982) and Jiang (1984) who also found, independently, that the measurement of dynamic strength is more reliable for determining a person's capability for manual lifting tasks.

A variation of the dynamic strength testing described above is used to determine the number of repetitions that can be performed, with or without external loading (Nummi et al., 1978; Troup et al., 1981; Griffin et al., 1984). A similar approach was adopted by Ayoub et al. (1978) who measured time (dynamic endurance). In this case, subjects were asked to lift a weight of approximately 25% of their isometric arm strength and were assigned a pace of lift. The total time elapsed was recorded and this

Table 2.4. Correlations between strengths and maximum acceptable weight of lift (Mital et al., 1986c)

	Maximum acceptable weight of lift			
Strength	Floor to 81 cm	81–152 cm	At 81 cm height	At 152 cm height
SJDS				
Floor to 81 cm	0.517	—	—	—
81—152 cm	—	0.672	—	—
At 81 cm height	—	—	0.533	—
At 152 cm height	—	—	—	0.574
Isometric				
Arm	0.356 (Maximum Value)			
Back	0.295 (Maximum Value)			
Composite	0.380 (Maximum Value)			
Shoulder	0.383 (Maximum Value)			

time, dynamic endurance, was successfully used to predict the maximum acceptable weight of lift (Ayoub *et al.*, 1978; Mital *et al.*, 1978; Mital and Ayoub, 1980) and isometric strengths (Mital and Ayoub, 1980; Mital and Manivasagan, 1984).

This review clearly indicates the importance of strength testing to determine materials handling capabilities of individuals and its role in the reduction of occupational low-back pain. The relationship between strengths and the maximum acceptable weights of lift and its implication in engineering design are discussed in Chapter 4. The measurement of static and dynamic strengths, their prediction and additivity are discussed in Chapter 7.

2.2.7 Training/Experience

Training in MMH has been advocated by many individuals over the years (Brown, 1971, 1974; Davies, 1978). The rationale is that worker preparation will lead to the avoidance or the successful survival of injuries. Generally, it is expected that after completion of training, a worker will have learned: (i) the correct method of load handling, and (ii) to recognize and cope with the hazards due to the load, the task or the environment. The training is also expected to lead to improvements in physical capability of workers and, thereby, a reduction in incidences of low-back pain and injury. This section deals only with the review of literature addressing the issue of physical fitness. Educational training and training programmes are dealt with in Chapter 8.

Physical fitness appears to improve the physical capabilities of individuals. Some of these improvements are: (i) increased isometric strength (Salter, 1955; Chapman and Troup, 1969; Astrand and Rodahl, 1977; Mital and Ayoub, 1981a; Kanehisa and Miyashita, 1983; Asfour *et al.*, 1984a) and isokinetic strength (Kanehisa and Miyashita, 1983), (ii) increased maximal oxygen uptake (Astrand, 1967a; Ekblom *et al.*, 1968; Chapman and Troup, 1969; Saltin *et al.*, 1969; Fox *et al.*, 1973, 1975; Astrand and Rodahl, 1977; Asfour *et al.*, 1984a), (iii) increased oxygen uptake during submaximal work (Ekblom *et al.*, 1968; Flint *et al.*, 1974), (iv) decreased heart rate at submaximal workload (Ekblom *et al.*, 1968; Frick *et al.*, 1963, 1967; Fox *et al.*, 1973, 1975), (v) increased lifting capability (Asfour *et al.*, 1984a), and (vi) increased maximum acceptable frequency of lift (Garg and Saxena, 1981, 1982; Mital and Asfour, 1983).

These findings are, however, not without controversy. Several researchers have reported an increase in mechanical efficiency at submaximal loads (Ekblom *et al.*, 1968) while others (Clausen *et al.*, 1970; Tzankoff *et al.*, 1972; Fox *et al.*, 1975) did not observe any change in energy cost, due to training, at submaximal loads. Similar conflicting findings have been reported for the effect of training on maximal heart rate. While Fox *et al.* (1973, 1975) and Saltin *et al.* (1968) noticed a slight decrease in maximal heart rate, Ekblom *et al.* (1968) and Maksud *et al.* (1972) did not. Mixed results were also obtained by Mital and Asfour (1983). The subjects in their study showed improvements in the maximum acceptable frequency of lift ranging from 0% to nearly 30%.

It appears that, overall, the effects of physical training are positive and lead to enhanced physical capabilities. Increased strength, for example, means enhancing the individual's capability when it is restricted by his strength (Yates et al., 1980; Jorgensen, 1970; Kroemer, 1974; Asfour et al., 1984a). Dehlin et al. (1978) and Karvonen et al. (1977) have also reported that exercise and physical training led to decreases in perception of exertion and reduced back pain. An increased cardiovascular capability, on the other hand, means reduced physiological stress (Flint et al., 1974; Cunningham and Hill, 1975; Asfour et al., 1984a). A recent report by Dehlin et al. (1981), however, indicated that physical training and ergonomic education programmes had little preventive effect on the perception of work and the subjective assessment of low-back problems such as pain and discomfort.

Table 2.5. *Frequency effects on the maximum acceptable weight of lift (Mital and Manivasagan, 1983a)*

Population	Frequency (lifts/min)	Weight (kg)	Difference (B − A) (kg)
Student (A)	2	21·39 (100%)	—
Industrial (B)	2	27·14 (100%)	5·75*
Student (A)	4	19·60 (91·60%)	—
Industrial (B)	4	25·81 (95·11%)	6·21*
Student (A)	6	19·06 (89·11%)	—
Industrial (B)	6	25·06 (92·34%)	6·00*

* Not significantly different from each other ($P \geq 0·10$).

Table 2.6. *Comparison of industrial and student population carrying capacities, in kg (Mital and Ilango, 1983a)*

Carrying distance	Student population		Industrial population		Level of significance
	$\bar{x}^*$	S†	$\bar{x}$	S	(P)
3·05	21·70	6·51	32·96	11·68	0·01
9·14	21·47	6·98	29·96	9·34	0·01

* $\bar{x}$ = mean; †S = standard deviation.

Table 2.7. *Lifting capacity (psychophysical estimates) of students as a percentage of industrial subjects (Mital, 1985f)*

Variable	Males		Females	
	$\bar{x}^*$	S†	$\bar{x}$	S
Box size (cm)				
30·46	82	81	93	99
45·72	81	70	93	75
60·96	81	72	94	75
Frequency (lifts/min)				
1	78	75	91	97
4	78	89	92	74
8	82	71	94	87
12	90	89	96	87
Height of lift				
Floor to knuckle	83	77	94	75
Knuckle to shoulder	78	68	91	87
Shoulder to reach	82	78	95	102

* $\bar{x}$ = mean; †S = standard deviation.

Mital and Manivasagan (1983a) determined that experienced male industrial workers could, on the average, lift 6 kg more than their inexperienced student counterparts (Table 2.5). Similarly Mital and Ilango (1983) and Mital (1985f) found that experience on the job does in fact increase MMH capability as expressed by maximum acceptable weights of carry and lift in the two respective studies (Tables 2.6 and 2.7). Further discussion on training is included in Chapter 8.

2.2.8 Physical Fitness

It is generally believed that physically fit individuals are less prone to injuries and incapacity for work (Doelan and Wright, 1979). A reduction in payroll costs has been reported by Cox et al. (1981) when employees participated in an exercise programme. Physically fit individuals are also likely to undertake more demanding jobs and might not absent themselves from work due to a minor injury. Their pain tolerance limits are usually higher (Brooke, 1967; Scott and Gijsbers, 1981) and they tend to report only serious injuries. Ironically, Cady et al. (1979) reported that among fire fighters in California, the most fit group suffered the most severe injuries which indicates that they take on tasks of greater risk than less fit workers. Lack of physical fitness or a history of back disorders are well known clues for identifying individuals who might have a recurrence of back or sciatic pain (Dillane et al., 1966; Rowe, 1971; Karvonen et al., 1980b; Nordgren et al., 1980; Troup et al., 1981; Lloyd and Troup, 1983).

2.2.9 Psychological Factors

The effects of psychological factors on maximal muscular performance and psychophysical MMH capacity are pervasive albeit not clearly understood. A review of some of these factors and their likely effect on strength performance is summarized in Table 2.8.

Selan (1986) has summarized some of the empirical evidence supporting those factors presented in Table 2.8. In an early study by Lehman et al. (1939), pervitin (1-phenyl-2-methylaminopropane) was administered to subjects prior to riding on a bicycle ergometer. Administration of the drug resulted in up to three-fold increases in task performance. In the absence of any definable facilitation of the physiological systems, the authors concluded that the drug created central excitation which led to improved performance. The authors theorized that the 'end point' of human performance occurs when the sum of all possible negative factors (e.g. fatigue, muscle pain) exceed, in the person's perception, the sum of all possible positive factors (i.e., motivation). The results of this study are largely in agreement with the findings of Alles and Feigen (1942) and Smith and Beecher (1959).

One of the most ambitious studies in this area was conducted by Ikai and Steinhaus (1961). Using a cable tensiometer, the tension exhibited by the right forearm flexors during a maximal voluntary contraction (MVC) was measured. MVC was measured under the following conditions: (1) after having a blank shot fired behind the subject, (2) having the subject shout during the exertion, (3) exerting while under hypnosis, (4) exerting while under a post-hypnotic suggestion, and (5) exerting 25 min

Table 2.8. *Psychological factors affecting maximum muscular strength (Kroemer and Marras, 1981)*

Factor	Likely effect
Feedback of results	Positive
Instructions on how to exert strength	Positive
Arousal of ego involvement, aspiration	Positive
Pharmaceutical agents (drugs)	Positive
Startling noise, subject's outcry	Positive
Hypnosis	Positive
Setting of goals, incentives	Positive or negative
Competition, contest	Positive or negative
Verbal encouragement	Positive or negative
Spectators	?
Deception by researcher	?
Fear of injury	Negative
Deception by subject	Negative

following the ingestion of 30 mg of amphetamine sulphate. Significant increases in flexor tension occurred under the after-shot condition, when shouting during the exertion and following the amphetamine sulphate ingestion. Under the hypnosis conditions (during and after) significant increases and decreases in tension occurred as a function of the nature of the suggestion given the subject. Ikai and Steinhaus (1961) explain their results in terms of Pavlovian classical conditioning. They assumed that a muscular exertion is normally restrained by the negative conditioned inhibitions associated with maximal contractions (i.e., pain, spasms). The shot and the shout served as external inhibitors, in the Pavlovian sense, and the amphetamine sulphate and hypnotic suggestions served to suppress these conditioned inhibitions.

Bakken (1983) looked directly at the effects of psychological factors on the psycho-physically-determined MAWL. Bakken designed and administered an attitude questionnaire to each of 12 subjects participating in a psychophysical lifting experiment. The data indicated that the greatest weight of lift variability reduction occurred with the sole use of attitude or PWC (Physical Work Capacity) variables (reduction of approximately 70%). Use of attitude as a predictor of maximum acceptable weight of load (MAWL) was superior to the use of strength measurement or Cybex-II type variables (approximately 50% reduction), or the sole use of climatic, anthropometric, or task parameter variables (approximately 30% reduction). Unfortunately, the use of the attitude questionnaire in this study was intended only as a preliminary assessment of the worth of attitude as a predictive tool. As such, no analyses were conducted to determine the 'content' of these attitudes, nor was the questionnaire formally validated or tested.

Fox (1982) performed a psychophysical study involving bi-manual lifting which implemented social facilitation. Social facilitation is defined as an improvement in the performance of well-learned tasks and a performance decrement in poorly learned tasks under certain conditions of social stimulation (Zajonc, 1965). Some social stimuli commonly associated with social facilitation effects include:

1. Presence/absence of experimenter during task;
2. Presence of audience (other than experimenter) during task;
3. Presence of co-actors (individuals working on some task separately from subject).

Fox determined the psychophysical MAWL for subjects lifting individually and with a partner. The hypothesis was that for a given bi-manual team, a team member would be lifting approximately one half of a weight of load equal to twice his individual MAWL. However, the results indicated that, for a given bi-manual team, the team member with the smaller individual MAWL was lifting 23·22 lbs over his individual MAWL while the team member with the larger individual MAWL was lifting 12·76 lbs over his individual MAWL. Table 2.9 presents the average MAWL selected by each team. Table 2.10 presents the bi-manual MAWLs if the weight of load selection were based on the lower individual MAWL. As the tables indicate, in some instances, the team was lifting over twice as much weight as might be expected based on the smaller individual MAWL.

Table 2.9. *Actual weights (in lbs) of lift selected by bi-manual teams (Fox, 1982)*

S#	1	2	3	4	5	6	7	8
1	—	117·5	100·5	129·0	113·8	113·3	118·3	119·5*
2		—	103·3	100·8	121·5	86·8	104·5	135·8
3			—	115·3	111·8	103·5	93·5	97·8
4				—	117·3	99·0	118·3	134·3
5					—	122·5	119·5	144·3
6						—	107·0	124·8
7							—	113·8
8								—

* Average of two trials.

Table 2.10. *Predicted bi-manual capacities based on individual MAWL (Fox, 1982)*

S#	1	2	3	4	5	6	7	8
1	—	63·0	55·0	73·0	80·0	66·5	82·0	94·5*
		64·0	64·0	76·0	82·0	67·0	87·0	96·0†
			55·0	64·0	63·0	63·0	63·0	63·0
2		—	64·0	64·0	64·0	64·0	64·0	64·0
				55·0	55·0	55·0	55·0	55·0
3			—	64·0	64·0	64·0	64·0	64·0
					73·0	73·0	73·0	73·0
4				—	76·0	67·0	76·0	76·0
						66·5	80·0	80·0
5					—	67·8	82·0	82·0
							66·5	66·5
6						—	67·0	67·0
								82·0
7							—	87·0
8								—

* Lower individual's average MAWL. † Lower individual's highest MAWL.

Fox (1982) also found that the subjects with the smaller individual MAWL had statistically significant higher heart rates (increase of approximately 4 beats/min) and per cent PWC (increase of approximately 2%) working bi-manually than working alone. Fox termed these differences of no practical significance. The subject with the higher individual MAWL showed no significant change in heart rate or per cent PWC when working alone or working bi-manually. A reanalysis of Fox's data reveals inconsistent heart rate changes whether the subject worked alone or bi-manually (Table 2.11). Looking at per cent PWC (Table 2.12), more consistent results were found. Six of eight subjects showed an overall increase in percent PWC when working bi-manually as opposed to lifting individually.

Drawing firm conclusions from Fox's (1982) study is difficult because it is noted that in the bi-manual condition a biomechanical advantage may have been introduced (type of box and lift) which contributed to the unexpectedly high bi-manual MAWL. However, Fox also noted that, in post-experiment interviews, subjects agreed that the bi-manual lifting was more interesting and less boring than individual lifting, and therefore increased drive (in social facilitation terms). This cannot be dismissed as a possible explanation of the results.

Table 2.11. Changes in heart rate when lifting alone or in teams (Fox, 1982)

Subject	Alone	Bi-manual	% Change
1	127·0	116·9	8 d*
2	111·2	117·7	6 i†
3	102·4	99·6	3 d
4	118·3	124·6	5 i
5	136·6	136·5	—
6	104·3	117·1	11 i
7	133·6	140·9	5 i
8	112·6	109·4	3 d
Average	118·3	120·3	2 i

* d = decrease; † i = increase.

Table 2.12. Changes in percent PWC when lifting alone or in teams (Fox, 1982)

Subject	Alone	Bi-manual	% Change
1	45·6	39·0	14 d*
2	23·2	25·7	10 i†
3	34·4	32·6	5 d
4	33·5	37·0	10 i
5	36·4	37·0	2 i
6	33·7	38·3	12 i
7	34·3	40·6	16 i
8	28·3	29·7	5 i
Average	33·7	35·0	4 i

* d = decrease; † i = increase.

2.3 The 'Task' Component

Elements which describe or define the MMH activity compromise the task component. Table 2.13 shows a listing of various task characteristics that have been recognized as factors which contribute to the MMH hazard (Herrin *et al.*, 1974; Chaffin and Ayoub, 1975). Some of these are related to the object being handled, such as shape,

Table 2.13. Task characteristics comprising the task component of the MMH system (Chaffin and Ayoub, 1975). Reprinted with permission of American Industrial Hygiene Association Journal

Load	Measure of mass, push/pull force requirements, mass moment of inertia
Dimensions	Measures of size of unit load, such as height, width, breadth when indicating the form of rectangular, cylindrical and spherical
Distribution of load	Measure of the location of the unit load c.g. with respect to the worker for one-hand and two-hand activities
Couplings	Measures of simple devices used to aid in grasping and manually manipulating the unit load, such as texture, handle size, shape and location and foot traction
Stability of load	Measures of load c.g. location consistency, as a concern in handling liquids and bulk materials
Workplace geometry	Measures of spatial properties of the task, such as movement distance, direction and extent of path, obstacles and nature of destination
Frequency/duration/pace	Measures of the time dimensions of the handling task including frequency, duration and required dynamics of activity over the short-term and long-term
Complexity	Measures of combined or compounding demands of the load, such as manipulation requirements of movement, objective of activity, precision of tolerance and number of kinetic components

size, load distribution and stability, while others, such as frequency, distance moved, duration and work place geometry, describe the task itself. Table 2.14 further breaks down the factors listed in Table 2.13 and, in addition, shows the major studies which have investigated designated task characteristics affecting material handling capabilities of workers. The approach used by investigators to study these task characteristics is also shown (there are basically three approaches, psychophysical, physiological and biomechanical, which have been used to quantify the effects of task characteristics on worker's MMH capability. These approaches are discussed in detail in Chapter 3). Table 2.14 shows net effects of the task characteristics (listed in Table 2.15) on operator responses. The net effects indicate the material handling capability of workers (see section 2.6).

Table 2.14. Net effect of task characteristics on operator response for MMH activities

Characteristic	Activities	Net effect*
Frequency (+)	All	Metabolic energy expenditure rate (+) Heart rate (+) Acceptable weight/force (−) Work-rate (+) Perceived exertion (+)
Task Duration	All	Metabolic energy expenditure rate (−) Heart rate (− or 0) Work-rate (−) Acceptable weight/force (−)
Object Size (+)	All	Metabolic energy expenditure rate (+) Heart rate (+) Acceptable weight/force (−) Spinal stresses (+)
Object Shape Collapsible Non-collapsible (volume +) Non-collapsible (volume 0)	Lifting, carrying	Acceptable weight/force (+) Acceptable weight/force (+) Acceptable weight/force (0)

Continued

Table 2.14. Continued

Characteristics	Activities	Net effect*
Couplings	All	Acceptable weight/force (+) Metabolic energy expenditure rate (−) Spinal stresses (−)
Object weight/force	All	Metabolic energy expenditure rate (+) Heart rate (+) Spinal stresses (+) Intra-abdominal pressure (+) Perceived exertion (+)
Load stability and distribution	Lifting, carrying	Acceptable weight/force (−) Metabolic energy expenditure rate (0) Heart rate (0) Perceived exertion (+)
Vertical lift height	Lifting, lowering	Acceptable weight (−) Metabolic energy expenditure rate (+) Heart rate (+) Perceived exertion (+)
	Carrying	Acceptable weight (+)
Height of force application/ starting point	Pulling, pushing lifting, lowering, carrying	Acceptable weight/force (+) Acceptable weight/force (−)
Distance travelled	Pushing, pulling carrying	Acceptable weight/force (+) Metabolic energy expenditure rate (+) Heart rate (0)
Speed/grade	Pushing, pulling, carrying	Acceptable weight/force (−) Metabolic energy expenditure rate (+) Heart rate (+)
Posture/technique	All except carrying	Free-style
Asymmetrical lifting	Lifting	Acceptable weight (−) Heart rate (0) Metabolic energy expenditure rate (0) Perceived rating (+) Intra-abdominal pressure (+) Spinal stress (+)

* Net effects: +, increase; −, decrease; ?, unknown or not clear; 0, no effect.

Table 2.15. Major task characteristics affecting MMH capabilities

Characteristic	Researcher(s)	Approach		
		Psychophysical	Physiological	Biomechanical
Frequency	Asfour (1980)	×	×	
	Asfour et al. (1984b; 1985)	×	×	
	Ayoub (1977)	×	×	
	Ayoub et al. (1980b; 1983b)	×		
	Ciriello and Snook (1983)	×		
	Foreman et al., (1984)	×		
	Garg (1976)		×	
	Garg and Saxena (1982)	×		
	Garg and Saxena (1978)		×	

Table 2.15. Continued

Characteristic	Researcher(s)	Approach		
		Psychophysical	Physiological	Biomechanical
	Karvonen and Ronnholm (1964)		×	
	Karwowski and Ayoub (1984a, b)	×		
	Khalil et al. (1985)		×	
	Legg and Pateman (1984)		×	
	Mital (1984a, b)	×	×	
	Mital and Asfour (1983)	×	×	
	Mital and Ayoub (1980; 1986)	×		
	Mital and Ayoub (1981a)		×	
	Mital and Okolie (1982)	×		
	Mital and Manivasagan (1983a)	×		
	Mital and Fard (1986)	×	×	
	Mital et al. (1985b)	×	×	
	Mital (1985f)	×	×	
	Petrofsky and Lind (1978a, b)		×	
	Ronnholm et al. (1962)		×	
	Samantha and Chatterjee (1981)		×	
	Snook (1978a, b)	×		
Task duration	Bonjer (1962)		×	
	Deivanayagam and Ayoub (1979)		×	
	Garg (1980)		×	
	Karwowski and Yates (1985)	×		
	Legg and Myles (1981)	×	×	
	Mital (1983c)	×	×	
	Snook (pers. comm.)	×		
Object size	Ayoub et al. (1980b)	×		
	Ciriello and Snook (1983)	×		
	Frievalds et al. (1984)	×		×
	Garg and Saxena (1980)	×		
	Haisman et al. (1972)		×	×
	Kromodihardjo and Mital (1985)	×		×
	Martin and Chaffin (1972)			×
	Mital (1984a, b)	×	×	
	Mital and Ayoub (1981a)		×	
	Mital and Fard (1986)	×	×	
	Mital and Kromodihardjo (1986)	×		×
	Mital and Manivasagan (1983b)	×		
	Morrissey and Liou (1984)		×	
	Tichauer (1971)			×
Object shape	Garg and Saxena (1980)	×		
	Haisman et al. (1972)		×	
	Mital and Okolie (1982)	×		
	Mital and Manivasagan (1983b)	×		
	Smith and Jiang (1984)	×	×	
Couplings	Bakken (1983)	×		
	Coury and Drury (1982)	×		
	Drury (1980)	×		
	Drury et al. (1982)		×	
	Drury et al. (1985)	×	×	×
	Frievalds et al. (1984)	×		×
	Garg and Saxena (1980)	×		
	Harrison and Malkin (1983)			×

Table 2.15. Continued

Characteristic	Researcher(s)	Approach		
		Psychophysical	Physiological	Biomechanical
	James (1983)			×
	Kromodihardjo and Mital (1985)	×		×
	Mital and Ayoub (1981a)		×	
	Mital and Kromodihardjo (1986)	×		×
	Perkins and Wilson (1983)			×
Object weight/force application (exertion)	Asfour (1980)	×	×	
	Asfour et al. (1984b; 1985)	×	×	
	Ayoub et al. (1980)	×	×	×
	Ayoub and McDaniel (1974)	×		×
	Chaffin et al. (1983)	×		×
	Ciriello and Snook (1983)	×		
	Davis and Stubbs (1980)	×		
	Frievalds et al. (1984)	×		×
	Garg (1976)		×	
	Garg and Chaffin (1975)			×
	Garg and Herrin (1979)		×	×
	Garg and Saxena (1979)	×	×	
	Garg and Saxena (1980)	×		
	Karwowski and Ayoub (1984b)	×		
	Kroemer (1969; 1974)	×		
	Kromodihardjo and Mital (1985)	×		×
	Martin and Chaffin (1972)			×
	Mital (1984a, b)	×	×	
	Mital et al. (1984).	×	×	
	Mital and Ayoub (1981a)		×	
	Mital and Fard (1986)	×	×	
	Mital and Kromodihardjo (1986)	×		×
	Mital and Manivasagan (1983a, b)	×		
	Mital and Okolie (1982)	×		
	Poulsen (1970; 1971; 1981)	×		
	Poulsen and Jorgensen (1971)	×		
	Snook (1978a, b)	×		
Load stability and distribution	Ayoub et al. (1979)	×		
	Mital (1986c)	×	×	
	Mital and Fard (1986)	×	×	
	Mital and Ilango (1983b)	×		
	Mital and Manivasagan (1983)	×		
	Mital and Okolie (1982)	×		
Vertical lift height/ height of force application	Astour et al. (1984b; 1985)	×	×	
	Ayoub and McDaniel (1974)	×		×
	Ayoub et al. (1980b)	×		
	Ciriello and Snook (1983)	×		
	Frederik (1959)		×	
	Garg (1976)		×	
	Garg and Chaffin (1975)			×
	Martin and Chaffin (1972)			×
	Mital (1984a, b)	×	×	
	Mital and Okolie (1982)	×		
	Snook (1978a, b)	×		

Table 2.15.   Continued

Characteristic	Researcher(s)	Psychophysical	Physiological	Biomechanical
Work place geometry	Boussenna *et al.* (1982)	×		
	Corlett and Bishop (1976)	×		
	Corlett *et al.* (1979)	×		
	Mital (1986c)	×	×	
	Ridd (1983)	×		×
Distance travelled/ slope and traction	Ciriello and Snook (1983)	×		
	Datta *et al.* (1978)		×	
	Gordon *et al.* (1983)		×	
	Mital and Manivasagan (1983b)	×		
	Goldman and Iampietro (1962)		×	
	Mital and Okolie (1982)	×		
	Pimental and Pandolf (1979)		×	
	Snook (1978a, b)	×		
	Grieve (1979)	×		
	Soule *et al.* (1978)		×	
	Ilmarinen and Louhevaara (1984)		×	
	Yousef *et al.* (1972)		×	
Posture/technique	Ayoub and McDaniel (1974)	×		×
	Brown (1976)	×		
	Chaffin *et al.* (1983)	×		×
	Datta and Ramanathan (1971)		×	
	Garg and Herrin (1979)		×	×
	Garg and Saxena (1979)	×	×	
	Kumar (1984)		×	
	Kumar and Magee (1982)		×	
	Leskinen *et al.* (1983a, b)			×
	Park (1973)			×
	Ridd (1983)	×		×
	Soule and Goldman (1969)		×	
	Troup *et al.* (1983)			×
Asymmetrical lifting/ carrying	Kromodihardjo and Mital (1985)	×		×
	Kumar (1980)		×	×
	Kumar and Magee (1982)		×	
	Kumar (1984)		×	
	Mital (1986c)	×	×	
	Mital and Ilango (1983)	×		
	Mital and Fard (1986)	×	×	
	Mital and Kromodihardjo (1986)	×		×

2.3.1 Frequency

Frequency, or pace, is associated with repetitive tasks. Both psychophysical and physiological approaches (see Chapter 3) have been used to investigate the effects of this task characteristic. Use of the biomechanical approach, however, has been limited to analysing infrequent tasks. Table 2.15 lists major studies which have investigated the effects of task pace on MMH capability, manual lifting in particular.

For a given task, the maximum weight of lift acceptable to an individual decreases non-linearly with an increase in pace (e.g., Garg and Saxena, 1979; Snook, 1978a, b; Ayoub et al., 1980b, 1983b; Ciriello and Snook, 1983; Mital and Manivasagan, 1983a; Asfour et al., 1984a, b; Foreman et al., 1984; Karwowski and Ayoub, 1984a, b; Mital, 1984a, b; Khalil et al., 1985). The metabolic energy expenditure rate also increases with an increase in the task pace (e.g., Garg, 1976; Ayoub, 1977; Garg and Saxena, 1979; Petrofsky and Lind, 1978a; Nag et al., 1979; Sanchez et al., 1979; Ayoub et al., 1980b; Asfour et al.,1984a, b; Mital, 1984a, b). Even though the maximum acceptable weight of lift decreases with pace, the total acceptable workload (kg-m/min), as several researchers have stated, increases (e.g., Mital et al., 1978; Mital and Ayoub, 1981a; Garg and Saxena, 1979; Mital and Okolie, 1982; Mital and Asfour, 1983). These studies also point out that, with respect to physical fatigue, task frequency is relatively more important than the characteristics of the load handled. Thus, the net effect is an increase in task demand with pace.

Recently, several predictive models have been developed that determine the effect of task pace on an individual's MMH capability. These have been based on either the psychophysical approach (e.g., Poulsen, 1970, 1971, 1981; Mital et al., 1978, 1984; Mital and Ayoub, 1980, 1986; Mital, 1983a, b) or the physiological approach (e.g. Mital et al., 1984; Garg, 1976; Asfour, 1980; Mital and Ayoub, 1981b). These models are discussed in more detail in Chapters 3 and 6.

Ayoub et al. (1980b) have developed various limiting combinations of load and frequency for manual lifting tasks based on subjective estimates of acceptable weight. Garg and Ayoub (1980) found that for high load (more than 15 kg), low frequency (less than 6 lifts/min) tasks, the physiological fatigue criterion (based on a limit of 5 kcal/min metabolic energy expenditure rate) appears to be more conservative than the psychophysical measure. In other words, the psychophysical approach would result in more liberal standards for low frequency, high load tasks, while the physiological approach will result in more liberal standards for high frequency, low load tasks (Figure 2.5).

Mital et al. (1984) and Mital (1983c) have suggested that the load frequency combinations determined by the psychophysical approach overestimate individual's lifting capabilities, and the weight estimates based on a trial period of 25–45 minutes cannot be sustained throughout the work shift. Mital and Shell (1984, 1985a) and Mital et al. (1984) also determined that the physiological fatigue criterion of 5 kcal/min of metabolic energy expenditure rate results in overexertion and excessive physiological fatigue. Based on psychophysical and metabolic energy expenditure rate data collected on 37 industrial male and 37 industrial female workers, Mital (1985a) plotted load–frequency combinations. Figures 2.6 and 2.7 show the relationships between load and frequency, based on the psychophysical and the physiological fatigue criteria, for males and females, respectively. Both the 33% and the 28–29% metabolic energy expenditure rate criteria (Mital, 1983c) were applied. At no point across the entire frequency range (1–12 lifts/min), did the physiological fatigue criteria provide conservative estimates. In addition, Mital (1985a) states that the psychophysical criterion provides a more accurate determination of MMH capabilities of individuals if it is corrected for the work duration (section 2.3.2).

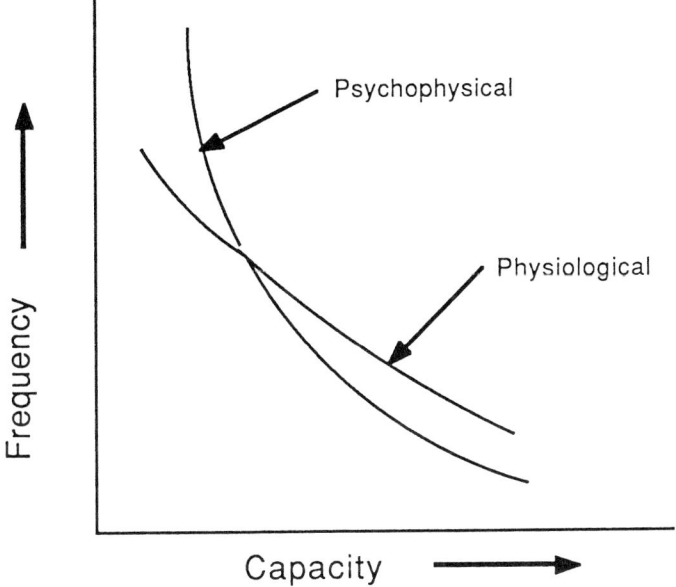

Figure 2.5. Psychophysical vs. physiological fatigue criterion *(based on Garg and Ayoub, 1980)*. From *Human Factors, Vol. 22* © *1980 by The Human Factors Society, Inc., and reproduced by permission*

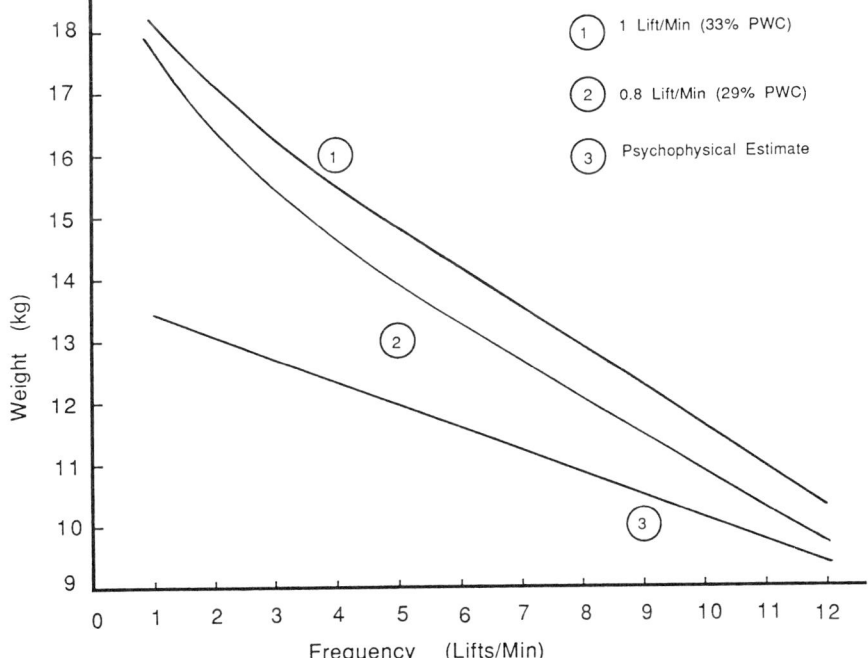

Figure 2.6. Relationship between weight of lift and lifting frequency based on psychophysical and physiological fatigue criteria (industrial males) (Mital, 1985a)

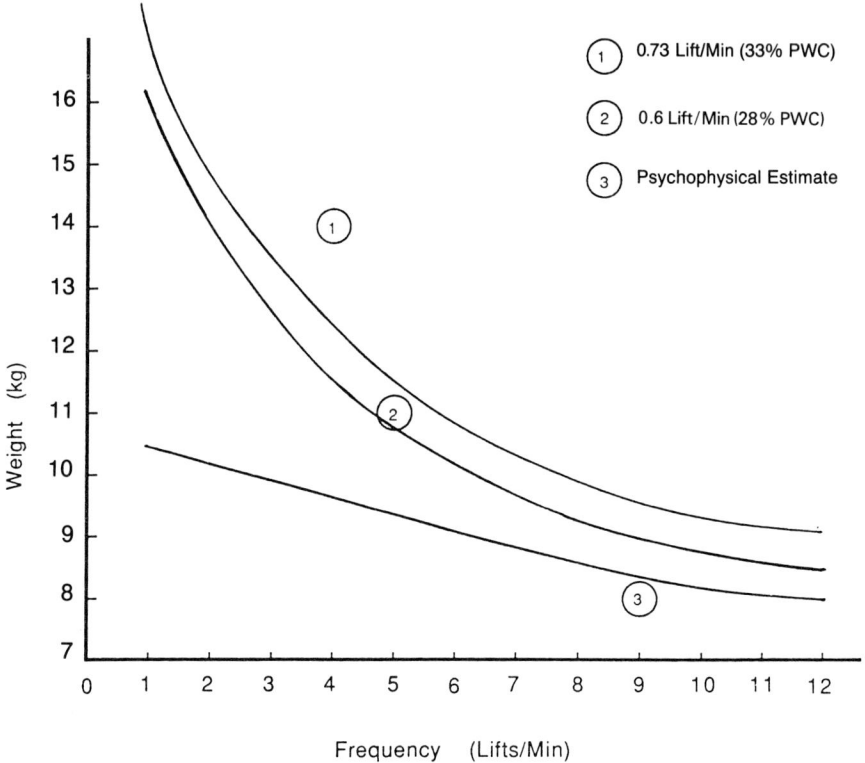

Figure 2.7. Relationship between weight of lift and lifting frequency based on psychophysical and physiological fatigue criteria (industrial females) (Mital, 1985a)

Maximum acceptable frequency of lift has been determined for one-handed (Garg and Saxena, 1982; Mital and Asfour, 1983) and two-handed (Snook and Ciriello, 1974a; Mital et al., 1985b) manual lifting tasks. The maximum acceptable frequency for one-handed manual lifting tasks has been determined to be 50% of the maximum frequency that can be sustained for a 4-min period. For two-handed lifting tasks, the pace most acceptable to workers has been found to be 4 lifts/min (Snook and Ciriello, 1974a). There is, however, some controversy about this number. In a simulation of palletizing and stacking tasks (Mital et al., 1985b), both males and females, when given a choice, selected a substantially higher frequency of lift (Table 2.16). The physiological responses, however, did indicate excessive metabolic energy expenditure rate both for males and females. When the pace was reduced and kept fixed (6 lifts/min), the physiological responses for males dropped to within acceptable levels. Females did not complain of excessive fatigue when lifting at the fixed pace, even though working at 42% of their aerobic capacity.

Table 2.16. Average ($\bar{x}$), standard deviation (S) and range (R) of responses at rest and at the steady state (Mital et al., 1985c)

Response	Machine-Paced			Self-Paced		
	$\bar{x}$	S	R	x	S	R
Resting heart rate (Beats/min)						
Males	72·7	8·6	60·0–90·0	72·7	8·6	60·0–90·0
Females	82·8	5·3	69·0–92·0	82·8	5·3	69·0–92·0
Resting oxygen uptake (kcal/min)						
Males	1·56	0·27	1·2–2·0	1·56	0·27	1·2–2·0
Females	1·67	0·47	0·9–2·5	1·67	0·47	0·9–2·5
Final heart rate						
Males	102·7	10·0	86·0–115·0	130·4	16·8	94·0–160·0
Females	122·3	8·3	104·0–132·0	134·0	8·4	121·0–147·0
Final oxygen uptake						
Males	4·2	0·5	3·4–5·8	8·6	1·7	5·7–12·2
Females	4·6	1·2	3·6–6·9	7·1	1·2	4·6–9·2
Final frequency						
Males	*	*	*	12·9	3·3	7·4–20·3
Females	*	*	*	9·7	3·4	6·1–18·1

* Fixed for machine pacing (6 containers/min).

2.3.2 Task Duration

How long a person can perform a given manual task depends on his endurance. Several studies have shown that as the task duration increases, the worker's energy expenditure level also increases, gradually (Bonjer, 1962; Deivanayagam and Ayoub, 1979; Garg, 1980). This increase may be due to the accumulative effect of the products of metabolism, changes in blood flow distribution, or deterioration in mechanical efficiency (Deivanayagam and Ayoub, 1979). There is also a decreased level of energy expenditure rate that can be maintained throughout the working time as the working time increases (Mital and Shell, 1984, 1985a).

Therefore, with increased task duration, either the task burden should be reduced, or appropriate rest allowances should be given. This is especially important since prolonged work days (more than 8h) are not uncommon in industry. Mital (1983c) determined that when given a choice, workers will reduce the weight of load with time. Karwowski and Yates (1986) arrived at similar conclusions. Figures 2.8 and 2.9 show how the weight declines with time for males and females, respectively (Mital, 1983c).

As evidenced by nearly constant heart rates, the workers reduced the weight to maintain the same levels of fatigue. If the weight is not reduced, the metabolic energy expenditure and heart rate will increase (Deivanayagam and Ayoub, 1979). An exception was noted by Snook (1985), who did not observe any reduction in load weight with a 4 hour work duration.

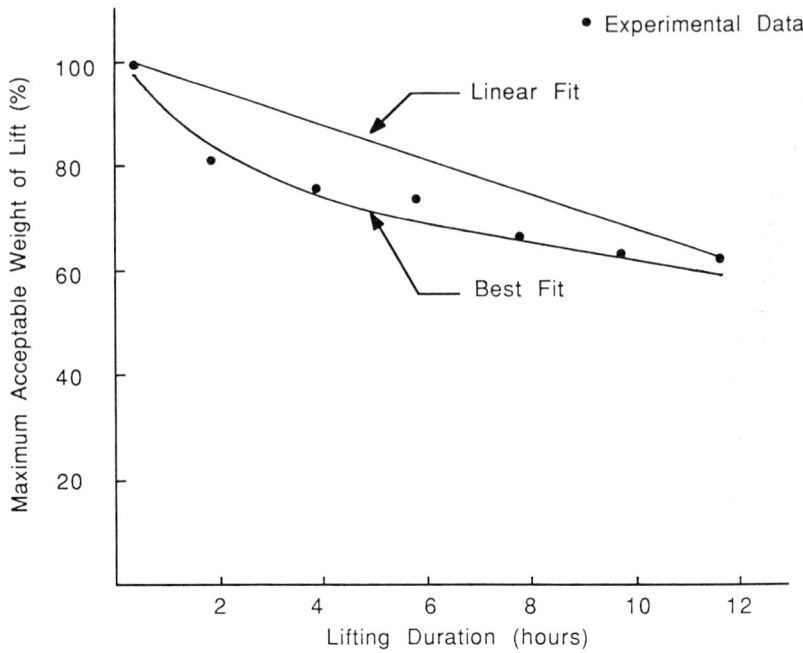

Figure 2.8. Change in maximum acceptable weight of lift with time for males (Mital, 1983c). From Human Factors, Vol. 25. © *1983 by* The Human Factors Society, Inc., *and reproduced by permission*

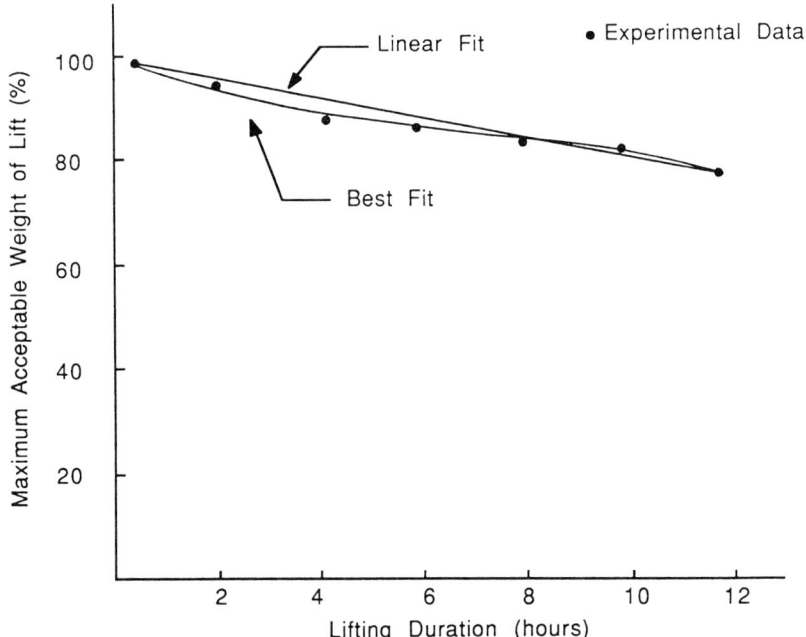

Figure 2.9. Change in maximum acceptable weight of lift with time for females (Mital, 1983c). From Human Factors, Vol. 25 © *1983 by* The Human Factors Society, Inc., *and reproduced by permission*

2.3.3 Object Size

The size (bulk) of the object to be handled has been considered in only a few studies (Table 2.15). From these, it appears that object size does make a difference depending on which measurement criterion (psychophysical, physiological or biomechanical) is used. Mital and Ayoub (1981a) reported that object bulk has no effect on the metabolic energy expenditure rate of an individual for lifting and lowering activities. Also, for manual carrying activities, Morrissey and Liou (1984) reported no significant change in either the metabolic energy expenditure rate or the heart rate. Haisman et al. (1972), on the other hand, reported an increase in metabolic cost as the size of a cart being pushed increases.

Garg and Saxena (1980) found that subjective estimates of the maximum acceptable weight of lift decrease with an increase in volume for rectangular containers (non-collapsible type), and increase with an increase in volume for bag type (collapsible) containers (also see section 2.3.4). This, however, was not found true for one-handed carrying of non-collapsible containers (Mital and Manivasagan, 1983b) in which the maximum acceptable weight of lift increased significantly with an increase in volume, while the heart rate did not change appreciably.

With regard to container dimensions, it has been found that the height of the container (object) does not affect the metabolic energy expended in lifting it (Mital and Ayoub, 1981a). Nevertheless, practical considerations, such as arm length and vision obstruction, should be used to determine container height.

Container width (dimension in the frontal plane) and length (dimension in the sagittal plane) do influence the lifting capability of an individual. This is reflected in terms of lower acceptable weights of lift (Garg and Saxena, 1980; Frievalds et al., 1984; Kromodihardjo and Mital, 1985; Mital, 1985c d; Mital, 1986b; Mital and Kromodihardjo, 1986), increased metabolic energy expenditure rate (Mital and Ayoub, 1981a; Mital, 1985; Mital, 1986b), and higher spinal stresses (Tichauer, 1971; Frievalds et al., 1984; Kromodihardjo and Mital, 1985; Mital and Kromodihardjo, 1986).

Figure 2.10 shows the change in oxygen uptake for given container dimensions. According to this, the container dimension in the sagittal plane should not exceed 50 cm. Any increase in container volume should be accomplished by first increasing its height, then its length (up to a limit of 50 cm) and finally its width. Biomechanically, however, the length of the container should be as small as possible (Tichauer, 1971; Frievalds et al., 1984; Kromodihardjo and Mital, 1985; Mital and Kromodihardjo, 1986). As the length increases, compressive and shear forces on the spine increase as well. Table 2.17 shows the relationship between the maximum acceptable weight of lift and the resulting compressive, shear and ground reaction forces (Mital and Kromodihardjo, 1986). These results are similar to those obtained by Frievalds et al. (1984). In this study, the lack of difference in biomechanical responses is explained by the fact that the subject reduced the weight as the container length increased. The reason for the weight reduction was, apparently, an attempt to maintain the same levels of spinal stresses. It is also noted that lifting strength decreases with the horizontal distance (Martin and Chaffin, 1972; also see section 2.3.8).

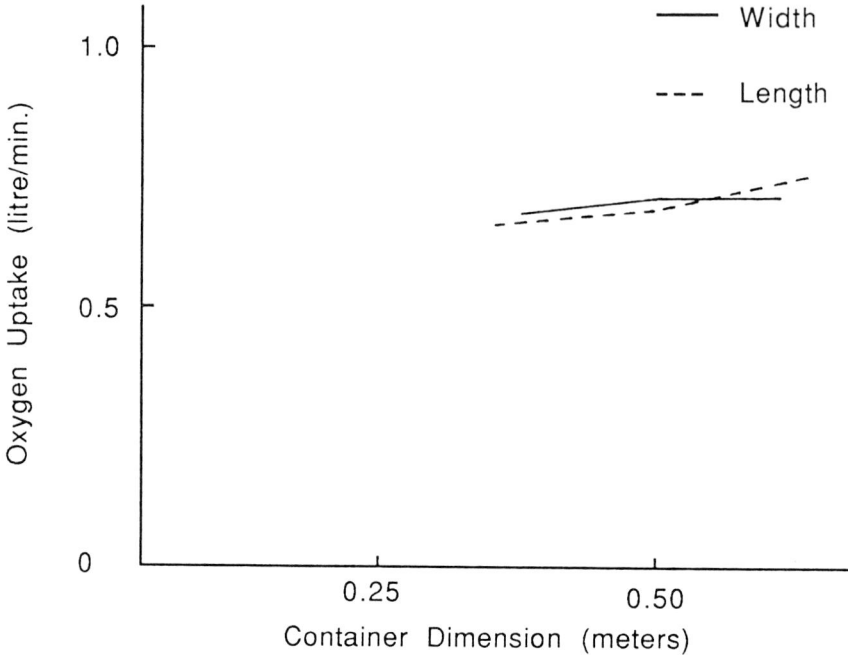

Figure 2.10. Change in oxygen uptake with variations in box width and box length (lifting) (Mital and Ayoub, 1981a). Reprinted with permission of American Industrial Hygiene Association Journal

Table 2.17. Change in psychophysical and biomechanical responses with box length (Mital and Kromodihardjo, 1984)

Box length (cm)	Maximum acceptable weight of lift (N)	Compressive force (N)	Shear force (N)	Ground reaction force (N)
30·48	297	3952	477	1145
45·72	284	4181	484	1085
60·96	249	4245	480	1055

It is clear that large size objects lend themselves to awkward and hazardous lifting. Figure 2.11 contains a set of curves showing the compressive force resulting from varying the distance of the load from the spine using back lift and leg lift techniques.

Tichauer (1973) derived an equation to express the relationship between the weight of an object and its moment arm to the lumbar spine in producing a biomechanical equivalent moment, ML, assuming body thickness to be 8 in.

$$ML = W\frac{L}{2} + 8''$$

Where W = weight of load in lbs; L = length of container midsaggital plane; ML = moment at lumbar spine. Table 2.18 shows the ML values produced by different loads with various lengths using the above equation.

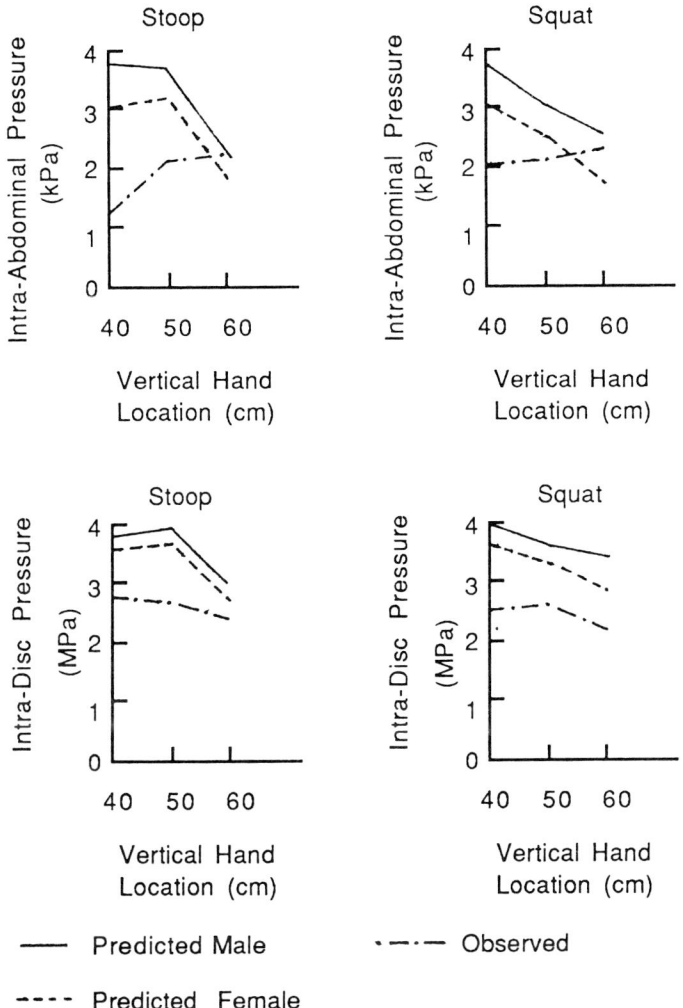

Figure 2.11. Comparison of observed and predicted intra-abdominal and intra-discal pressures vs. hand location (*Andersson* et al., *1985*)

Table 2.18. Static holding task

True weight (lbs)	Object length (in)	Approximate moment at lumbar spine (lbs-in)
5	7·5	60
6·4	30·0	150
10·0	7·5	120
14·75	7·5	175
12·3	30·0	285
17·0	30·0	380

As Table 2.18 indicates, a lighter load can create a larger lumbar moment than a heavier load with a shorter object length. From a handling ease and safety viewpoint, it is suggested that containers or objects be as small and compact as practically possible. Bulky or irregularly shaped objects should be handled mechanically.

2.3.4 Object Shape

Studies investigating the effect of container shape are relatively few (Haisman et al., 1972; Garg and Saxena, 1980; Mital and Okolie, 1982; Mital and Manivasagan, 1983b; Smith and Jiang, 1984). Haisman et al. (1972) reported that the metabolic energy expenditure of individuals is influenced by the shape of the hand cart they push. With regard to maximum acceptable weight of lift, Garg and Saxena (1980) reported that a greater amount of weight can be lifted using bag containers (collapsible) than using boxes (non-collapsible). Smith and Jiang (1984) not only confirmed these findings, but also found that this greater amount of load is lifted at only a slightly higher physiological cost (Table 2.19).

Table 2.19. Comparison of the maximum acceptable weight of lift between collapsible (bag) and non-collapsible (box) containers (Smith and Jiang, 1984)

Container	Load (kg)	Oxygen uptake (l/min)	Heart rate (beats/min)
Bag	24·26	1·50	129·7
Box	22·05	1·45	121·3

Figure 2.12. Shapes of containers used in one-handed carrying tasks (Mital and Manivasagan, 1983b)

More specifically, Mital and Okolie (1982) reported that 18% more weight is handled in collapsible containers than in non-collapsible containers. They also found that the shape of non-collapsible containers does not influence the maximum weight of liquid acceptable for carrying activities. Although the weight of liquid carried ranged from 24·4–24·8 kg, subjects indicated their preference for carrying liquid in cylindrical containers, followed by barrel and rectangular shapes.

The finding that the shape of non-collapsible containers does not influence carrying capabilities (Mital and Okolie, 1982), however, was not confirmed by Mital and Manivasagan (1983b) for one-handed carrying tasks. Using four different container shapes (Figure 2.12), they determined the maximum acceptable weight of carry. Subjects selected weights that, on the average, resulted in a heart rate of 100 beats/min. Table 2.20 shows the acceptable weight of carry and the corresponding pulse rate for the four container shapes which are approximately the same volume. Figure 2.12 shows the four container shapes.

Table 2.20. *Acceptable weight of carry and pulse for containers of various shapes (Figure 2.12) used in one-handed tasks (Mital and Manivasagan, 1983b)*

Shape	Load (kg)	Pulse (beats/min)
Plastic bucket	9·67	99·13
Galvanized iron bucket	9·12	101·15
Tool box	10·66	100·81
Radiator can	10·16	98·66

It is evident from this limited number of studies that there is no general agreement regarding the effect of container shape on MMH capabilities. What is clear, however, is that collapsible containers (bags) are easier to handle probably because their functional dimensions change during handling due to their flexibility.

2.3.5 Couplings

These are handles used to grip objects or shoes worn to prevent slipping. Figure 2.13 shows these two couplings in a MMH system.

All three approaches — psychophysical, physiological and biomechanical — have been used to study this variable in manual lifting tasks, and the findings are consistent. That is, handling loads with handles is safer and less stressful. According to Bakken (1983), lifting loads in boxes without handles results in about 15% decrease in the maximum acceptable weight of lift. Garg and Saxena (1980) confirmed this when they observed a decline, ranging from 4–11·5%, in the maximum weights acceptable to their subjects for lifting boxes without handles.

Mital and Ayoub (1981a) reported that lifting boxes without handles requires additional metabolic energy (0·17 kcal/min). Although Frievalds *et al.* (1984), Kromodihardjo and Mital (1985) and Mital and Kromodihardjo (1986) found that lifting containers without handles results in lower compressive forces, this does not suggest that handling containers without handles is safer. On the contrary, individuals are much more cautious of the hazards when lifting loads without handles and,

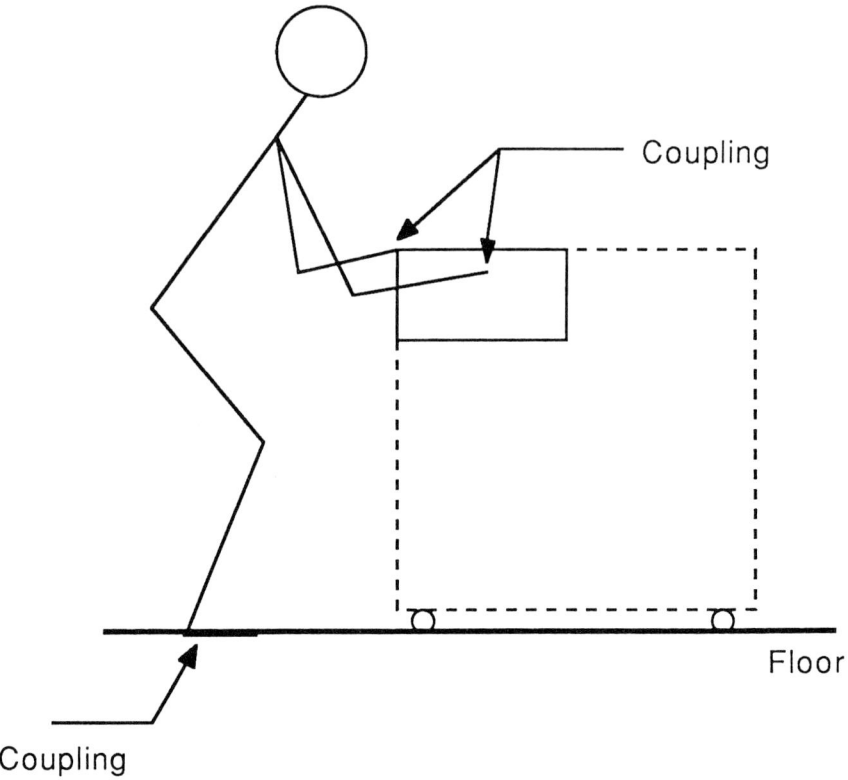

Figure 2.13. *Coupling between the worker and the object and floor*

therefore, if given a choice, choose a lower weight. The reduction in compression, shear and ground reaction forces, in the case of no handles, is primarily due to the fact that individuals lift less weight. When compression and shear forces are normalized by the weight lifted, this becomes very clear.

Having determined that handles are warranted, at least in manual lifting activities, the next step is to determine their optimal location and shape. In terms of shape, many of the studies that have been conducted relate to hand-tools rather MMH activities in which vertical loads are supported (Drury, 1980). Container designs and warehousing needs do not permit handles other than the cut-out type. Handles, 115 mm long and 25–38 mm wide (or in diameter, in the case of cylindrical handles for carts, etc.), with 30–50 mm clearance all round are favoured (Drury, 1980; Coury and Drury, 1982; Drury and Pizatella, 1983). The angle of the handle should be 0°–10° ulnar (Drury *et al.*, 1985).

As observed by Mital and Ayoub (1981a) and Drury *et al.* (1982), in most cases when handling containers without handles, the containers are held at diagonally opposite ends. The heart rate and rating of perceived exertion (RPE) values are lowest when one hand is under the load, centrally, and the other on the leading edge, at least half way up. Combinations of these positions (hands at diagonally opposite ends, one hand under the load in the middle and the other on the top of the leading edge, or one hand under the load in the middle and the other half way on the leading edge) are also feasible and permit both horizontal and vertical stability (Drury and Pizatella, 1983).

Slipping can start whenever kinetic friction (friction during walking) is less than 0·3 (ratio of horizontal to vertical force) (Perkins and Wilson, 1983). James (1983) has recommended similar values (0·24–0·39). The necessity to shorten stride on a slippery floor has also been demonstrated. In general, since slippery floors have a low coefficient of friction, the step length should be reduced to arrest or prevent slipping. Table 2.21 shows the relationship between step length and the coefficient of friction that must be maintained to avoid slipping.

Table 2.21. Relationship between the coefficient of friction and step length (James, 1983)

Coefficient of friction	Step length (cm)
0	0
0·087	15
0·18	30
0·27	45
0·36	59
0·47	73
0·58	86

To determine floor conditions, the portable skid tester designed by the Road Research Laboratory (Giles et al., 1964) could be used. Table 2.22 shows the relationship between the skid tester reading and floor condition (Greater London Council, 1971).

Table 2.22. Safe limits using the skid resistance tester (Greater London Council, 1971)

Skid resistance reading	Floor condition
Below 19	Dangerous
20–39	Marginal
40–74	Satisfactory
Above 75	Excellent

In order to avoid slipping, it is necessary to maintain a strong coupling between the individual and the floor (a static coefficient of friction of at least 0·5). This is primarily accomplished by wearing shoes. While sole patterns, heels and profiled floors help in gripping the floor, no shoe or floor material can provide complete protection against slipping accidents if mud, oil or other liquids are present on the floor and/or the shoes. Water spillage relief surfaces should be used to reduce the chance of slipping (Harrison and Malkin, 1983). Floor design and material should allow foot comfort and easy cleaning. In the case of hard floors, shoe materials should be selected to provide comfort. Table 2.23 provides general guidelines for floor and shoe material.

Table 2.23. Guidelines for floor and shoe material (National Safety News, 1974)

Floor		Surface Condition of the Floor		
		Dry	Wet	Greasy
Ceramic tile	Recommended	Plastic (vinyl) and neoprene Supple rubber (white non-hardened), preferably with shallow tread	Plastic (dense vinyl) and neoprene Non-crepe and non-hardened rubber, preferably with pronounced tread Leathers	Plastic (except dense vinyl) Vegetable tanned leather
	Not recommended	Leathers (especially vegetable tanned) Hardened rubber Crepe rubber	Hardened rubber Crepe rubber	Plastic (dense vinyl only) and neoprene Leathers (expect that advised above) All types of rubbers
Synthetic silcrete slab and corundum concrete slab	Recommended	Plastic (vinyl) and neoprene Supple rubber (white, non-hardened)	Plastic (dense vinyl) and neoprene Non-crepe and non-hardened rubber preferably with pronounced tread Leathers	Plastic (dense vinyl) Hardened rubber Leathers
	Not recommended	Leathers (especially vegetable tanned) Hardened rubber Crepe rubber	Hardened rubber Crepe rubber	Non-hardened rubber Neoprene
Concrete	Recommended	Plastic (vinyl) and neoprene Supple rubber (white, non-hardened)	Plastic (dense vinyl) and neoprene Non-crepe rubber and non-hardened rubber, preferably with pronounced tread Leathers	Plastic (dense vinyl) Hardened rubber Leathers

Variables in Manual Materials Handling

Surface					
Roughened concrete	Not recommended	Leathers (especially vegetable tanned) Hardened rubber Crepe rubber	Hardened rubber Crepe rubber Supple rubber	Supple or non-hardened rubber Neoprene	
	Recommended	Plastic (vinyl) Supple rubber (white, non-hardened) Crepe rubber Chrome retanned leather	Plastic (dense vinyl) and neoprene Non-crepe and non-hardened rubber, preferably with pronounced tread Crepe rubber Leathers	Plastic (dense vinyl) Hardened rubber Leathers	
Wood	Not recommended	Vegetable tanned leather Hardened rubber	Hardened rubber	Non-hardened rubber Neoprene	
	Recommended	Plastic (vinyl) and neoprene Supple rubber (white, non-hardened), preferably with shallow tread	Plastic (dense vinyl) and neoprene Non-crepe and non-hardened rubber, preferably with pronounced tread Leathers	Hardened rubber Leather	
	Not recommended	Leathers (especially vegetable tanned) Hardened rubber Crepe rubber	Hardened rubber Crepe rubber	Plastic (dense vinyl) and neoprene Non-hardened rubber	

2.3.6 Object Weight/Force Application (Exertion)

The magnitude of weight that can be handled or the force that can be exerted, up to a large extent, determines the MMH capability of an individual. Table 2.14 lists the major studies that have investigated this task characteristic. Recommendations range from data norms to elaborate prediction models for infrequent and repetitive tasks. The recommendations for infrequent MMH tasks are based mainly on the biomechanical and the psychophysical approaches. For repetitive MMH tasks, recommendations are based on the physiological and the psychophysical fatigue criteria. The various weight/force data norms are reviewed extensively in Chapter 4. The state of the art recommendations are also made. Models for predicting individual's MMH capability, for either frequent or infrequent tasks, are discussed in Chapters 3 and 6.

2.3.7 Load Distribution and Stability

Load distribution and stability affect the location of the centre of gravity in handling asymmetrical objects or containers filled with liquids. To date, very little research has been done on the effects of load distribution and stability on worker's MMH capability.

When the load centre of gravity falls along the side of the body (asymmetric load), it causes a lateral bending moment on the lumbar column resulting in rotation of each vertebra on its adjacent vertebra. Torsional resistance to such rotation is generally provided by the spinal column discs. Weakening or degeneration of a disc would most likely result in a higher magnitude of strain at the lumbar column. Torsional stress due to asymmetric loading reduces the maximum acceptable weight of lift or carry (Ayoub *et al.*, 1979; Mital and Okolie, 1982; Mital and Ilango, 1983; Mital and Manivasagan, 1983a; Mital, 1986c; Mital and Fard, 1986). Ayoub *et al.* (1979) have shown that a lateral offset of about 10 cm in the centre of gravity decreases lifting capability by up to 10%. Mital and Manivasagan (1983a) investigated three different centre of gravity locations: 0, 12·7 and 25·4 cm from the midsagittal plane in the frontal plane. The maximum acceptable weight of lift decreased by 3·1% when the offset was 12·7 cm. A further decrease of 4·5% was observed when the offset was increased from 12·7 to 25·4 cm. Mital and Fard (1986) also noticed a decrease of 4% in lifting capability when the centre of gravity offset was 10·16 cm. The lifting capability declined by 10·45% when the offset was 20·32 cm. No changes in physiological costs, however, were detected (Table 2.24). This indicated that subjects reduced the maximum acceptable weight of lift, due to increasing torsional stresses, to maintain the same physiological burden.

It was further observed that lifting an asymmetrical load with 20·32 cm centre of gravity offset was as demanding as lifting a symmetrical, but considerably longer box (Table 2.24).

If the load is positioned such that the worker must reach for it, the stresses on the spine and on some particular muscle groups are increased. There was an effect on the electromyographic activity (EMG) level of the deltoid muscle when reach was at a maximum (Habes *et al.*, 1985). Such situations as reaching over obstacles or across a table surface can produce large compression forces even when lifting small loads.

Table 2.24. *Maximum acceptable weights of lift and resulting physiological responses when lifting symmetrical and asymmetrical containers (Mital and Fard, 1986)*

Box (cm) (length × width)	Weight (kg)	Oxygen uptake (l/min)	Heart rate (beats/min)
60·96 × 30·48 (symmetrical)	15·94*	0·69	105·18
30·48 × 35·56 (symmetrical)	17·79	0·64	102·23
30·48 × 35·56 (10·16 cm offset)	17·08	0·64	103·4
30·48 × 35·56 (20·32 cm offset)	15.93*	0·65	102·76

All heart rates and oxygen uptake values are not different from each other ($P \geq 0·10$).
* Not different at the 10%, or greater, level of significance.

The height of the lift also affects the stress on the spine. Lifting from the floor is more stressful than lifting from table height. Data shows that the greatest compressive and shear forces occur during the first few fractions of a second of a lift from the floor (Mital and Kromodihadjo, 1986; Figures 2.14 (a) and (b)). This is true for both symmetrical and asymmetrical lifting.

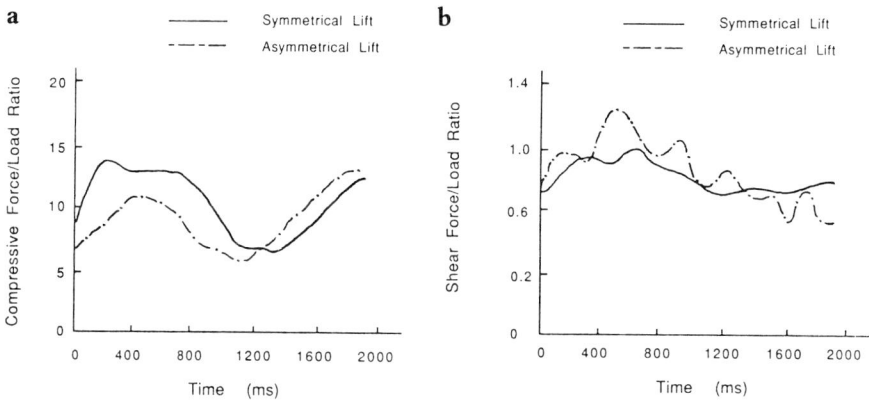

Figure 2.14. *Time history of (a) normalized compressive force and (b) normalized shear force for symmetric and asymmetric lift (Mital and Kromodihardjo, 1986)*

For manual carrying activities also, the capability declined proportionately when the centre of gravity was offset (Mital and Ilango, 1983; Mital, 1986c). The centre of gravity is constantly offset when handling liquids.

When handling liquids, the problem of stability is less acute if the container is full, but becomes serious when a partially filled container is handled. Then, when the container is moved, the liquid inside also moves resulting in a continuously changing centre of gravity. This moving centre of gravity results in high–low torques at the couplings and torsional stresses at the lumbar column. Mital and Okolie (1982) concluded that for liquid carrying activities, capacity may be reduced by as much as 31% due to the continuously changing centre of gravity. They also found that restricting the movement of liquid in the container, by using partitions, significantly enhances the carrying capability, by up to 10%.

2.3.8 Vertical Lift Height

It has been shown that MMH capability is a function of vertical distance of lift. An individual's lifting capability decreases with the vertical height or vertical distance (e.g., Martin and Chaffin, 1972; Garg and Chaffin, 1975; Ayoub et al., 1978; Mital et al., 1978; Snook, 1978a, b; Ciriello and Snook, 1983; Asfour et al., 1984a, b; Mital, 1984a, b; Asfour et al., 1985). The reduction in lifting capability may be as much as 30% (Ayoub et al., 1978) as the vertical height of lift increases from 0·76–1·65 m. Similar results have been obtained by others (Ciriello and Snook, 1983; Snook, 1978a, b; Mital, 1984a, b; Mital et al., 1984). The metabolic energy expenditure rate increases if the weight lifted across various vertical distances remains unchanged (Garg, 1976; Mital and Ayoub, 1981a). The lifting strength, as predicted by Martin and Chaffin (1972), either first decreases then increases, or first increases then decreases, depending upon the horizontal distance between the ankles and load. Garg and Chaffin (1975) have indicated that the overall lifting strength is maximum when the elbow angle is approximately 60°. The strength declines for lower or higher elbow angles.

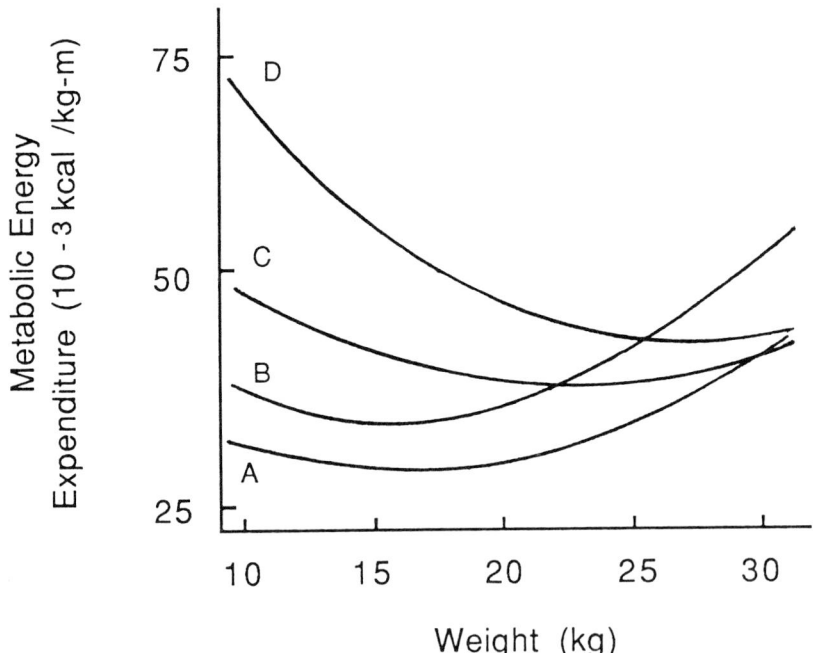

Figure 2.15. The relationship between lifting height (A-100 to 150 cm, B-150 to 200 cm, C-50 to 100 cm, D-0 to 50 cm), weight and metabolic energy expenditure (Frederick, 1959)

Mital and Okolie (1982), Snook (1978a, b) and Ciriello and Snook (1983) observed that for manual carrying activities, the capability increases as the carrying height decreases; more weight is carried at the knuckle height compared to at the elbow height. This seems to partially contradict strength simulation results (Martin and Chaffin, 1972; Garg and Chaffin, 1975), but supports the strength data of Chaffin (1974).

The starting point of lift or lower is also important. For the same vertical distance, lifting capability can decrease with a change in the starting point of lift. The decrease in capacity may be as large as 23% when the starting point of lift is moved from the knuckle to shoulder height.

According to Frederick (1959), the height range between 1 and 1·5 m is least expensive in terms of metabolic energy expenditure (Figure 2.15).

For pushing and pulling activities, the height of force application influences worker capabilities somewhat differently. In general, the force exertion capability of individuals increases with an increase of the height for pushing activities. This trend is reversed for pulling activities (see Chapter 5 for details).

2.3.9 Workplace Geometry

It has been suggested that workplace geometry may contribute to the hazards of MMH, and, while the practical effects of restrictions in working space on posture, movement and working capacity are generally recognized, little specific information is available relating these hazards to MMH capabilities. Drury (1985) lists a number of changes in the working patterns imposed by the lack of adequate space and the need for increased accuracy of placement.

Ridd (1983) determined that intra-abdominal pressures are generally increased by frontal and lateral barriers as well as by restricted head room during lifting. Postural constraints created by relative position of hands and feet, or dictated by the workplace layout, increase the perceived discomfort ratings (Corlett and Bishop, 1976; Corlett et al., 1979; Boussenna et al., 1982).

The results of Mital (1986c) indicate that the effect of restricted space on manual carrying capabilities is perhaps more profound than the effect of any other task characteristic. Both males and females suffered a 13% decline in their carrying capabilities when they carried loads through a 56-cm wide passage. In cases of load carriage through open and confined spaces, heart rates were not different from each other. Subjects were observed reorienting the load and the body while carrying loads through the narrow passage. This reorientation and postural instability resulted in slower and cautious movement, and was the reason for lower capabilities. Additional body and load movements, however, compensated for the decline in heart rate due to the lower amount of weight carried. Table 2.15 shows the average weight and heart rate for load carriage through open and confined spaces.

Loading, unloading and maintenance are some common activities performed in confined spaces. Since no other specific information is currently available that relates potential workplace geometry hazards to an individual's capabilities, educating employees in safe procedures appears to be a logical recourse.

Table 2.25. Effect of space restriction on manual carrying capabilities (Mital, 1986c)

Space	Sex	Weight (kg)	Heart rate (beats/min)
Open	Males	28·7	96·7
	Females	18·6	95·8
Confined	Males	24·9	96·0
	Females	16·1	95·7

2.3.10 Distance Travelled/Slope and Traction

Distance travelled/slope and traction are relevant mainly to pushing, pulling and carrying activities. Research using the psychophysical approach has shown that although the maximum acceptable weight of carry decreases with distance travelled, the acceptable work-rate (kg-m/min) increases (Snook, 1978a, b; Mital and Okolie, 1982; Ciriello and Snook, 1983). The metabolic energy expenditure rate shows an increase with the work-rate for carrying/walking activities (Datta *et al.*, 1978; Soule *et al.*, 1978).

For one-handed carrying activities, the acceptable weight of carry decreases with distance (Mital and Manivasagan, 1983b). This decrease may be as much as 16% when the carrying distance increases from 30·48–91·44 m.

The metabolic energy expenditure rate also increases with speed (Morrissey and Liou, 1984) and grade (Goldman and Iampietro, 1962; Yousef *et al.*, 1972; Pimental and Pandolf, 1979; Gordon *et al.*, 1983; Ilmarinen and Louhevaara, 1984). Walking downgrade with loads is less expensive than walking upgrade with loads (Yousef *et al.*, 1972; Pimental and Pandolf, 1979). The heart rate also increases with an increase of slope (Grieve, 1979). The increase in metabolic energy expenditure in physiological studies of manual carrying activities appears to indicate that a considerable proportion of the increase is either due to the load itself or the walking speed. However, the exact proportion is difficult to determine in many studies, the effect of distance is confounded either with the load or the carrying speed (Datta *et al.*, 1978; Soule *et al.*, 1978; Morrissey and Liou, 1984).

The force exertion or walking capability is known to be affected by the terrain. It makes a difference if the handling is performed on soft snow, hard concrete or grass, for example. The coefficient of friction changes with terrain (Kroemer, 1969, 1974; Pimental and Pandolf, 1979; Grieve, 1979) and so does the metabolic cost (Pimental and Pandolf, 1979). It is not known, however, how much the metabolic energy expenditure rate changes with terrain (also see section 2.3.5).

2.3.11 Posture/Technique

Posture is defined as the configuration the body assumes to initiate an activity. Different material handling activities require different body postures. The body, however, may assume different configurations for the same activity. For instance, loads can be lifted in stoop (straight legs), squat (straight back) or free-style (semi-squat) postures.

The state of the art suggests that for manual lifting tasks, the free-style posture is the best. Between stoop and squat postures, squat posture is biomechanically less stressful (Garg and Herrin, 1979; Leskinen *et al.* 1983a, b), but the stoop posture leads to lower metabolic energy expenditure (Garg and Herrin, 1979; Kumar and Magee, 1982; Kumar, 1984). Among all the three postures, the free-style posture is considered least stressful (Brown, 1976; Garg and Saxena, 1979; Kumar and Magee, 1982; Kumar, 1984) or least tiring. Metabolically, however, there is some contradiction. According to Brown (1976) and Garg and Saxena (1979), the free-style method of lifting is least expensive. Kumar (1980), however, found the stoop method of lifting to be less expensive.

When considering compression on the lumbar spine during the performance of a lifting task, the recommended posture for lifting has traditionally been the squat or the 'leg lift'. The stoop or 'back lift' approach has been shunned as it is believed to present greater risk of injury during lifting. In general, this traditional view of lifting posture is appropriate, although some important exceptions exist.

Data which supports the squat posture as safer have used loads which can be lifted between the knees, which is the most common type of industrial lifting. Some loads are simply too large to fit between the knees and cannot be lifted in the squat posture. The leg lift, in this case, will actually cause an increase in stress since the worker must reach in front of the knees to grasp the load, thereby increasing the distance from the spine to the load. Another problem with the squat posture is the strength the leg muscles require. The squat posture requires the worker to lift the weight of the load and his own body on each lift. Many workers simply do not have the strength in their legs to lift their body-weight plus the load. Therefore, the squat posture may reduce the compressive stress on the spine but increase the stress to other parts of the musculoskeletal system.

The stoop posture places greater stress on the back muscles, which is the prime force in producing compression on the spine. However, for loads too large to pass between the knees, the stoop posture can reduce the spine-to-load distance by allowing the worker to place the trunk over the load (Figure 2.16). But in such a lifting posture, the arms might have to be extended in order to reach the load. Thus, the load moment arm on the spine is reduced, but a high torque will be produced on the shoulder and might hyperflex the lumbar column, causing another type of musculoskeletal problem. The back lift has the advantage of not requiring the worker to lift the weight of his own body when performing a lift. Figure 2.16 displays the difference between a stoop posture and squat posture that produces less compression on the spine. Garg and Herrin (1979), in studying these lifting postures, reported that the compressive force on the lumbosacral disc is equal to, or slightly lower, for loads greater than 5 kg when the load is lifted using the stoop posture.

Some loads which must be lifted are too large to fit between the knees and too tall to permit bending over them. Neither posture is ideal in this case; the stress to the spine will be large either way. Therefore, rather than advocating a specific posture, any posture which minimizes the spine-to-load distance while not requiring more strength than the worker can produce is probably the optimum posture.

When trunk rotations are involved in the body posture during the lifting task, the

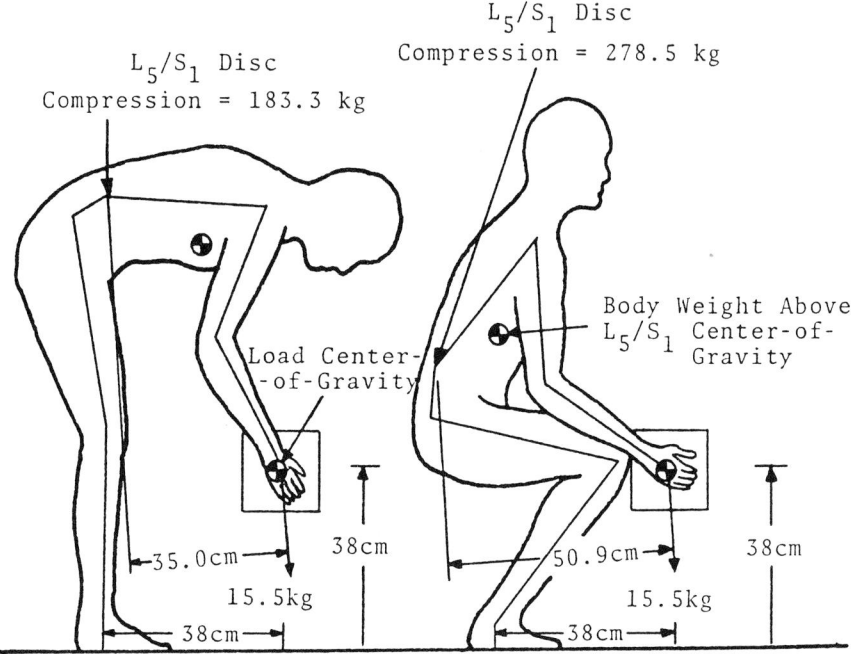

Figure 2.16. Low-back compression associated with two lifting postures (Park, 1973)

intra-abdominal pressure (IAP) and intradiscal pressure parallel increases of angle of rotation. Anderson et al. (1985) developed a biomechanical model to predict the lumbar load including compression, moment load and stress and strain of spinal connective tissues. In the sensitivity analysis, they compared the predicted values of IAP and disc pressure versus the degree of trunk rotation from the reported model with the observed values from Andersson et al. (1976). The results are shown in Figure 2.17.

It is clear that biomechanically the squat posture is generally favoured, while subjective ratings of fatigue and metabolic energy expenditure favour the free-style method of lifting.

Between static (holding) and dynamic conditions, it is rather clear that endurance for static conditions is greater as reflected by torque exertion (Marras et al., 1984) and holding time (Evans et al., 1983). Kumar and Davis (1983), however, did not find any differences in the intra-abdominal pressure between the static and dynamic postures. This is contradictory to the findings of Marras et al. (1984).

The physiological costs are affected by non-erect postures. Morrissey et al. (1983), for example, have shown that a 60% erect posture is physiologically more demanding than a normally erect or 80% erect posture (Table 2.26). For seated postures, lumbar and/or thoracic support results in lower values of intra-abdominal pressure (Boudrifa and Davis, 1984).

The force exertion capability is also affected by posture (Ayoub and McDaniel, 1974; Davis and Stubbs, 1980; Chaffin et al., 1983). For pushing and pulling activities, the

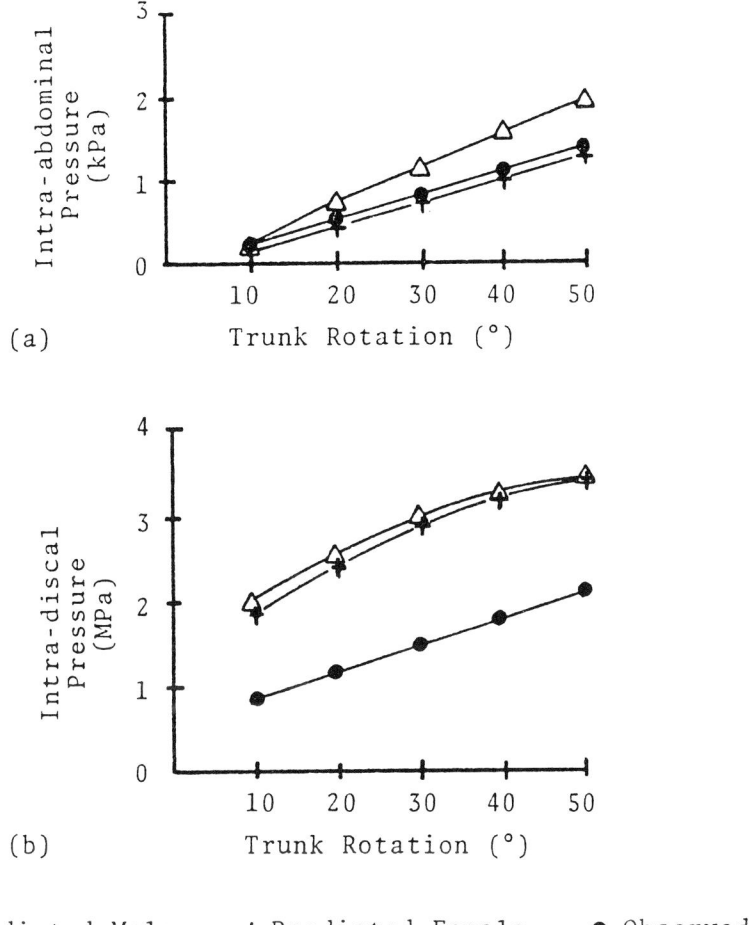

Figure 2.17. *The effects of trunk rotation on (a) IAP and (b) intra-discal pressure for a 200 N load. Observed values are from Andersson et al. (1976), predicted values are from Andersson et al. (1985). (Adapted from Andersson et al., 1985)*

Table 2.26. *Physiological responses to shovelling in different postures (Morrissey et al., 1983)*

Posture	Heart rate (beats/min)	Oxygen uptake (l/min)	Ventilation volume (l/min)
Normal erect	127·6	1·38	50·9
80% of erect	131·1	1·47	56·0
60% of erect	139·4	1·56	67·4
Kneeling posture	128·4	1·29	57·1

force exertion is affected by handle location, foot placement, and pushing/pulling height (Ayoub and McDaniel; 1974; Chaffin *et al.*, 1983).

With regard to carrying activities, carrying load on the head is least expensive in terms of metabolic cost (Soule and Goldman, 1969; Datta et al., 1978). Carrying loads by hand in a suitcase grip, however, is quite demanding.

For tasks that must be carried out manually, but are beyond the capability of one operator, load sharing or use of multiple workers should be considered. If using bi-manual lifting, leg-lifting (squat) posture should be employed (Bendix and Eid, 1983).

2.3.12 Asymmetrical Lifting/Carrying

Lifting or carrying asymmetrical objects asymmetrically is the rule rather than an exception in industrial settings, and yet only a few investigations dealing with this problem have been undertaken (Kumar, 1980, 1984; Kumar and Magee, 1982; Mital and Ilango, 1983; Mital and Manivasagan, 1983a; Kromodihardjo and Mital, 1985; Mital and Fard, 1986; Mital and Kromodihardjo, 1986). From all these investigations, the following facts have clearly emerged:

(i) Lifting asymmetrical objects is more stressful than lifting symmetrical objects as indicated by reduced lifting/carrying capabilities, increased intra-abdominal and intra-discal pressures and electromyographic activity of erector spinae and external obliques.
(ii) The metabolic energy expenditure rate of the individual is not affected significantly by the plane of the activities.

The decline in lifting capability is approximately 8·5% (Mital and Fard, 1986). Since lower weight is usually accepted for lifting, the resulting compressive and ground reaction forces are also low (Mital and Kromodihardjo, 1986). The shear forces, however, are substantially higher due to torsional stresses on the spine. It appears that for asymmetrical activities, shear tolerance capability might very well be the limiting factor.

2.4 The 'Environment' Component

Those MMH system characteristics which define or influence the physical working conditions (environment) are considered environment components. It has been known that the working environment affects individuals' performance. How it translates in terms of MMH capability is, however, not known for all environmental characteristics. The environmental characteristics discussed in this section are temperature and humidity, noise, illumination, vibration and altitude.

2.4.1 Temperature and Humidity

It is well known that heat load (combinations of air temperature, velocity, relative humidity and radiant heat) influences a person's physiological and psychological behaviour. When the heat stress exceeds the person's heat tolerance capacity, adverse heath effects occur (Dukes-Dobos, 1981). At lower levels of heat stress, even though there is no risk of health damage, people feel discomfort. The state of discomfort is responsible for reduced work-rate (Henschel, 1971), increased irritability, carelessness, a feeling of fatigue (Pepler, 1963), and increased accident rates (Belding et al., 1961; Brouha, 1967; Edholm, 1967).

Since hot and humid conditions are frequently encountered by workers in steel mills, foundries, mines, chemical industries, forestry, construction sites and other similar work places, it is essential to know the influence of temperature and humidity on the MMH capability of workers.

According to Brouha (1967), the ability to reach a high level of metabolic energy expenditure (oxygen uptake) and to maintain the body in the state of thermal homeostasis are essential requirements for efficient working in hot and temperate climates. If it becomes clear that the metabolic energy expenditure rate of an individual is not affected by the heat load, the working efficiency then would simply be a function of the person's ability to get rid of the excess heat quickly. The effect of heat on oxygen uptake, however, is not clear. Several studies have reported no change (Brouha, 1967; Kamon and Belding, 1971; Snook and Ciriello, 1974b) while others have reported a decline (Sengupta et al., 1979; Suzuki, 1980) or rise (Consolazio et al., 1963; Frye and Kamon, 1981).

In the context of MMH, higher temperatures and humidity results in elevated heart rates (Brouha, 1967; Kamon and Belding, 1971; Givoni and Goldman, 1973; Kamon et al., 1978; Kamon, 1979; Krajewski et al., 1979; Sengupta et al., 1979) and increased rectal temperatures (Kamon and Belding, 1971; Givoni and Goldman, 1972; Snook and Ciriello, 1974b). The physiological responses to humid and dry heat are different for men and women. In general, women show longer tolerance limits and relatively lower heart rates and rectal temperatures as compared to men (Avellini et al., 1980; Shapiro et al., 1980).

A decline in lifting capability of individuals was observed by Snook and Ciriello (1974b) at a WBGT(Wet Bulb Globe Temperature) of 27°C, compared to a WBGT of 17·2°C. Specifically, the lifting capability declined by 20%, pushing capability by 16%, and carrying capability by 11%.

Based on data available in the literature, Smith and Ramsey (1982) have developed a classification of lifting tasks according to metabolic workload (Table 2.27). Threshold WBGT levels, based on these workload classifications, are also provided (Table 2.28).

Table 2.27. Classification of lifting tasks according to metabolic workload (Smith and Ramsey, 1982). Reproduced with permission of Institute of Industrial Engineers

Height	Weight (kg)	Light (100–199 kcal/h)	Moderate (200–299 kcal/h)	Heavy (300–399 kcal/h)	Very heavy (400 + kcal/h)
Floor to waist	4.54	< 10*	11–15	16–20	> 21
	11·35	< 5	6–10	11–14	> 15
	20·43	< 2	3–5	6–8	> 9
Floor to head	4·54	< 3	9–12	13–16	> 17
	11·35	< 4	5–8	9–11	> 12
Waist to head	4.54	< 20	—	—	—
	11·35	< 10	10–16	17–19	—

* Lifts/min.

Table 2.28. Threshold WBGT levels* (Smith and Ramsey, 1982). Reproduced with permission of Institute of Industrial Engineers

Metabolic costs (see Table 2.20)	Threshold WBGT† (°C)
Light work	30
Moderate work	28
Heavy work	26
Very heavy work	25

* Two hour time weighted average values.
† Adult male, acclimatized, physically fit, normally clothed, good nutrition.

Procedures suggested by Kamon (1979) and Krajewski *et al.* (1979) should be used to design work–rest cycles for MMH tasks that are performed in hot conditions.

The effects of cold weather are not the same as the effects of climate. Powell *et al.* (1971) have shown that, in the cold, hand injuries can be reduced by wearing gloves though at the cost of stability and grip strength (Hertzberg, 1955). Williamson *et al.* (1984) have shown an inverse relationship between the age of the operator and the time spent in cold. It is, therefore, unwise to insist upon a rigid work/recovery routine for all operators. However, an exposure time of 40% of the total work-cycle time would not cause an unacceptable risk or discomfort. Exposure times may be increased by the use of heavier clothing and increased physical activity. Heavier clothing, however, restricts physical movement.

2.4.2 Noise, Illumination and Vibration

No studies have been conducted in which noise, illumination and vibration have been investigated in the context of the MMH capabilities of individuals. These factors generally create physiological and safety problems. For instance, excessive noise can lead to constriction of blood vessels, changes in blood circulation and upsetting the sense of balance (Glorig, 1971). Vibration, on the other hand, increases oxygen uptake (Duffner *et al.*, 1962) and cardiac output (Hood *et al.*, 1966). The problems of discomfort (glare) or visibility may become acute due to illumination. The exact effects of noise, vibration and illumination on the MMH capabilities of individuals, however, are not known.

2.4.3 Altitude

As elevation increases, the maximal oxygen uptake decreases. At about 3800 m, the maximal oxygen uptake is approximately 80% of the maximal oxygen uptake at 230 m. Increased oxygen debt, longer time period to reach a steady state of oxygen consumption and ventilation, and higher heart rates are some of the symptoms associated with submaximal work at high altitudes (Billings *et al.*, 1971). However, there is insufficient evidence to draw any meaningful conclusions regarding the effect of altitude on the MMH capability of an individual. It appears, though, that the capability will decrease due to oxygen deficiency, for example, but how much and in what proportion is not known.

2.5 Interactive Effects of System Components

The previous sections (2.2–2.4) described the effects of worker, task and environmental characteristics on the MMH capability of an individual when considered separately. In reality, however, several of these are present at any one time and collectively pose a given type of hazard. The net effect can be quite complex. Figures 2.18 and 2.19 are examples of the interactive effects when only three task characteristics are present (Ayoub et al., 1978; Asfour, 1980).

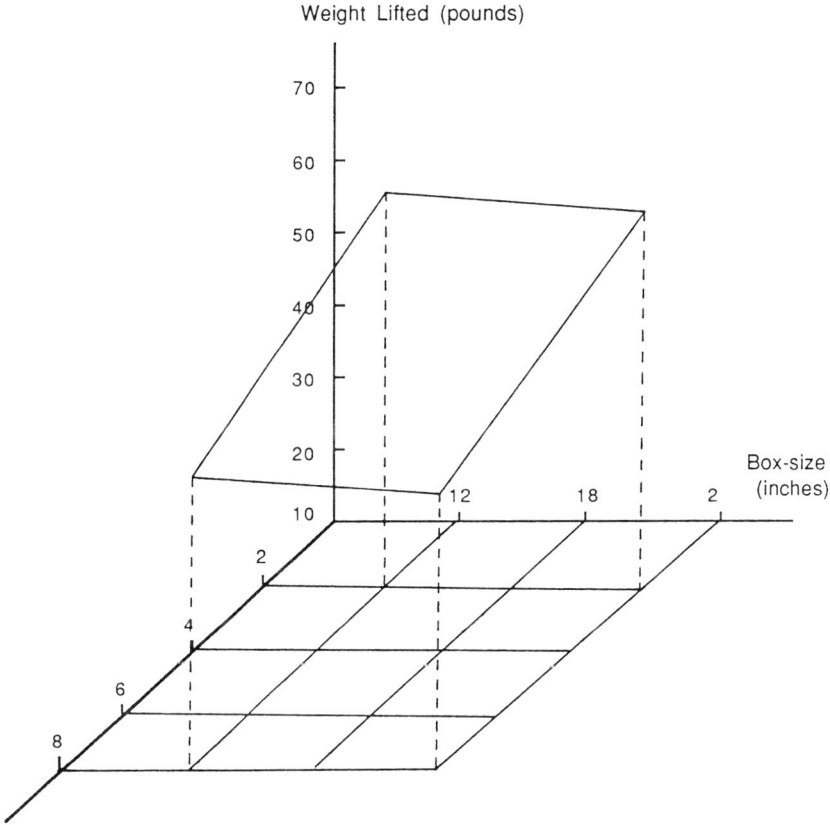

Figure 2.18. *Interactive effects of box-size and frequency (Ayoub et al., 1978)*

Figure 2.19 shows the interactive effects of box-size dimension in the sagittal plane and lifting frequency on the maximum acceptable weight of lift (Ayoub et al., 1978) using a three-dimensional projection. When there are more than two factors involved, it becomes almost impossible to portray the effect in this fashion. Figure 2.19 is an example depicting the effects of lifting height, lifting frequency and load lifted on

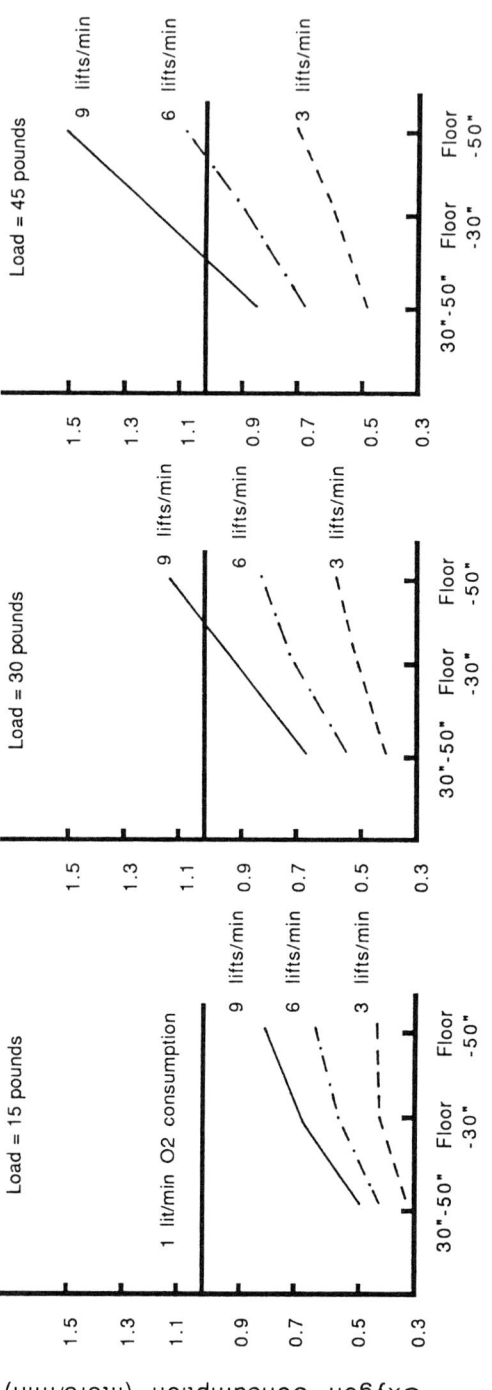

Figure 2.19. Height of lift, frequency of lift, and load lifted interaction (Asfour, 1980)

oxygen uptake. When third or higher order interactions are involved, the net influence can be determined only with the help of regression models. Chapters 3 and 6, for instance, describe several models which help evaluate the net effect of more than three variables on responses.

2.6 System response measures

The response of an MMH system can be ascertained by measuring one or more of several well recognized response measures. Frequently, the selection of a response measure(s) is a function of the approach used (psychophysical, physiological or biomechanical).

The previous review has shown that the following response measures can be used reliably to estimate the demands or stresses of an MMH activity:

(i) Strength or force exertion capability (static or dynamic);
(ii) Compressive and shear forces acting on the spine;
(iii) Ground reaction forces;
(iv) Metabolic energy expenditure rate;
(v) Respiratory demands (expressed in terms of volume of air or percentage of the physical work capacity);
(vi) Heart rate or pulse rate;
(vii) Workload perception (ratio scales, interval scales, ranking scales, rating scales, etc.);
(viii) Intra-abdominal pressure; and
(ix) Endurance time or number of cycles endured.

The utility of these responses measures in selecting design criteria or job design is discussed in subsequent chapters.

Chapter 3
Design Criteria

3.1 Introduction

3.1.1 Stress/Strain Concept

Man in the workplace is affected by two types of forces: forces from the immediate physical environment, and biomechanical forces of his body. These forces, fundamental to the basic principle of ergonomics, are described by the laws of Newtonian mechanics and the biological laws of life. In this context, the biological laws of life describe the musculoskeletal and physiological systems and their response to the demands of the task (Tichauer, 1978). Thus, the stress/strain concept involves external forces and/or effort imposed upon the worker which potentially produces a strain on the worker's musculoskeletal and physiological systems. It is the goal of the ergonomist to reduce stress sufficiently to minimize musculoskeletal and/or physiological strain.

3.1.2 MMH Stresses

Applying the stress/strain concept to MMH, three approaches are used: biomechanical, physiological and psychophysical. The biomechanical approach studies the musculoskeletal structure such that the physical, or mechanical, limits of the individual are determined. The physiological approach studies the circulatory responses and the human body's metabolic response to various loads. The psychophysical approach establishes acceptable lifting weights to the individual. That is, the individual subjectively quantifies his tolerance of stress (NIOSH, 1981). This chapter discusses each of the approaches in more detail.

3.2 The Biomechanical Approach

3.2.1 Biomechanical Analysis for MMH

3.2.1.1 Definitions and Applications of Biomechanics

Biomechanics has been defined by Contini and Drillis (1966) as 'the science which investigates the effect of internal and external forces on human and animal bodies in movement and at "rest" '. Winter (1979) defined biomechanics of human movement as 'the interdiscipline which describes, analyzes, and assesses human movements.' Frankel and Nordin (1980) defined biomechanics as the discipline which 'uses laws of physics and engineering concepts to describe motion undergone by the various body segments and the forces acting on these body parts during normal daily activities.' Given these definitions, several disciplines contribute directly to biomechanics as shown in Figure 3.1. Biomechanics, in turn, can be divided into general and applied biomechanics.

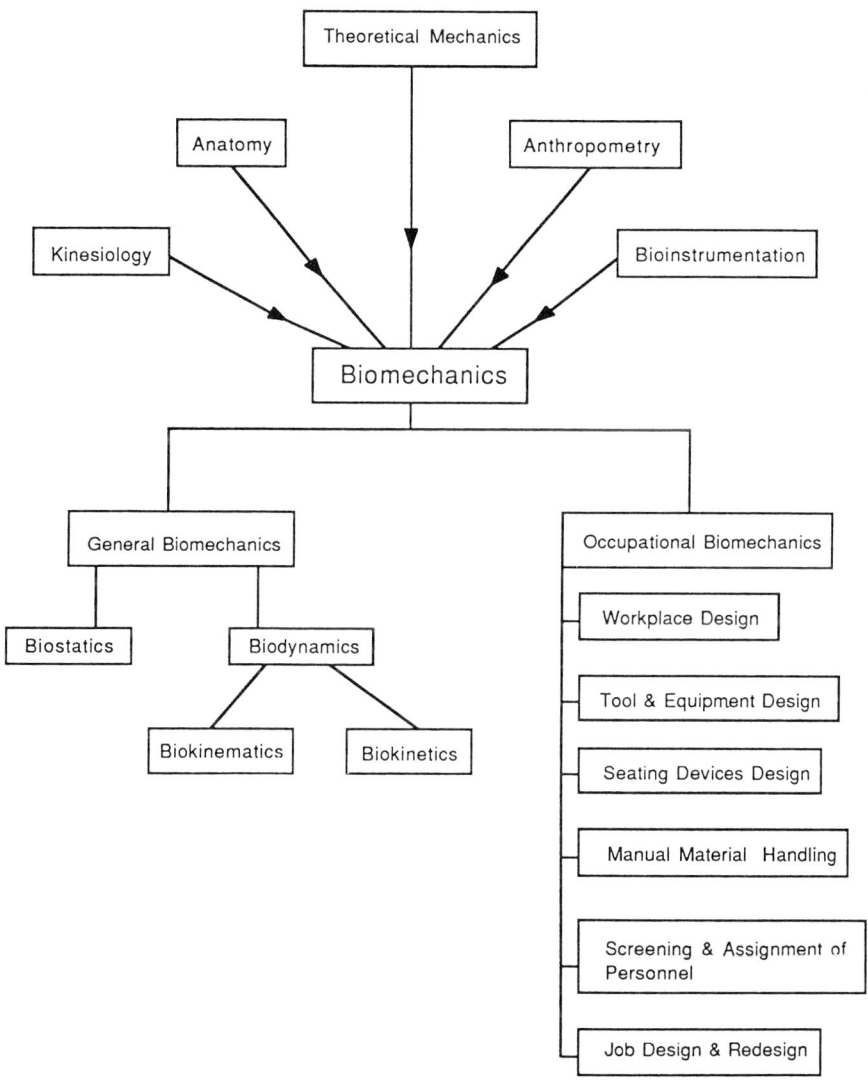

Figure 3.1. *A schematic diagram of biomechanics, modified from Contini and Drillis (1966)*

General biomechanics is concerned with the fundamental laws and rules governing organic bodies at rest or in motion. Biostatics considers those situations in which only analysis of bodies at rest or bodies moving in a straight line at uniform velocity (i.e., no acceleration generated, thus no force yield) is involved. Biodynamics is concerned with the description of the movement of the body in time without consideration of the forces involved (kinematics) and motion caused by forces acting on the body (kinetics). Both internal and external forces are included in the kinetic analysis of motion.

Occupational biomechanics is a division of applied biomechanics which involves applying the principles of biomechanics towards work in improving everyday activities, especially dealing with human disorders and performance limitations which exist at present in a variety of manual tasks in industry. Occupational biomechanics can be defined as 'the study of the physical interaction of workers with their tools, machines, and materials so as to enhance the worker's performance while minimizing the risk of musculoskeletal disorders' (Chaffin and Andersson, 1984).

3.2.1.2 The Body as a System of Levers

Biomechanics is based on disciplines. These include anthropometry, engineering science, bioinstrumentation, and kinesiology. Therefore, one can expect that occupational biomechanics is highly empirical. This is particularly true because biomeasurements to be made relate to human body motions and functions and, therefore, some criteria must be met in making measurements for application purposes. Because of the complexity of measurements and the need for safety in developing design criteria, the methodologies of modelling have been used extensively. Through modelling with necessary simplifications and assumptions, a system or an industrial task can be designed and evaluated without the need for comprehensive experimentation and elaborate data collection and analysis. In studying a particular system through modelling, if the model does not simulate or predict a system's behaviour well enough, we can change some input parameters, or part of the model itself, to study the complex nature of the real system. The following sections will show some examples of basic biomechanical modelling to estimate the mechanical stresses on the body.

In order to make biomechanical analysis possible, the body is viewed as a system of links and connecting joints. In the biomechanical model, each of the links is the same length and possesses the same mass and moment of inertia as their corresponding human segments. The mass is considered to be concentrated at a single point on the link, the centre of mass (CM). Different numbers of links are used by researchers in different biomechanical models. Muth (1976) proposed a five-link biomechanical model using optimization techniques, assuming no existing wrist joint. On the other hand, Ayoub and El-Bassoussi (1978) used six body links in their dynamic biomechanical model. A seven-link model is reported by Chaffin and Andersson (1984), allowing for spinal flexion. Figure 3.2 shows the body represented as a system of links. It is worth noting that in the use of biomechanical analysis and modelling, the torso is often considered as a simple one-link or two-link system. Such simplication may be needed to facilitate the use of mechanics, and can also be justified on the basis that the inclusion of additional torso links does not contribute towards understanding the mechanical trauma resulting from the different positions assumed by different workers of varying body size. It is anticipated that biomechanical analysis will become more sophisticated as we are able to better define task requirements and work postural demands.

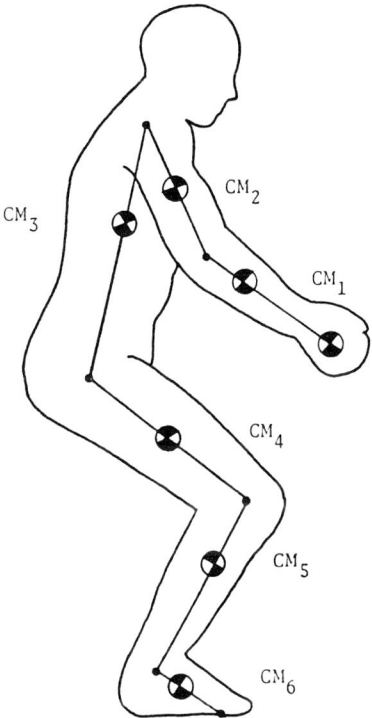

Figure 3.2. The body as a system of six links (modified from Chaffin and Andersson, 1986)

The body segment links are rotated around the joint by the action of the skeletal muscles. Muscle attachments are close to the joint where they act allowing the small contractile distance of the muscle to be transformed into large movements of the link. However, the short distance between the joint and muscle attachment means load or resistances at the distal end of the link have a mechanical advantage over the muscles. As a result, large muscle forces are required for relatively small loads.

3.2.1.3 Stress on the Musculoskeletal System

In order to estimate the mechanical stress imposed on the body while at rest or in motion when performing occupational tasks, the analyst must utilize the various mechanical properties of the body segments to perform the appropriate mechanical analysis. It is ncessary to make some simplifications and assumptions concerning the human body system. To accomplish these mechanical analyses, biomechanical models of various degrees of sophistication are developed. In the following sections, some simple biomechanical link segment models are displayed to estimate stresses on the various parts of the body while handling loads.

Forces are vector quantities with four characteristics: magnitude, direction, line of action and point of application. Three types of forces constitute all the forces acting on the total body system (Winter, 1979):

(a) *Gravitational forces*, acting downward through the centre of mass of each segment with magnitude equal to mass times gravitational acceleration.
(b) *Ground reaction or external forces*, due to applied workload and body segment weights.
(c) *Muscle forces*, expressed in terms of net muscle moment acting at a joint. Some other forces such as joint friction and forces within the muscle also contribute to the net moment.

3.2.1.3.1 STATIC ANALYSIS

Static analysis is used to study the rotational moments and forces acting on the human body when no movement is involved. Many physical activities can be analysed as if they were executed statically, even when they involve movement. Dynamic considerations are important mechanically only when a motion involves significant linear or angular accelerations. If this dynamic situation is not the case, the static analysis techniques are quite useful for studying the static and quasi-static or, equivalently, quasi-isometric physical activities.

3.2.1.3.2 ANALYSIS OF ONE-SEGMENT LINK

To study the stresses imposing on the musculoskeletal system, we start the analysis by examining the forearm free body diagram in Figure 3.3, assuming no significant wrist joint exists. Consider the case in which an 8 kg load is held in the hands. The load acting at the hand produces a torque at the elbow as does the weight of the forearm and hand. The involved muscles' contractile activities then produce the necessary torque to counterbalance these torques. Since the weight is held in the hands with no body movement involved, the static analysis situation is assumed.

In a static equilibrium situation, according to Newton's law, the sum of the torques about the point of rotation must sum to zero. Likewise, the sums of the vertical and horizontal forces must also sum to zero. In mathematical form, these two static equilibrium conditions can be stated as:

$$\Sigma \text{ Moments } (M) = 0; \Sigma \text{ Forces } (F_x \text{ or } F_y) = 0.$$

Let the weight of load be 8 kg in Figure 3.3, assuming symmetric lifting, 4 kg in each hand. Using a male 1·78 m in height, the values of D_1 and D_a are found to be 0·36 m and 0·15 m, respectively. If the subject weighs 75 kg, then W_a is estimated to be 1·65 kg. Then the reactive force and torque acting at the elbow joint, due to the load held in the hand and weight of forearm, can be calculated as:

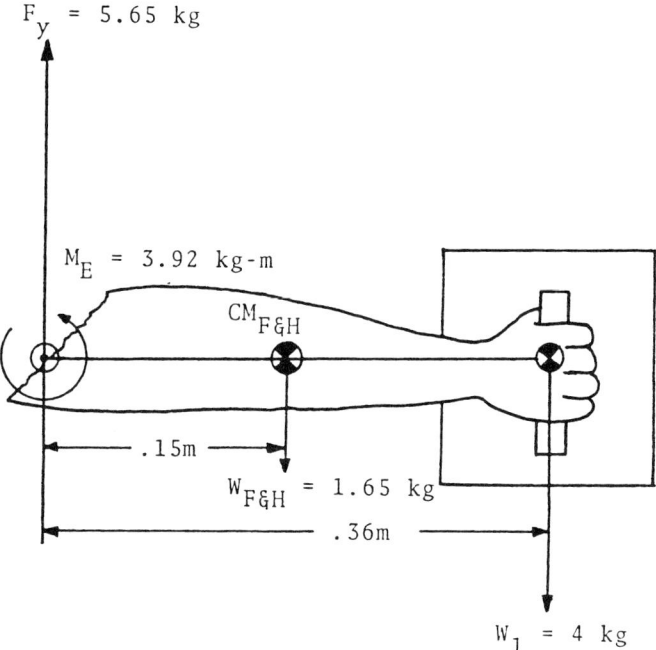

Figure 3.3. *Free body diagram of forearm and hand in horizontal position holding 4 kg load*

$$Y_t = 0 = -W_1 - W_a + F_y \text{ (Force equilibrium)}$$
$$= -4\,\text{kg} - 1\cdot 65\,\text{kg} + F_y$$
$$F_y = 5\cdot 65\,\text{kg}.$$

where W_1 = the weight of load; W_a = the weight of forearm and hand; F_x = the reactive force at elbow joint in the x-direction; F_y = the reactive force at elbow joint in the y-direction; Y_t = the force in Y direction; X_t = the force in X direction.

Since there is no force developed in the horizontal direction, the reactive force at the elbow joint in the X direction should equal zero.

$$X_t = 0$$

and

$$F_x = 0.$$

The reactive torque at the elbow joint is the torque necessary to counteract those forces produced by the load weight and body segment weight multiplying by their corresponding distances from point of force application to centre of rotation, i.e., the elbow joint.

The following equation should meet this condition.

$$\Sigma M = 0 = -W_1(D_1) - W_a(D_a) + M_E$$
$$= -4\,\text{kg}\,(0\cdot 36\,\text{m}) - 1\cdot 65\,\text{kg}\,(0\cdot 15\,\text{m}) + M_E$$
$$M_E = 3\cdot 92\,\text{kg}-\text{m}$$

where D_1 = the length of the link; D_a = the distance from elbow to the link CM; M_E = the reactive moment at elbow joint.

Suppose the forearm is no longer held in a horizontal position but with an angle as shown in Figure 3.4. The reactive force acting at the elbow joint will still remain the same, but the reactive moment is reduced. This is due to the reduction of the moment arm actually used, d_1, as shown in Figure 3.4. The revised value will be calculated as

$$M = 0 = -W_1(D_1)(\cos a) - W_a(D_a)(\cos a) + M1_E$$

or

$$M1_E = (\cos a) \times M_E.$$

Thus, the load and the weight of the arm has an additive effect on the elbow moment, with its maximum moment value when the arm is horizontal and a minimum effect when the arm is vertical.

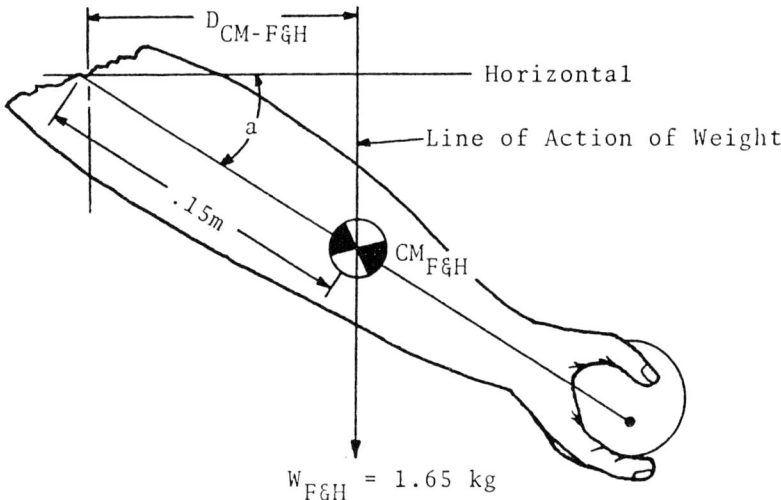

Figure 3.4. *One-link model, when segment link is at an angle to horizontal line, showing the reduction of lever arm*

3.2.1.3.3 ANALYSIS OF TWO LINKS

The effect of weight of load applied on the hand and the effects of accumulated body segment weights will be transmitted from the hand through connecting body segment links to the feet, where the reaction force takes place. Such forces and torque transmissions can be examined by studying the two-link body segment model in Figure 3.5, representing the human upper extremity. First, we can treat this two-link model as two separate one-link systems and perform exactly the same force and torque analyses, as introduced previously, on the forearm. Under the static equilibrium condition, these reactive forces and the torque should yield equal amounts but opposite directions at the same point of application, which is the distal end of upper arm link.

Again, the particular muscle group activity should produce force and torque at the shoulder joint large enough to counteract the force and torque effects due to the body segment weight itself, and reactive force and torque from the previous link. Note that the forces and torque have additive effects on the succeeding links. Using the same assumed subject, the weight of the upper arm is determined to be 2·1 kg, with length 0·33 m. The distance for centre of mass is 0·13 m from the centre of rotation (shoulder joint). The reactive force and moment acting on shoulder joint can then be calculated as

$$Y_t$$
$$= -2\cdot1\,\text{kg} - 5\cdot65\,\text{kg} + R_s$$
$$R_s = 7\cdot75\,\text{kg}$$

and

$$\Sigma M = 0 = -W_u(D_2) - R_e(D_u) - M_E + M_s$$
$$= -2\cdot1\,\text{kg} \times (0\cdot13\,\text{m}) - 5\cdot65\,\text{kg}\,(0\cdot33\,\text{m}) - 3\cdot92\,\text{kg-m} + M_s$$
$$M_s = 5\cdot86\,\text{kg-m}$$

where W_u = the weight of upper arm; R_e = the reactive force transmitted from forearm link; R_s = the reactive force transmitted from shoulder link; D_2 = the distance from the shoulder to the link CM; D_u = the length of the link; M_E = the moment transmitted from previous link; M_u = the reactive torque at shoulder joint; M_s = the moment acting on the shoulder.

In the case where the posture of the arm is altered, this will have a great effect on the moments at the elbow and shoulder, but will have no effect on the external reactive forces. The previous analyses of the two-link model will remain in effect, except the

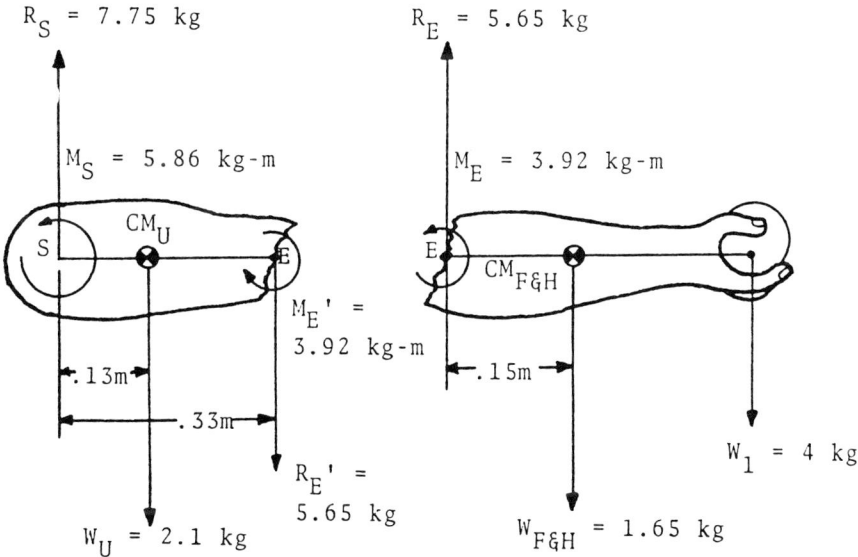

Figure 3.5. *Two-link static planar analysis of upper arm and forearm-hand segments held horizontal*

vertical distance, from point of rotation to the action line of force, will be used in the calculations of moment rather than the length of the segments themselves. A typical representation of the two-link model at angles from the horizontal is shown in Figure 3.6.

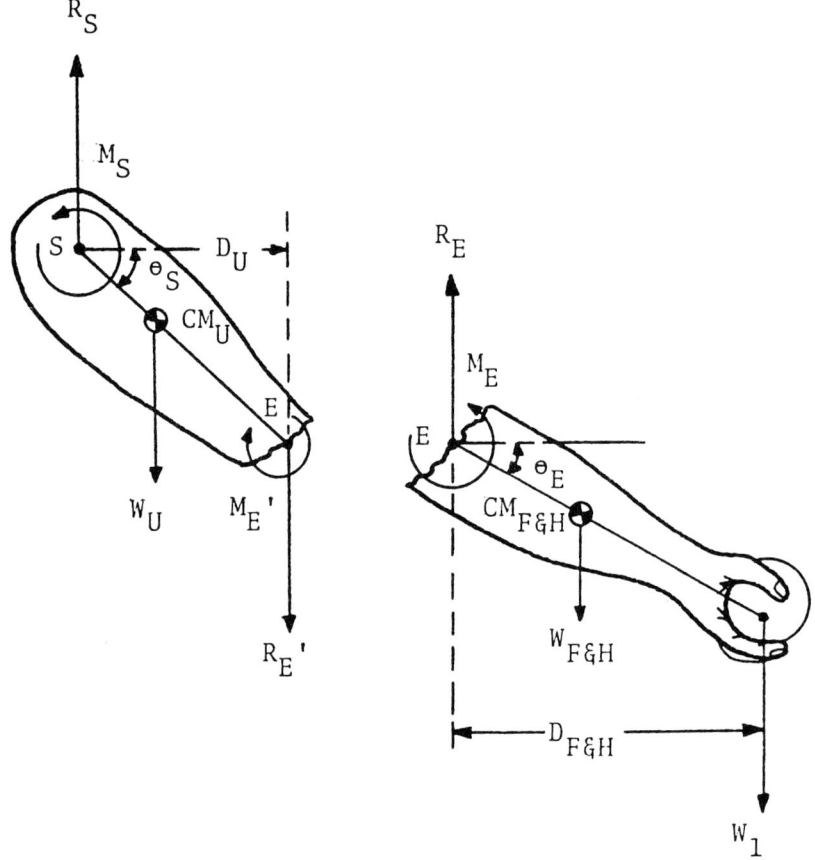

Figure 3.6. *Two-link static planar analysis of upper arm and forearm-hand segments including angle effect. The actual lever arms used are $D_{F\&H}$ and D_U, for forearm and hand and upper arm, respectively. Forces acting downwards and moments acting clockwise are negative, angles measured clockwise from right horizontal axis*

3.2.1.3.4 ANALYSIS OF MULTIPLE LINKS

Following the sequence of body segment links, we can use the same method of analysis to calculate the reactive force and torque at each joint. If any body segment forms an angle from the horizontal line, the calculated values of reactive forces will still remain the same, but the moment will be changed due to the increased or decreased actual moment arm in effect. The ankle joint, which is the last joint in the body-link model, would bear the entire body weight and applied external load and the resulted accumulative torque. Figure 3.7 shows the effects of three different postures on the reactive moments and forces. The ground reactive forces are the same in all three body postures, but the reactive torques are drastically different. It is necessary to note the convention of force and torque in the above analyses. If the resultant moment acts counterclockwise it will be termed positive in a Cartesian coordinate system, otherwise it will be termed negative. The forces acting upward and toward the right will be termed positive, other forces will be termed negative.

3.2.1.3.5 ANALYSIS OF INTERNAL FORCES

Besides the 'load moments' introduced in the proceeding sections, we might be interested in looking at models where internal muscle forces are involved. This is represented in Figure 3.8, which is very similar to Figure 3.3, except an internal muscle force provided by the biceps brachii is introduced.

Using the previous anthropometric data for the subject, the resulting moment equation can be expressed as follows:

$$\Sigma M = 0 = -W_1(D_1) - W_a(D_a) + F_m(D_m)$$

where ΣM = the sum of the moments about the elbow; F_m = the force due to muscular contraction; D_m = the distance from the elbow to the point of muscle action on the link.

Substituting those previous values into the above equation and isolating the unknowns produces:

$$\Sigma M = 0 = -4\,\text{kg}(0\cdot 36\,\text{m}) - 1\cdot 65\,\text{kg}(0\cdot 15\,\text{m}) + F_m(D_m)$$
$$F_m(D_m) = 1\cdot 69\,\text{kg} - \text{m}.$$

The torque of the muscle contraction must equal $1\cdot 69\,\text{kg}-\text{m}$ for the link to maintain static balance. If the value for D_m is assumed to be $0\cdot 05\,\text{m}$ the magnitude of the muscle force can be determined.

$$F_m(0\cdot 5\,\text{m}) = 1\cdot 69\,\text{kg} - \text{m}$$
$$F_m = 33\cdot 75\,\text{kg}.$$

Thus, the muscles must exert a force of $33\cdot 75\,\text{kg}$ to balance the $4\,\text{kg}$ load being held in the hand.

Next the horizontal and vertical forces can be determined. Since there are no horizontal forces acting in this example, the horizontal force, F_x, drops out of the analysis. Substituting the known values into the following equation and isolating the unknown, F_y, produces the vertical force on the elbow.

$$Y_t = 0 = -W_1 - W_a + F_y + F_m$$
$$= -4\,\text{kg} - 1\cdot 65\,\text{kg} + F_y + 33\cdot 75\,\text{kg}$$
$$F_y = -28\cdot 1\,\text{kg}.$$

The bones of the forearm are drawn upward by the muscular force because the weights of the load, forearm and hand only partially offset the vertical pull of the muscles. Therefore, the bone of the upper arm must press down reactively to hold the system in balance.

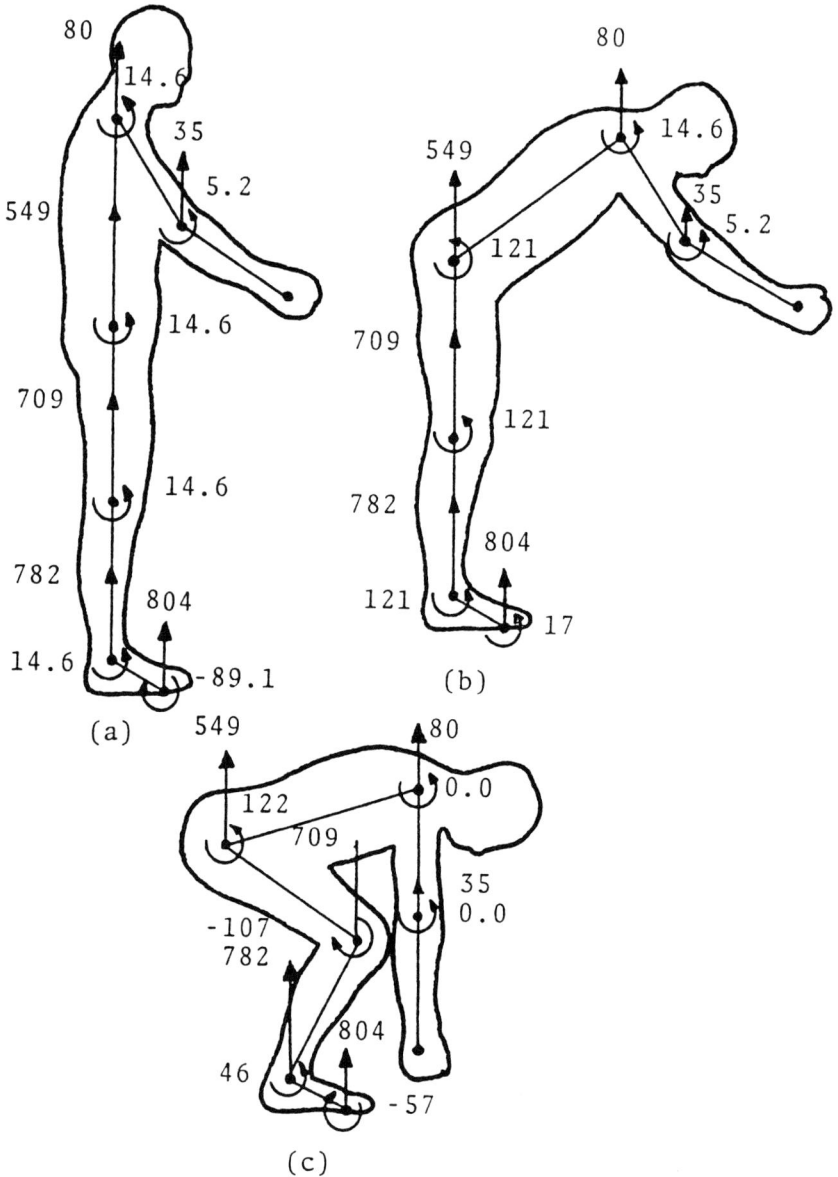

Figure 3.7. Reactive moments (Nm) and forces (N) for three different postures using average male anthropometric data (no load in hands) (from Chaffin and Andersson, 1984)

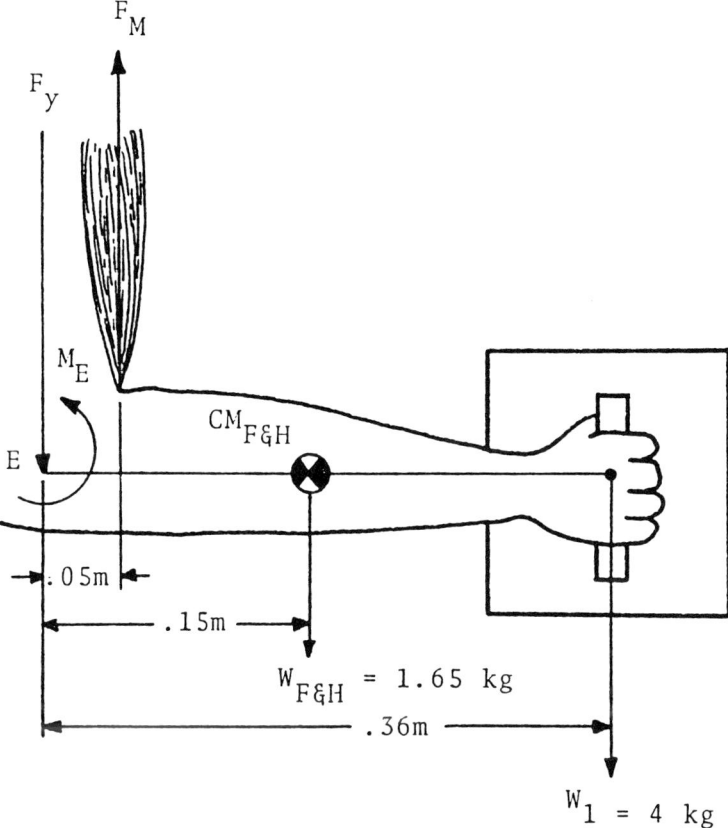

Figure 3.8. Analysis of internal muscle force F_M and joint reaction force R created by weight of forearm, hand and load W_1 held horizontal

In industrial tasks, workers frequently handle loads in the range of 20 kg. This would correspond to 10 kg supported by each hand. Using this value for W_1, the muscle contractile force and the vertical force on the elbow will be

$$M = 0 = -10 \text{ kg}(0 \cdot 36 \text{ m}) - 1 \cdot 65 \text{ kg}(0 \cdot 15 \text{ m}) + F_m(0 \cdot 05 \text{ m})$$
$$F_m = 76 \cdot 95 \text{ kg}$$
$$Y_t = 0 = -10 \text{ kg} - 1 \cdot 65 \text{ kg} + F_y + 76 \cdot 95 \text{ kg}$$
$$F_y = -88 \cdot 6 \text{ kg}.$$

Thus one can see that the forces sustained by the body holding a common industrial load can be quite large. One of the most important assumptions made in the estimation of internal forces is that a singular muscle action accounts for the internal force. This is not quite true, since body postural changes involve different muscle group activities, and usually several muscles share the moment requirement at a joint. Models accounting for multiple internal muscle forces will be discussed in detail in a later section.

3.2.1.4 Dynamic Analysis

Most industrial activities are dynamic in nature, i.e., they involve body movement. Body movement is often assessed using photographs at various time intervals. Picture taking of MMH activities is integral to the development of kinetic biomechanical models; whereby the use of cinematography enables researches to record human body movement. The recordings, from which body movements are derived, often produce quantitatively different results due to differences in film speed (Mital and Kromodihardjo, 1984; Kromodihardjo and Mital, 1986a). Different film speed criteria have been offered, and thus chosen by researchers. Choices of film speed are, however, infrequently accompanied by explanation or reason. Kromodihardjo and Mital (1986a) offer a criteria for film speed, which has been balanced for accuracy and cost.

Figures 3.9 and 3.10 delineate sagittal lifting, and lifting and twisting body motions with film speeds of 17 f/s, 125 f/s and 500 f/s of subjects lifting a 11·36 kg box from the floor to a table (0·91 m high). Acceleration peak differences should be noted. The relationship between film speeds is shown in Figure 3.11. The calculation of this relationship is based on the assumption that data collected at 500 f/s is error free, relatively speaking.

According to the results plotted on Figure 3.10, there is a sharp decrease in the percentage of error until the filming speed of 125 f/s. Film of 125 f/s is 98% accurate (Kromodihardjo and Mital, 1986a). Since there is only a slight decrease of error percentage and an increase of film cost beyond 125 f/s, a film speed of 125 f/s is recommended.

In a biomechanical analysis, body dynamics need to be considered only when the inertial forces and inertial moments produced are of magnitudes that are significant when compared with the force and moments needed for equilibrium (Shultz and Andersson, 1981). When the body movement speed is large enough to impose significant stress on the body, dynamic modelling techniques should be used to study the force and moment load on each segment link. The example used to illustrate the static analysis will now be used to demonstrate the analysis of dynamic activities in the sagittal plane, hence maintaining planar motion rather than three-dimensional motions. In addition to the values already given for the example, three new kinematic variables must be defined for the motion of the link to be included in the analysis. They are the angular acceleration, W, the angular velocity, V, and the angle, a, between the link and horizontal axis. In the static example the links forming the upper arm and lower arm were placed at right angles to each other and in the vertical and horizontal directions for simplification. The diagram in Figure 3.12 shows the lower arm elevated by the value of angle a.

The forearm and hand link is essentially rotating about the elbow joint and, therefore, at any point in time during the motion the angular acceleration, angular velocity and the angle between the horizontal axis and the forearm link can be identified. Let the angular acceleration be defined as 15 rad/s^2, the angular velocity as 1·5 rad/s, and the angle as 40°. The other values used in the static example will remain the same for this dynamic example.

Several intermediate values must be computed before the actual calculation of the torque and forces can be performed. The masses of the load and forearm link must be

Design Criteria

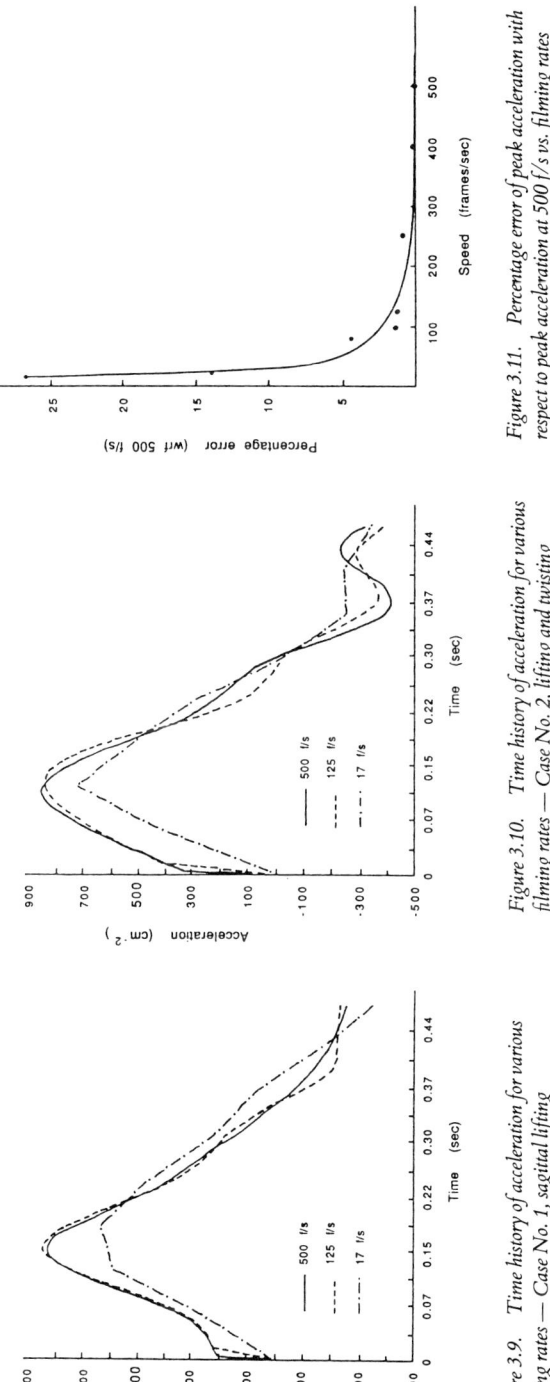

Figure 3.9. Time history of acceleration for various filming rates — Case No. 1, sagittal lifting

Figure 3.10. Time history of acceleration for various filming rates — Case No. 2, lifting and twisting

Figure 3.11. Percentage error of peak acceleration with respect to peak acceleration at 500 f/s vs. filming rates

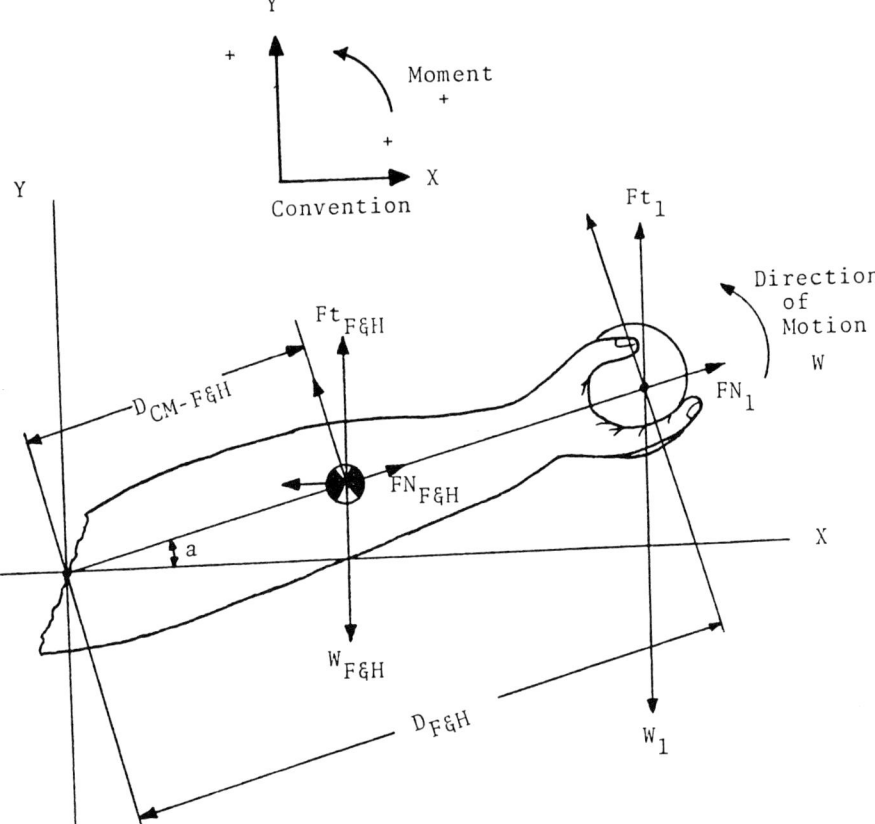

Figure 3.12. *The variables involved in a dynamic analysis of lifting a load in the hand and the sign convention used in the analysis*

determined as well as the inertial forces acting normally and tangentially to the forearm.

First, the mass values for the weight of the load and the forearm link are determined by dividing them by the acceleration due to gravity.

$$W_1 = M_1 g$$
$$W_a = M_a g$$

where W_1 = weight of the load; W_a = weight of the forearm and hand link; M_1 = mass of the load; M_a = mass of the forearm and hand link; g = acceleration due to gravity (9·8 m/s²).

Substituting the weights and acceleration due to gravity and solving for the masses:

$$4 \text{ kg} = M_1 (9·8 \text{ m/s}^2)$$
$$1·65 \text{ kg} = M_a (9·8 \text{ m/s}^2)$$
$$M_1 = 0·41 \text{ kg} - \text{s}^2/\text{m}$$
$$M_a = 0·17 \text{ kg} - \text{s}^2/\text{m}.$$

The inertial forces are determined using the masses just obtained and the angular acceleration and velocity. These inertial forces can be described by the tangential and normal forces, which are orthogonal to each other. When an object is moving around a point, the centre of rotation, with angular velocity, V, and radius of rotation, r, the tangential and normal acceleration can be calculated by the following equations:

$$a_N = V^2 r \quad \text{(in radial direction)}$$
$$a_T = w\, r \quad \text{(tangent to motion curve).}$$

where a_N = normal acceleration due to rotation; a_T = tangential acceleration due to rotation; w = angular acceleration due to rotation. Thus, the tangential forces are computed in following equations:

$$F_{ta} = M_a\, a_{Ta} = M_a(D_a)\, w$$
$$F_{A1} = M_1\, a_{T1} = M_1(D_1)\, w$$

where F_{ta} = tangential inertial force at link CM; F_{t1} = tangential inertial force at load; M_a = mass of the forearm link; M_1 = mass of the load; D_a = distance of the link CM from the elbow; D_1 = distance of the load from the elbow; a_{Ta} = tangential acceleration at centre of gravity; a_{T1} = tangential acceleration at point of load application.

The radius of rotation and the lever arm necessary to form torque is zero. The following equation expresses the dynamic moment equilibrium about the elbow joint:

$$M = 0 = M_E + (-M\,\text{static}) + (-M\,\text{dynamic tangential})$$
$$+ (-M\,\text{dynamic rotational}).$$

The moment of inertia, I_0, about CM for forearm and hand, can be calculated by:

$$I_0 = M_a\, P^2$$

where P is the radius of gyration and M_a is the mass of the forearm and hand. Values for P can be found in literature on anthropometry. The radius of gyration and segment length described from either the proximal or distal end for each link are obtained from Winter (1979). Using these data, the P value about CM for forearm and hand is found as $0{\cdot}168$ m (i.e., $0{\cdot}468 \times 0{\cdot}36$ m). Thus

$$I_0 = (0{\cdot}17\,\text{kg} - \text{s}^2\text{m}) \times (0{\cdot}36\,\text{m})^2$$
$$= 0{\cdot}022\,\text{kg} - \text{s}^2\text{m}.$$

Referring to the moment equilibrium equation about the elbow joint,

$$\begin{aligned}
M_E &= M\,\text{static} + M\,\text{dynamic tangential} + M\,\text{dynamic rotational} \\
&= M_1 g\,(\cos a)\, r + M_a g\,(\cos a)\, r \quad \text{(static)} \\
&\quad + F_{ta}(D_a) + F_{t1}(D_1) \quad \text{(dynamic)} \\
&\quad + I_0\, w \quad \text{(rotational)} \\
&= 1{\cdot}65\,\text{kg}\,(0{\cdot}15\,\text{m})\,(\cos 40) + 4\,\text{kg}\,(0{\cdot}36\,\text{m})\,(\cos 40) \\
&\quad + 0{\cdot}383\,\text{kg}\,(0{\cdot}15\,\text{m}) + 2{\cdot}21\,\text{kg}\,(0{\cdot}36\,\text{m}) \\
&\quad + (0{\cdot}22\,\text{kg} - \text{s}^2\text{m})\,(15\,\text{rad/s} \times \text{s}) \\
&= 2{\cdot}48\,\text{kg} - \text{m}.
\end{aligned}$$

Comparing this dynamic reactive torque with the one under the static situation

(1·69 kg-m) we can see the magnitude increases by 46·75% if the forearm and link move upward with a speed of 1·5 rad/s.

If the muscle strength moments are of interest, the moment equilibrium equation needs to be modified. Referring to Figure 3·2 for the model and variables name, the modified equation becomes:

$$M = 0 = \text{muscle} + (-M \text{ static}) + (-M \text{ dynamic tangential}) \\ (-M \text{ dynamic rotational}).$$

Substituting and solving:

$$F_{ta} = 0·17 \text{ kg} - \text{s}^2/\text{m} \ (0·15 \text{ m}) \text{ rad/s}^2$$
$$F_{t1} = 0·41 \text{ kg} - \text{s}^2/\text{m} \ (0·36 \text{ m}) \ 15 \text{ rad/s}^2$$
$$F_{ta} = 0·383 \text{ kg}$$
$$F_{t1} = 2·214 \text{ kg}.$$

The normal forces are computed in a similar way in the following equations.

$$F_{na} = M_a \ (D_a) \ v \times v$$
$$F_{n1} = M_1 \ (D_1) \ v \times v$$

where F_{na} = normal inertial force at link CM; F_{n1} = normal inertial force at load; M_a = mass of the forearm link; M_1 = mass of the load; D_a = distance of the link CM from the elbow; D_1 = distance of the load from the elbow; v = angular velocity.

Substituting and solving:

$$F_{na} = 0·17 \text{ kg} - \text{s}^2/\text{m} \ (0·15 \text{ m}) \\ \times (1·5 \text{ rad/s} \times 1·5 \text{ rad/s})$$
$$F_{n1} = 0·41 \text{ kg} - \text{s}^2/\text{m} \ (0·36 \text{ m}) \\ \times (1·5 \text{ rad/s} \times 1·5 \text{ rad/s})$$
$$F_{na} = 0·057 \text{ kg}$$
$$F_{n1} = 0·332 \text{ kg}.$$

With these intermediate computations complete, the torques can be calculated. The sum of the torques about the elbow must be equal to zero for the system to be in equilibrium. The reactive torque about the elbow joint, M_E, must have the equal value but opposite direction to those torques due to the weight of the body segment and applied load, and effects of dynamic motion and inertia. Those torques consist of (1) torques and forces described in the static analysis section, (2) torques developed by all tangential inertial forces, and (3) torques resulting from the effects of rotation by multiplying the moment of inertia about CM of the segment link with instantaneous angular acceleration W. Note here, the normal inertial forces do not contribute to the formation of the reactive torque at the elbow joint, since it acts alone. In other words, the involved muscle group equivalent must generate an equal amount of torque to counterbalance all other torques.

$$M = 0 = F_m (\cos a) D_m - W_a (\cos a) D_a - W_1 (\cos a) D_1 \\ - F_{ta} (D_a) - F_{t1} (D_1) - I_0 W$$

where M = the sum of the torques at the elbow; F_m = force due to muscular contraction; D_m = distance of point of muscular action from elbow; W_a = weight of the forearm link; W_1 = weight of the load; a = angle of forearm link from horizontal; F_{ta} = tangential inertial force of forearm link; F_{t1} = tangential inertial force of load; D_a = distance of link CM from the elbow; D_1 = distance of the load from the elbow.

Only the muscular forces are unknown. By solving the above equation we can find the value of F_m.

$$M = 0 = F_m(0 \cdot 05 \text{ m}) \cos 40 - 1 \cdot 65 \text{ kg} (0 \cdot 15 \text{ m}) \cos 40$$
$$- 4 \text{ kg} (0 \cdot 36 \text{ m}) \cos 40 - 0 \cdot 303 \text{ kg} (0 \cdot 15 \text{ m}) - 2 \cdot 214 \text{ kg} (0 \cdot 36 \text{ m})$$
$$- (0 \cdot 22 \text{ kg} - \text{s}^2 \text{m}) (15 \text{ rad/s}^2).$$
$$F_m (0 \cdot 05 \text{ m}) (\cos 40) = 2 \cdot 48 \text{ kg} - \text{m}$$
$$F_m = 64 \cdot 75 \text{ kg}.$$

Based on the above calculation, the muscle must exert 64·75 kg of force to generate torque to equilibrate all other external torques. Comparing this value with the muscle force under the static condition (33·75 kg) there is approximately a 92% increase under the dynamic situation, even though the lever arm is reduced by an angle from horizontal in the dynamic situation. This is due to effects of motion and the mechanical disadvantage of the short muscle lever arm.

With the muscular force determined, the inertial forces, and the angle of the limb relative to the horizontal, the vertical and horizontal reactive forces can now be found. As before the sum of the forces must equal zero for the system to be in equilibrium.

$$X_t = 0 = F_x - F_{ta} (\sin a) - F_{t1} (\sin a)$$
$$- F_{na} (\cos a) - F_{n1} (\cos a)$$
$$Y_t = 0 = F_y - W_1 - W_a + F_m - F_{t1} (\cos a) - F_{ta} (\cos a)$$
$$+ F_{na} (\cos a) + F_{n1} (\cos a)$$

where X_t = total of forces in the horizontal direction; Y_t = total of forces in the vertical direction; F_x = horizontal reactive force; F_y = vertical reactive force; F_{ta} = tangential inertial force at the link CM; F_{t1} = tangential inertial force at the load; F_{na} = normal inertial force at the link CM; F_{n1} = normal inertial force at the load; W_a = weight of the forearm link; W_1 = weight of the load; F_m = force due to muscular contraction; a = angle of the forearm link to horizontal.

Substituting and solving for the reactive forces:

$$X_t = 0 = F_x - 0 \cdot 383 \text{ kg} (\sin 40) - 2 \cdot 214 \text{ kg} (\sin 40)$$
$$- 0 \cdot 057 \text{ kg} (\cos 40) - 0 \cdot 332 \text{ kg} (\cos 40)$$
$$Y_t = 0 = F_y - 4 \text{ kg} - 1 \cdot 65 \text{ kg} + 64 \cdot 75 \text{ kg} - 2 \cdot 214 \text{ kg} (\cos 40)$$
$$- 0 \cdot 383 \text{ kg} (\cos 40) + 0 \cdot 057 \text{ kg} (\sin 40) + 0 \cdot 332 \text{ kg} (\sin 40)$$
$$F_x = 1 \cdot 46 \text{ kg}$$
$$F_y = -57 \cdot 35 \text{ kg}.$$

In the model where no muscle strength analysis is involved, the calculation of the reactive force at the elbow joint remains unchanged. On the other hand, the Y component of the reactive force will be computed by dropping the muscle force component from the equation. That is:

$$Y_t = 0 = F_y - 4\,\text{kg} - 1\cdot 65\,\text{kg} - 2\cdot 21\,\text{kg}\,(\cos 40)$$
$$- 0\cdot 383\,\text{kg}\,(\cos 40) + 0\cdot 057\,\text{kg}\,(\sin 40) + 0\cdot 332\,\text{kg}\,(\sin 40)$$
$$F_y = 7\cdot 14\,\text{kg}.$$

Therefore, for the 4 kg load being moved by the arm, there is a vertical reactive force of 7·14 kg acting on the elbow joint. For the muscle strength model, a muscular force of 64·75 kg is required. The reactive forces at the elbow are 57·36 kg in a downward direction and 1·46 kg toward the body of the lifter. For comparison, the muscle force and reactive forces are solved for the condition where $a = 0°$ for both the dynamic and static situation. The results of these computations are shown in Table 3.1.

Table 3.1. Forces acting on the elbow joint under static and dynamic conditions using a 4 kg load.

Conditions	Vertical force (kg)	Horizontal force (kg)	Muscle force (kg)
Static model	3·75	—	—
	−28·1	—	33·75
Dynamic model	7·14	1·46	—
	−57·35	1·46	64·75

3.2.1.5 Three-dimensional Modelling

The models introduced in previous sections are related to single-link models or multiple-link models involving lifting in the saggital plane. In most industrial tasks, such as pushing, pulling and lifting with one hand, asymmetric loading of the human body occurs quite often. Mital and Kromodhardjo (1986) state that asymmetrical lifting leads to lower compressive and ground reaction forces. However, these results do not suggest that asymmetrical lifting is more desirable, but rather that workers lift lower weights at lower accelerations. It is necessary to utilize three-dimensional analysis techniques to depict and solve the forces and moment in such asymmetric loading systems. Under a static equilibrium situation, the following equations with reference to three orthogonal axes at each joint must be held:

$$F_x = 0 \quad F_y = 0 \quad F_z = 0$$
$$M_x = 0 \quad M_y = 0 \quad M_z = 0$$

In this case, the forces acting in the Z direction and the moment referring to the X and Y axes, M_x and M_y, respectively, are introduced into the model. A basic understanding of vector algebra is necessary before proceeding with further analyses. Some representative papers describing such techniques are found in Schultz and Andersson (1981), Chaffin and Andersson (1984) and Schultz (1986). The following example shows the moment generated about the lumbosacral junction of the spinal column, due to the holding of a 10-kg weight in the right hand. To simplify the problem, all the body segment weights are ignored. An orthogonal-axis reference system with its original set at the L5/S1 spinal junction and the coordinate of weight held in space is displayed in Figure 3.13. Before the value of the moment is calculated, three component vectors must be identified. First, the force vector of the weight must be defined. The weight force acts on a point at coordinate 35, 50, 20 (x, y, z axes, respectively) and downward to the ground, with a magnitude of 10 kg. By finding the unit vector representing the action line of force, the direction of the weight of load can be defined. Simply picking up the second reference point along the line of action (for example, a point with coordinate 35, 20, 20 below the first point), the direction of this force can be represented as:

$$D = (35-35)i + (20-50)j + (20-20)k$$
$$= -30j$$

$$D_u = \frac{-30j}{\sqrt{(35-35)^2 + (20-50)^2 + (20-20)^2}}$$

where D = the vector from point one to point two; D_u = the unit vector computed from vector D.

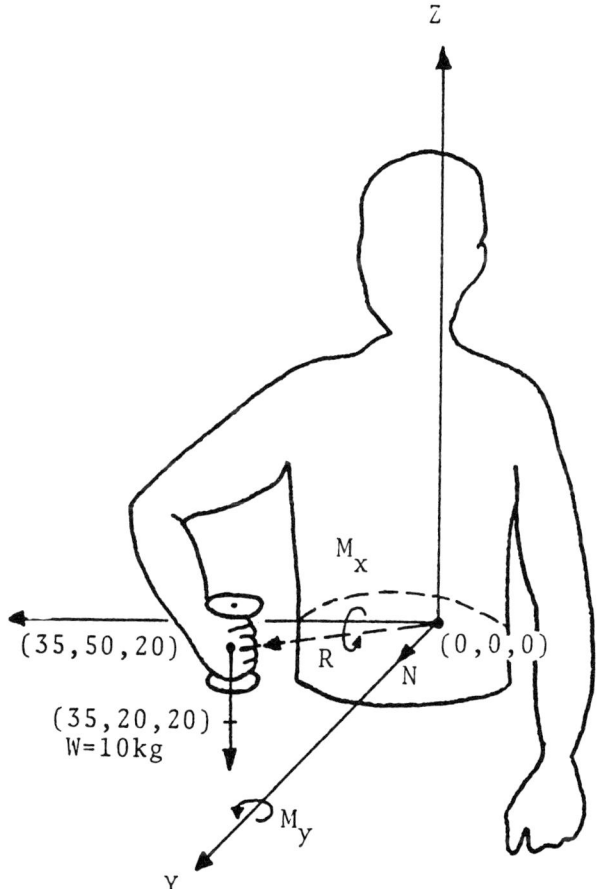

Figure 3.13. The moments acting at a certain point of the spinal column; refer to either X or Y axis due to the holding of 10 kg weight in the right hand (modified from Schultz, 1986)

Here, $-j$ indicates that the direction is acting downward and along the Y-axis. Any force acting downward to the ground will have a direction unit vector, $-j$. The direction of any force in space can be determined in the same way. Multiplying the magnitude of the force with the direction unit vector, D_u, the force vector of weight of load can be expressed as:

$$W_1 = 10\,(-j) = -10j \text{ (kg)}.$$

The second component vector defined is the unit vector representing the axis of interest. Assuming we want to compute the moment referring to the Y-axis, the unit vector representing the Y-axis will simply be N, j. Finally, the third component vector is the vector running from any point on the reference axis of interest to any point on the force vector. This vector, r, can be decided from the known two pairs of points on F and N as

$$Y = (35 - 0)\,i + (50 - 0)\,j + (20 - 0)\,k$$
$$= 35\,i + 50\,j + 20\,k.$$

The moment M_y about the y-axis is now computed as the 'triple scalar product' of r $(W_1 \times D_u)$. The concept of triple scalar product can be found in advanced engineering mathematics books such as O'Neal (1983). The value of moment M_y, taking the absolute value form, is computed from the determinant of the following matrix.

$$M_y = r\,W_1 \times D_u$$
$$= (35\,i + 50\,j + 20\,k)\,(-10\,j) \times (-j)$$
$$= \begin{vmatrix} 35 & 50 & 20 \\ 0 & -10 & 0 \\ 0 & -1 & 0 \end{vmatrix}$$
$$= 35\,(-10)\,(0) - (0)\,(-1) - 50\,(0)\,(0) - (0)\,(0)$$
$$+ 20\,(0) - 1) - (0)\,(-10)$$
$$= -350 \text{ (kg} - \text{cm)}$$
$$= 3 \cdot 5 \text{ (kg} - \text{m)}.$$

The direction of this moment is determined by the 'right-hand rule' as shown in Figure 3.14. The 'right-hand rule' states 'if you place your right hand so that the fingers curl in the direction of rotation from F to G, then the thumb points toward the direction of generated moment' (O'Neill, 1983). In other words, the line of action of the moment is defined as operating along a line perpendicular to the plane in which the force and moment arm exist. In the above example, the direction of the moment acts along the positive y-axis. Similarly, the moment about the x-axis acts toward the negative x-axis direction with magnitude of 5 kg-m.

3.2.1.6 Stress on the Lumbosacral Spine

Manual material handling tasks are inherent in many different jobs in industry today. Most MMH activities involve the lifting, pushing and pulling of loads in various body postures. The performance of such tasks exposes the worker to a variety of biomechanical hazards. The external load is applied to the hands, and the effects of

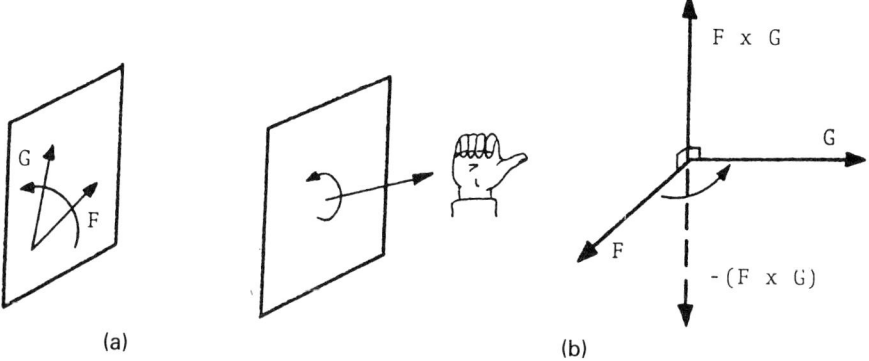

Figure 3.14. (a) *The right-hand rule for* F × G. (b) *The direction of cross product* F × G *and its orthogonal relationship with F and G (adapted from* Advanced Engineering Mathematics *by P. V. O'Neal, 1983, Wadsworth Inc., with permission of the publisher)*

external load and weights of body segments, through the body segments in sequence, are transmitted to the feet. In this 'body segment chain', the trunk has been identified as the weakest part in the chain, especially in the lumbosacral junction.

Lifting of moderate to heavy loads can create excessive mechanical stresses on various components of the musculoskeletal system. Such biomechanical stresses on the lumbosacral disc have been found to be high enough, when lifting compact loads of about 15·9 kg, that they would probably exceed the structural strength of the cartilage endplate or annulus fibrosis in people who have a pre-existing weakness in these structures (Chaffin, 1977). Loads on the lumbar spine should be kept as light as possible for several reasons: pre-existing lumbar spine conditions can be aggravated by heavy loads; workers with back pain lose more days from work when their jobs involve heavy loads; and it is suspected that heavy loads have a role in causing back pain (Schultz and Andersson, 1981).

Epidemiological data from Great Britain, the USA and Canada are consistent in the observation that 25% of all compensatory industrial injuries are of the back and are frequently associated with MMH (Snook, 1978a, b). Due to the escalating cost and litigation resulting from many cases of back injuries, much effort has been spent on biomechanical model development involving low-back kinetics, such as analysis of net moment, joint reaction forces, disc compression, nucleus pressure and vertebral stress. The following sections will introduce biomechanical models concerning the spinal load when performing physical activities either in the sagittal plane or in three-dimensional space.

3.2.1.6.1 Symmetric Lumbar Loading

Biomechanical models concerning the loads on the lumbar spine when performing physical tasks in the sagittal plane have been widely investigated and studied. Large load moments will act on the hip joint, as well as the lumbosacral junction, when lifting heavy loads. Thus, the internal forces of the trunk muscles must be relatively large to stabilize the spine. Due to these large muscle forces, significant compression and sheer forces have been identified on the L5/S1 disc. Morris *et al.* (1961) proposed a simple static sagittal-plane model of the lumbar spine during lifting, with both the back muscle and abdominal pressure being introduced into the computation of compression force on the spine. Kromodihardjo and Mital (1986b) apply the laws of mechanics to compute spinal stresses at the L5/S1 level in two steps. First, net forces and moments at the transverse plane at the L5/S1 level are measured using kinetics and Newton's Law. Secondly, net forces and moments of the L5/S1 joint are computed using Schultz and Andersson's (1981) internal trunk model. The muscle forces required become larger as the joint of rotation is further removed from the load. Thus, the muscle forces at the base of the spine become very large. Figure 3.15 shows the forces acting at the base of the spine during the holding of a load with the hands.

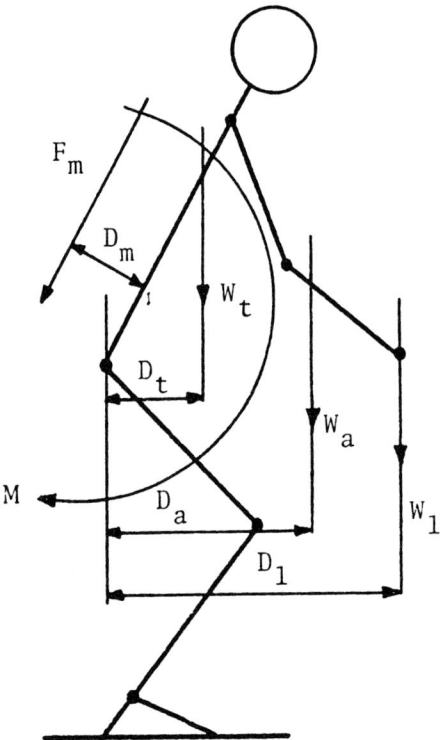

Figure 3.15. The forces, moment arms and angles involved in analysis of the stress on the back

The torque at the base of the spine can be computed as it was for the elbow. Then the vertical and horizontal forces and finally the compression and shear forces acting on the surface of the disc at the base of the spine can be determined.

$$M = 0 = F_m(D_m) - W_1(D_1) - W_a(D_a) - W_t(D_t)$$

where F_m = the force due to muscular contraction; D_m = the distance of the action of the muscle from the spine; M = the torque acting at the base of the spine; W_1 = the weight of the load; D_1 = the distance of the load from the spine; W_a = the weight of the arms and hands; D_a = the distance of the arms from the spine; W_t = weight of the trunk, head and neck; D_t = the distance of the CM of the trunk from the spine.

Using the same subject and load for this example, the moment at the base of the spine can be found. Substituting the values for the variables and solving the equation will yield the torque required of the muscles to balance the load in the hands.

$$M = 0 = 8 \text{ kg } (0 \cdot 36 \text{ m}) + 8 \cdot 5 \text{ kg } (0 \cdot 46 \text{ m})$$
$$+ 40 \cdot 8 \text{ kg } (0 \cdot 18 \text{ m}) - F_m (D_m)$$
$$F_m (D_m) = 14 \cdot 1 \text{ (kg} - \text{m)}.$$

If the distance of the action of the muscles from the base of the spine is taken to be 0·05 m then the muscular force required to balance the torque of the load can be found.

$$F_m (0 \cdot 05 \text{ m}) = 14 \cdot 1 \text{ (kg} - \text{m)}$$
$$F_m = 282 \text{ kg}.$$

As can be seen, the force of the muscle must be very large to balance the holding of a load of only 8 kg.

The compressive force is the more critical value in the back as this seems to be the source of many of the back problems associated with lifting. Combining the muscular force with the component of the weights acting on the base of the spine which are

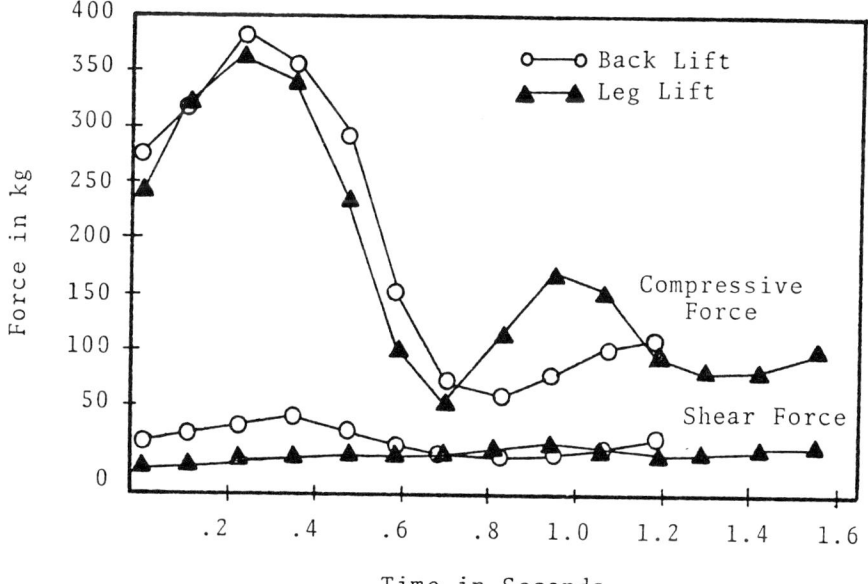

Figure 3.16. The compressive force over the duration of a floor to knuckle lift (based on Ayoub and El-Bassoussi, 1978)

normal to the surface of the disc will yield the estimate of the compressive force on the spine. The shear force is estimated from the component of the weights supported at the base of the spine which is parallel to the surface of the disc.

$$C = F_m + (W_1 + W_a + W_t)(\sin a)$$
$$S = (W_1 + W_a + W_t)(\cos a)$$

Again the effect of lifting a load rather than holding the load causes the compressive load to increase significantly depending on the body configuration. However, dynamic analysis requires additional computations. Figure 3.16 contains a typical profile of compressive force versus time during a lift from floor to knuckle height (76 cm above the floor).

3.2.1.6.2 ASYMMETRIC LUMBAR LOADING

Symmetric load handling, wherein the load is held with both hands and is moved in the midsagittal plane, is believed to be the most common method of handling a heavy load, based on both biomechanical models and experimental strength studies. The stresses on the musculoskeletal system are equalized bilaterally during symmetric load handling. The case of asymmetric handling, in which the load is handled by one hand or on the side of the body, causes not only a lateral bending moment on the lumbar column but, because of lordotic curvature of the column, produces a rotation of each vertebra on its adjacent vertebra. The effects of this vertebral rotation can result in a significant injury potential, due to the reducing torsional resistance strength of the annulus fibrosus, in the process of disc degeneration. In addition to these rotational effects, unbalanced loading on back muscles can produce highly concentrated stress which can overstrain a specific muscle or several muscles required to stabilize the column (NIOSH, 1981).

Kumar (1980) reported that asymmetric exertions appear to be more hazardous to the musculoskeletal system than symmetric exertions. Unfortunately, the studies of asymmetric load handling are sparsely performed, because of the experimental and biomechanical modelling complexities associated with three-dimensional force analysis (Chaffin and Andersson, 1984). A trunk model reported by Schultz and Andersson (1981) and Schultz et al. (1982) is able to study the biomechanics of asymmetric lifting. The model includes the major trunk muscle groups, and intra-abdominal pressure. Figure 3.17 shows the three-dimensional trunk model for this study. In the model, the body is visualized as being divided into upper and lower parts by an imaginary transverse cutting plane. The cutting plane is passed through the level of the lumbar spine at which the loads are to be determined. Based on the requirement that all body segments superior to this plane at the third lumbar level of the spine must remain in equilibrium during performance of the task, the reactive forces and moments acting on this particular vertebral disc are calculated due to weight of load and weights of body segments. Then the internal muscle forces are estimated, which are supplied by the muscle groups of the lower part of the body necessary to counterbalance those external forces and moments, as well as the compression and shear forces. The solutions are made based on different assumptions and simplifications, using techniques such as optimization under different but reasonable criteria.

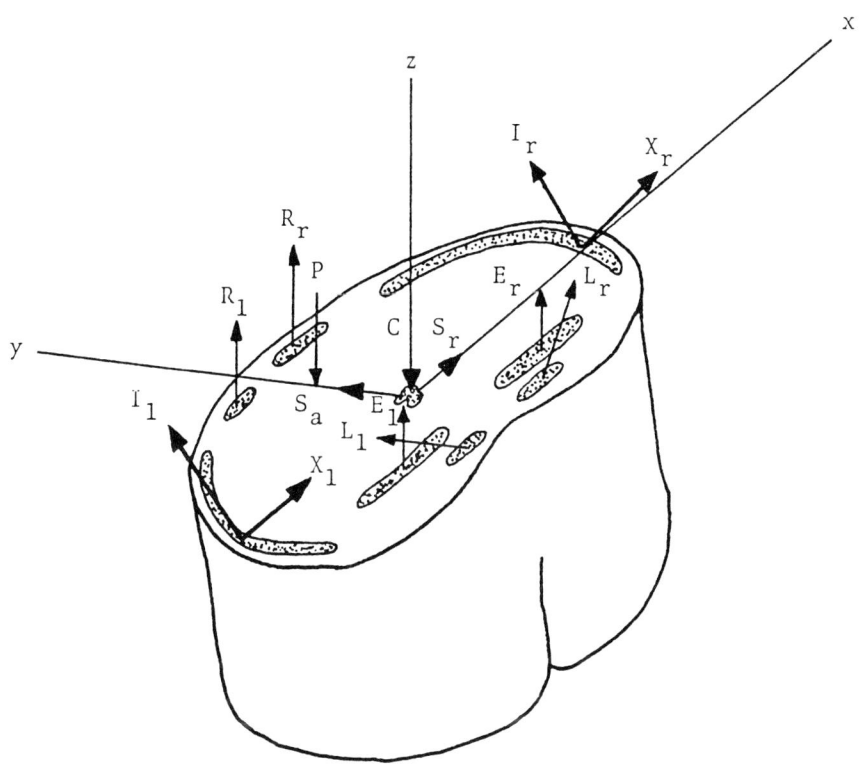

Figure 3.17. Three dimensional torso model (from Schultz and Andersson, 1981)
R_r = Right Rectus Abdominus
R_l = Left Rectus Abdominus
L_r = Right Latissimus Dorsi
L_l = Left Latissimus Dorsi
I_r = Right Internal Oblique
I_l = Left Internal Oblique
X_r = Right External Oblique
X_l = Left External Oblique
E_r = Right Erector
E_l = Left Erector
P = Intra Abdominal Pressure
C = Compression
S_r = Right Lateral Shear
S_a = Anterior Shear

3.2.1.6.3 THE ROLE OF INTRA-ABDOMINAL PRESSURE IN SPINAL LOADING

When a heavy load is being lifted the muscles of the trunk, and especially the abdominal muscles, are stressed and pressure rises in the abdominal cavity (Davis, 1956; Bartelink, 1957; Morris *et al.*, 1961). The intra-abdominal pressure (IAP) acting upward on the diaphragm and downwards on the pelvic floor can be thought of as an extensor of the trunk (Bartelink, 1957; Davis, 1959a; Morris *et al.*, 1961). It is believed that this IAP helps to relieve some part of the load applied on the spine by producing an extension moment. This extension moment contributes to the torque necessary for equilibrium of the trunk, thus the torque produced by the trunk muscle contraction is reduced and, in turn, the stress on the vertebral column is reduced.

Morris et al, (1961) reported that the calculated compressive force of about 30% on the lumbosacral level and about 50% on the lower thoracic portion, could be sustained by this IAP during lifting a load. Figure 3.18 shows the concept of how the IAP assists in the reduction of spinal load. The IAP measures can be used as a measure of the load imposed on the spinal column (Davis, 1959a).

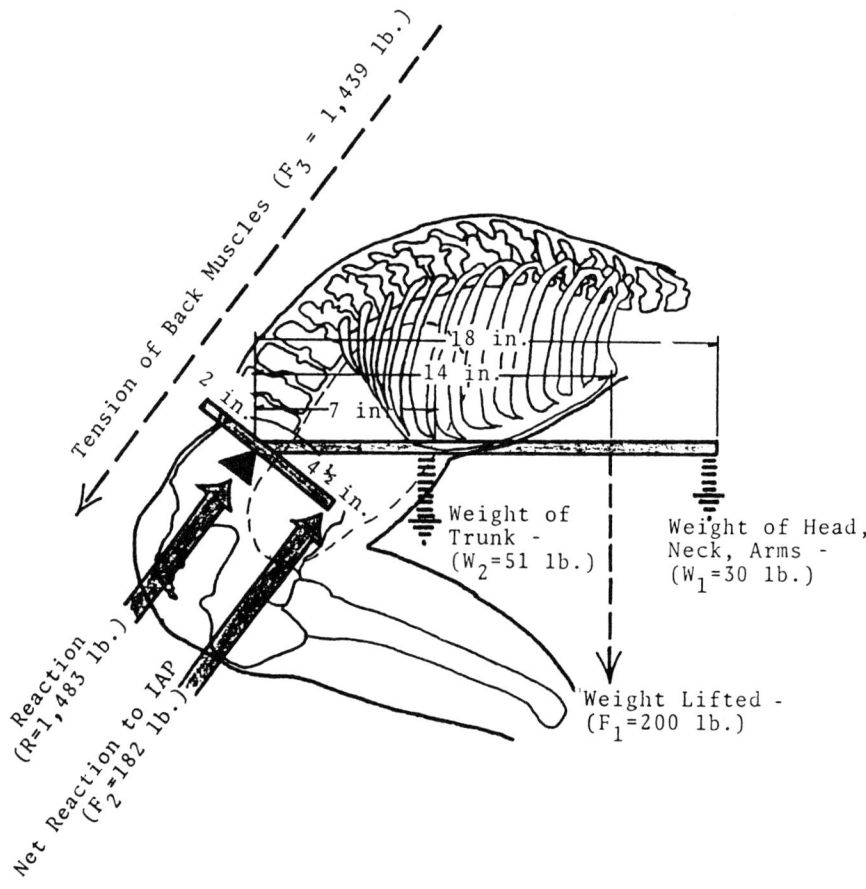

Figure 3.18. Intra-abdominal pressure acting as an extensor of the trunk to assist in the reduction of spinal load (modified from Morris, 1961)

By swallowing a capsule containing a pressure-sensitive element and a radio transmitter, this pressure can be measured. It has been reported by Davis and Stubb (1977), that the risk of back trouble increases when heavy loads are lifted if the intra-abdominal pressures exceed 100 mmHg. Davis (1956) reported the pattern of IAP response. During lifting a weight, the IAP rises abruptly to its peak value during the accelerative phase of the lift, and rapidly falls to a pressure above the resting level, while the load is held in a given position. The peak IAP at the start of a lift has been shown to be a function of the weight of the load, of its upward acceleration and of the angle of the trunk's forward flexion between 10° and 60° (after Mairiaux et al., 1984). Chaffin (1969) estimates the IAP from the data of Morris et al. (1961) using the following equation:

$$P = 10^{-4} [0 \cdot 6516 - 0 \cdot 005447 \text{ (hip angle)}] \text{ (hip torque)}^{1 \cdot 8}$$

where P = abdominal pressure in mm Hg with a max of 150 mm Hg; hip angle is in degrees from the erect position; hip torque is in kg – cm.

Mairiaux et al. (1984) studied the relation between IAP and lumbar moment when lifting weights in the erect posture. Results showed that IAP was well correlated with the lumbar moment by:

$$Y = 0 \cdot 079 X - 1 \cdot 127 \quad (r^2 = 0 \cdot 75)$$

where X = the angle between the trunk and the upper leg; Y = the predicted IAP. They suggested that the IAP measurement may be used as an index of spinal stress in lifting tasks when the body is in erect position. Troup et al. (1983) compared the IAP increase, hip torque and lumbar vertebral compression in different lifting techniques. The IAP data were compared with calculated peak values of lifting velocity, lumbosacral compression and hip torque, and with the integral of lumbosacral compression over time. They reported that the variation in peak IAPs was considerable and found no consistent relationship between IAP and any other variables. But, when the different lifting and lowering techniques were considered, significant differences in IAP were found. IAP was found to be less in the flexed trunk position than in a posture nearer to the vertical.

Marras et al. (1985) found that the IAP was primarily a function of angle and a weak indication of torque. Schultz et al. (1982), in examining isometric tasks, observed a poor correlation between IAP and measured intradiscal pressures, but felt that measurements of IAP may be useful as load indicators in tasks of brief duration. On the contrary, Ortengren et al. (1981), in studying the linear relationships between lumbar disc pressure and myoelectric back muscle activity and IAP, found the myoelectric signal and IAP to be good predictors for the disc pressure.

It is widely believed that IAP relieves the lumbar spine of compression. Recent literature, however, casts a doubt on this role of IAP. More accurate determination of the diaphragm area (299 cm^2) suggests that it is inadequate to generate sufficient IAP to alleviate spinal compression (Leskinen et al., 1983a,b). Also, according to Grillner et al. (1978), increases in IAP have to be concomitant with increased abdominal muscle activity. This imposes a direct compression penalty to the spine. Flexion moment resulting from increased abdominal muscle activity (pulling on the ribs and pelvis attachment points) also adds, indirectly, to the spinal compression. The net effect of extensor activity, due to IAP, and flexion moment is a net increase in spinal compression (McGill and Norman, 1985). Nachemson et al. (1986) also observed increases in spinal compression with IAP.

The contradiction in literature dealing with the role of IAP in manual lifting needs to be resolved. In the meantime, the role of IAP in reducing spinal compression must be reassessed with caution.

3.2.1.6.4 Use of EMG in Spinal Loading

The electrical activity of a muscle can be recorded with the help of an amplifier, a technique known as electromyography (EMG). This method has been widely accepted and used in both laboratory and field studies related to the prediction of muscle tension when performing physical acts. The relationship of EMG activity to muscle force is dependent on several factors and appears to be monotonic in the sense that an increase in signal amplitude signifies an increase in the muscle force, but is non-linear under many circumstances (Andersson and Ortengren, 1984, Chaffin and Andersson, 1984). These relationships are found to be true under isometric conditions, but there is a great uncertainty about the validity of the relationship under dynamic conditions (Chaffin and Andersson, 1984).

The myoelectric signals from back muscles in different static postures, with and without external loading, have been found responding similarly and consistently to changes in the trunk moment even when the trunk is under flexion and rotation (Andersson et al., 1976, 1977). They reported that the myoelectric activity increased when the angle of flexion increased and when the external load was increased at a fixed angle of flexion. During asymmetric loading, comparatively higher activity was found on the contralateral side in the lumbar region and on the ipsilateral side in the thoracic region. The responding EMG of trunk muscles is reported to correlate to trunk moment linearly. Thus, it is possible to use back muscle myoelectric activity at lumbar levels as a predictor of lumbar disc pressure (Ortengren et al., 1975, 1981). Analysis of myoelectric activity patterns gives useful guidance about how to reduce body loading in heavy work situations, and also permits quantitative evaluation of such improvements (Andersson and Ortengren, 1984).

3.2.1.7 Factors Affecting Biomechanical Stress on MMH

3.2.1.7.1 Tissue Characteristics

In performing heavy physical activities, or even in daily life, human lumbar spines have large loads impressed on them. These loads cause displacements and intradiscal pressure changes due to flexion, extension, right and left lateral bending and torsion moment. The compression force sustained by the lumbar spine could be up to 11 times the bodyweight in daily activities (Miller, 1986). An understanding of the mechanical behaviour of the human spine is useful in assessing many spinal pathologies and also provides a basis for the load limits on the spinal column in biomechanical assessment of physical activities. A lumbar motion segment, which consists of two adjoining vertebrae and their intervening ligamentous tissue, is a basic mechanical unit of the spine (Schultz et al., 1979; Nachemson et al., 1979).

Since it is difficult to perform tests on the lumbar vertebrae *in vivo*, in most experimental studies, human cadaver lumbar motion segments are used to study the influences on the mechanical behaviour of geometry, age, sex, disc level and degree of degeneration, for example. Care must be taken in the applications of these experimental results on the biomechanical evaluations of MMH jobs. Shultz et al. (1979), found significant intradiscal pressure increases resulting from bending or torsion. They reported that the interpretation of *in vivo* intradiscal pressure increases should be made with caution, since such pressure increases may result largely from the compressive load increase.

Compressive Strength of Disc

Various sources in the literature report failure of the vertebral bodies before failure of the intervertebral discs. Armstrong (1965) described the failure of disc cartilage endplates that distribute the compression loads to vertebral body, instead of the failure of the discs themselves. Brown *et al.* (1957), Perey (1957) and Hutton and Adams (1982) reported failure of the endplate or the vertebrae before the annulus fibrosus of the intervertebral disc failed. Hickey and Hukins (1980) developed a model of the intervertebral disc which predicted fracture of the endplates before the annulus would fail.

Based on this research, the intervertebral disc appears to be stronger in its resistance to failure under compression than the vertebral body. Therefore, the compressive strength of the vertebral body itself would seem to be the critical variable in determining the compressive limits of the spine.

Compressive Strength of Vertebra

A large variation is reported in the mechanical properties regarding the ultimate compressive strength of the vertebrae. In many cases, the variation within an individual's data is larger than the variation between variables being tested (Berkson *et al.*, 1979; Miller *et al.*, 1986). This individual variation is probably not of much importance to the overall mechanical behaviour of the trunk, and probably pre-existing variation in individual from a given activity. Part of the wide variation in the sensitivities of different individuals to low-back syndrome might be explained (Schultz *et al.*, 1979). However, even with the large variation trends observable in the data, the implication is that several variables are associated with the strength of the vertebrae.

The compression rate shows definite effects in the ultimate compressive strength of the vertebrae. Higher compression rates are associated with greater ultimate compressive strength (Evans and Lissner, 1959; Hutton *et al.*, 1979; Kazarian and Graves, 1977).

Differences in the ultimate compressive strength also appears to be affected by the sex of the individual. Studies by Hutton *et al* (1979), Hansson *et al.* (1980), and Hutton and Adams (1982) have found the ultimate compressive strength of females to be from 49–70% that of males.

A positive relationship appears to exist between the ultimate compressive strength and the level of the vertebrae (Kazarian and Graves, 1977; Hansson *et al.*, 1980; Hutton and Adams, 1982). Vertebrae at lower levels seem to exhibit greater ultimate compressive strength.

The cross-sectional area also appears to have an effect on the ultimate compressive strength (Kazarian and Graves, 1977; Hansson *et al.*, 1980; Hutton and Adams, 1982). Vertebrae having a larger cross-sectional area show a tendency to have higher ultimate compressive strength.

Another apparent factor in the ultimate compressive strength of a vertebra is the age of the individual. Hutton *et al.* (1979), Hansson *et al.* (1980) and Hutton and Adams

(1982) all reported data which showed a trend for younger persons to have higher ultimate compressive strengths. This trend is not as strong as that observed for some of the other variables, but it is none the less still present.

Another variable which affects the ultimate compressive strength of the vertebrae is the bone mineral content (Hansson et al., 1980). Bone mineral content was observed to increase with descending vertebral levels. Differences in the bone mineral content were found between males and females. Also, the data reported by Hansson et al. (1980) shows a decrease in bone mineral content with increasing age.

In general, the data seems to point to the effects of age, cross-sectional area, and bone mineral content as prime factors to be considered in determining the ultimate compressive strength of vertebrae. The vertebral level and sex differences seem to be correlated with the differences in cross-section area. Sex and age differences also may be explained by the bone mineral content. Other variables may be influential in determining the ultimate compressive strength. The activity level of the individual may have pronounced effects on the strength of the vertebrae. Hutton et al. (1980) suggested that people involved in activities which stress the vertebrae may develop stronger, heavier vertebrae as a result. Along with this is the weight of the individual. Heavier individuals may stress the vertebrae more resulting in stronger vertebrae. Other factors such as diet or illness may affect the ultimate compressive strength too. Also, heavier people tend to be larger and, therefore, would have larger vertebrae or possibly hypertrophied vertebrae due to greater stress from supporting more body-weight.

3.2.1.8 Task Characteristics

3.2.1.8.1 LOAD SIZE AND SHAPE
The size of the load, as described above, can cause the spine to load distance to be large. The shape of the load may also prevent the worker from bringing the load as close to the body also producing large spine to load distances. Neither of these conditions, large size or awkward shape, lends itself to optimum lifting postures or reduced stress to the spine. Figure 3.19 contains a set of curves showing the compressive force resulting from varying the distance of the load from the spine using a back lift technique. The shape of the load can also prevent symmetric lifting of the load adding a significant lateral torque to the spine. The possible damage to the spine from this kind of lifting is not known but the overall stress is larger. In addition, the shape of the load may contribute to an instability of the load causing abrupt changes in the relation of the load centre of mass relative to the spine. These sudden changes could produce compressive spikes which would be dangerous to the tissue of the spine.

Tichauer (1973) derived an equation to express the relationship between the weight of an object and its moment arm to the lumbar spine in producing a biomechanical equivalent moment, ML, assuming body thickness to be 8 in.

$$ML = 8 + \frac{\text{object length}}{2} \times \text{true weight}$$

where ML = moment at lumbar spine.

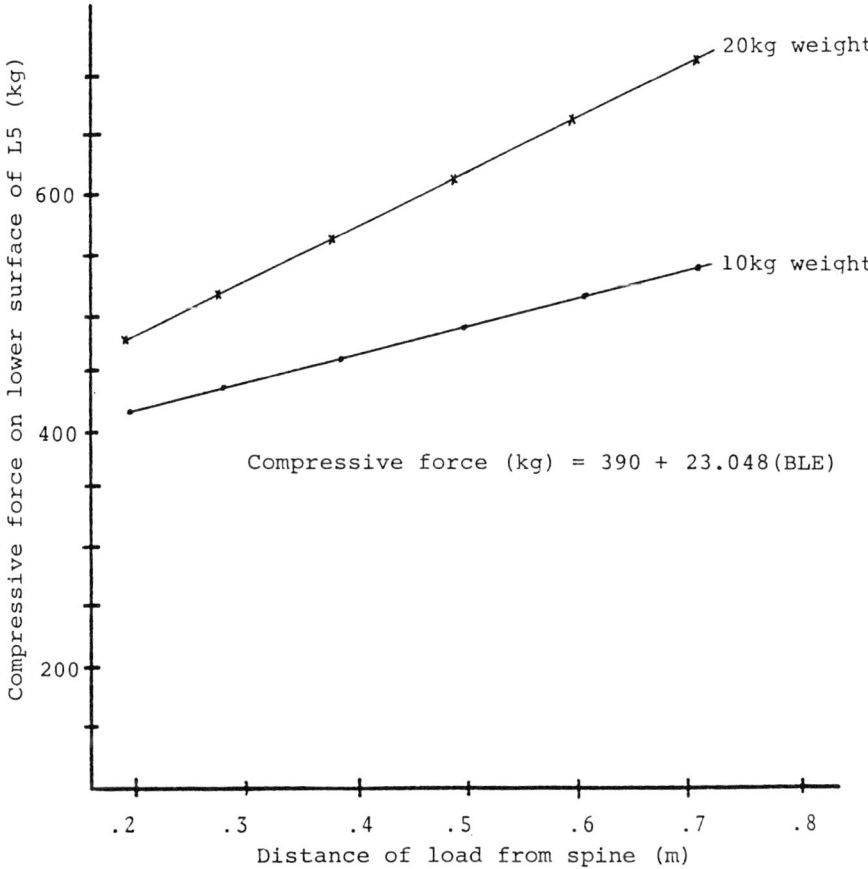

Figure 3.19. Effect of distance of load on compressive force; back lift technique

Table 3.2 shows the *ML* values produced by different loads with various length using the above equation.

As Table 3.2 indicates, a lighter load can create a larger lumbar moment than a heavier load with a shorter object length.

Table 3.2. Static holding task

True weight (lbs)	Object length (in)	Approximate moment at lumbar spine (lbs-in)
5	7·5	60
6·4	30·0	150
10·0	7·5	120
14·75	7·5	175
12·3	30·0	285
17·0	30·0	380

3.2.1.8.2 WEIGHT OF LOAD

By far the most obvious factor in determining lumbosacral stress is the weight of that load. Habes et al. (1985) confirmed that the most fatiguing task variable was weight based on the measured EMG for the muscle group involved in lifting tasks. Greater weight means greater compressive force on the spine. The effects of different weights in the hands on the IAP and intradiscal pressure, again, were displayed from the studies from Andersson et al. (1985) in Figure 2.11. Even when held close to the spine, large weights can place a great amount of stress on the spine.

3.2.1.8.3 POSITION OF LOAD RELATIVE TO THE SPINE

If the load is positioned such that the worker must reach for it, the stresses on the spine and on particular muscle groups is increased. There was an effect on the EMG level of the deltoid muscle when reach was at a maximum (Habes et al., 1985). Such situations as reaching over obstacles or across a table surface can produce large compression forces even when lifting small loads. The height of the lift also affects the stress on the spine. Lifting from the floor is more stressful than lifting from table height. Data shows that the greatest compressive force occurs during the first few fractions of a second of a lift from the floor. Figure 3.20 (a) and (b) show the compression and shear force profiles, respectively, of the L5/S1 and L4/L5 discs over time when lifting a 10 kg weight from floor to knuckle height in the sagittal plane (Ayoub and Chen, 1986).

The biomechanical effects of lifting to higher levels have not been clearly established. Research is underway which will contribute to understanding the biomechanics of lifting to shoulder levels.

3.2.1.9 Biomechanical Design Criterion

3.2.1.9.1 DESIGN CRITERION

There is a potential hazard in MMH tasks. The biomechanical design criteria for these physical activities deal with force loading on the musculoskeletal system as predictors of tissue tolerance and muscle sprains and strains (Garg and Herrin, 1979). The objective of many studies, using biomechanical criteria, is to reduce work-related injuries of the musculoskeletal system, and to increase work efficiency. Because the clinical data indicate that the back is a major area injured from lifting loads, it has been proposed that acceptable load limits be based on a percentage of back muscle strength.

Asmussen et al. (1965) have recommended weight limits based on 40% of isometric back strength with the indication that those limits may apply to frequent lifting activities. Poulsen and Jorgensen (1971) have also suggested weight limits for repeated lifting, based on 50% of isometric back strength. Garg and Ayoub (1980) reported allowable weights for single and repeated lifting based on isometric back strength, as shown in Table 3.3.

Though there is considerable disagreement as to the cause of low-back pain, the compressive force acting on the L5/S1 disc has been accepted as the primary measure of stress to the low back during lifting (Garg, 1979). The critical biomechanical factor in lifting loads is the compressive strength of the vertebrae. The major factors which appear to determine the compressive strength of the vertebrae are the age of the

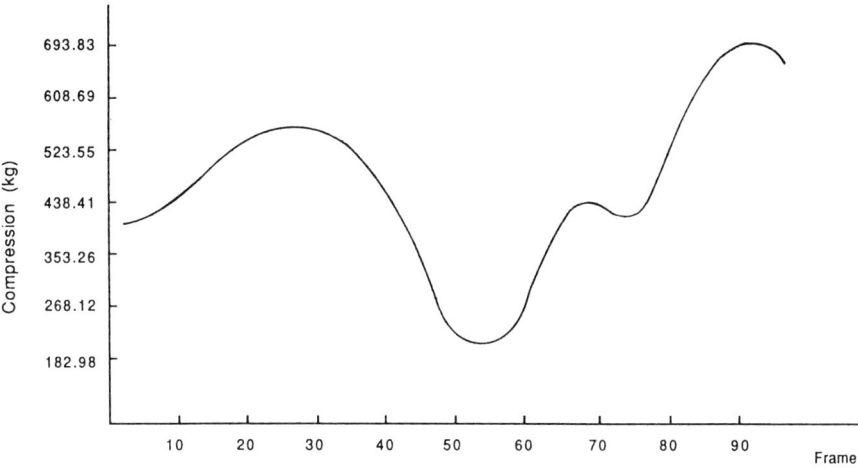

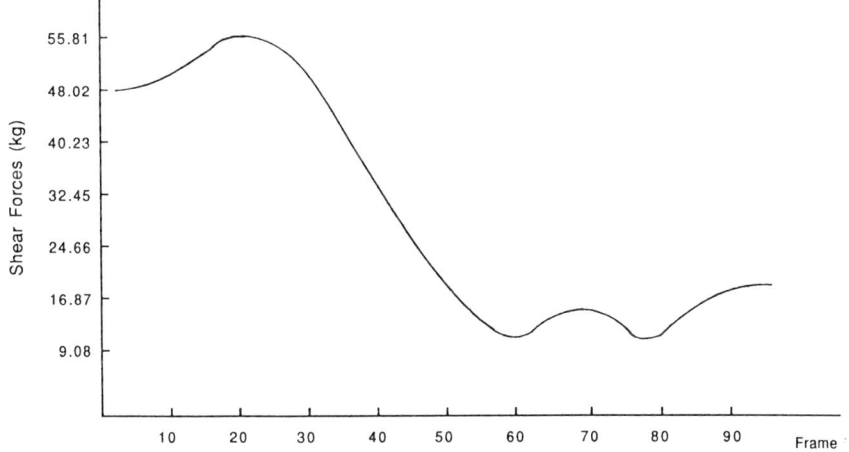

Figure 3.20. (a) *Compression force profile and* (b) *shear force profile on lumbosacral junctions when lifting from floor to knuckle height in sagittal plane. Time interval for every 10 frames stands for 0·2 s (Ayoub and Chen, 1986)*

worker, the cross-sectional area of the vertebrae, and the bone mineral content of the worker. Therefore, critical limits should take into account these factors when establishing the relative level of stress of MMH tasks.

The age of the worker can be easily established. The cross-sectional area can be determined through the use of X-rays, but estimates based on body size can be used more easily. The bone mineral content can be measured using the dual photon absorbtiometry method of Hansson *et al.* (1980), but such equipment is not convenient for

Table 3.3. Critical limits of compressive strength (kg) of the vertebrae (from Ayoub et al., 1983)

Age	General body size					
	Male			Female		
	n^*	$\bar{x}$†	S‡	n	$\bar{x}$	S
21–35	10	925·4	328·2			
36–50	6	668·4	234·6	19	406·6	105·7
51–70	13	377·8	119·8	28	306·9	103·0

*n = sample size; †$\bar{x}$ = mean; ‡S = standard deviation.

most situations. Therefore, the critical values presented in Table 3.3 based on age groups and body size can be used.

Chaffin and Park (1973) illustrated the observed incidence rates for the low-back compressive forces on the L5/S1 disc based on an industry study of 400 workers, as shown in Figure 3.21. From this statistical incidence rate and the strength data of the vertebrae from Table 3.3, it was concluded that any industrial job creating 650 kg compressive force on the spine will put workers in the above 35 age group at greater risk of

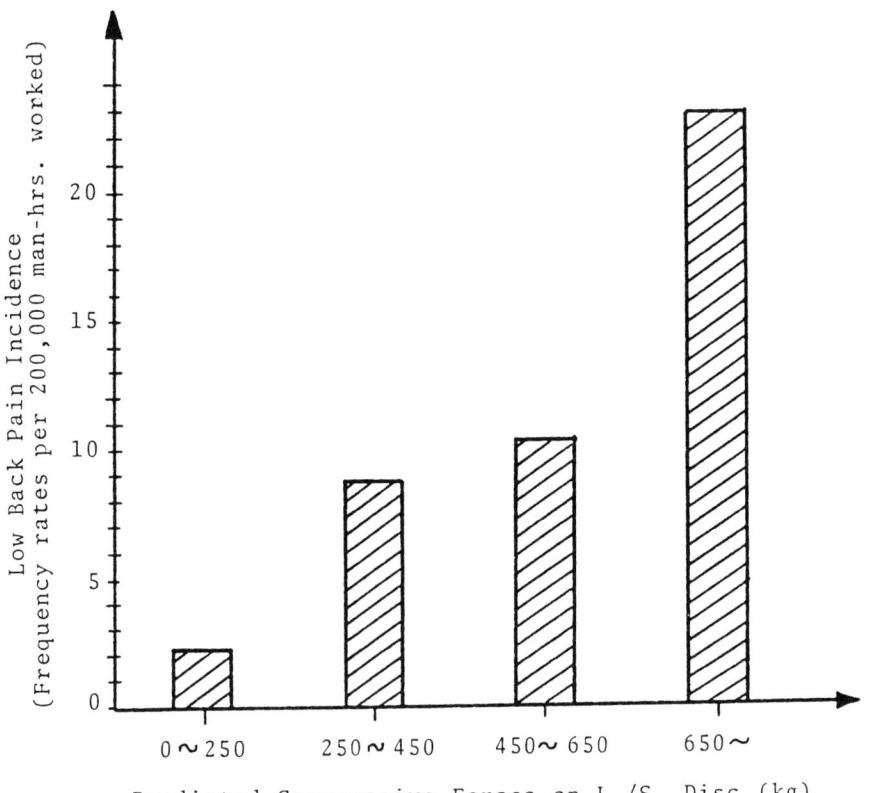

Figure 3.21. Relation between lower back pain and compressive force (Chaffin and Park, 1973). Reproduced with permission of American Industrial Hygiene Association Journal

low-back injuries. From the above observations, the muscle strength of workers and the estimated compressive force on the spine produced by a certain job, provide the biomechanical criteria for the evaluation.

3.2.1.9.2 MODELS USING THE BIOMECHANICAL CRITERIA
Several models are available for use in determining lifting capacity based on biomechanical criterion. El-Bassoussi (1974) developed a biomechanical model and a set of prediction equations based on the data of the model. The model equations predict the compressive force on the surfaces of the L4/L5 and L5/S1 discs as a function of the biomechanical equivalent (*BLE*). The *BLE* is a function of the weight of the load (W) and its distance from the base of the spine (D).

$$BLE = W(D).$$

Several assumptions are made with the *BLE*. First, the centre of mass of the load is assumed to be located at the midpoint of the length of the load. Secondly, the load will be pressed against the body while lifting. The third assumption is that the distance of the load from the spine does not vary during the lift. The values of D can be computed from the following equation:

$$D = (L/2 + 20) \text{ cm}$$

where L is the width of the load, and the thickness of the body from posterior surface to the base of spine is assumed to be 20 cm.

Thus, given these assumptions, the *BLE* allows a particular load to be compared with another load using a common measure.

El-Bassoussi (1974) developed both leg and back lift predictor equations. These equations (shown in Table 3.4) are linear functions of the *BLE* to predict the compressive force.

Freivalds et al. (1984) designed a dynamic biomechanical model consisting of seven rigid links joined at six articulations which evaluates job-related stresses imposed on the worker. This model, however, accounted for less than 20% variance of the data.

Kromodihardjo and Mital (1986b) and Mital and Kromodihardjo (1986) offer a three-dimensional model from which to determine spinal stresses generated during MMH activities based on the biomechanical criterion. The filming speed used by Kromodihardjo and Mital (1986b) and Mital and Kromodihardjo (1986) was 125 f/s compared to the film speed used by El-Bassoussi's (1974) which was 16 f/s, and the film speed used by Freivalds et al. (1984) which was 20 f/s. It was hoped that, since filming rate does affect acceleration profiles (Mital and Kromodihardjo, 1984; Kromodihardjo and Mital, 1986a), more accurate predictions could be made. However, Kromodihardjo and Mital's (1986b) three-dimensional dynamic model only explains 40% of the variance in the data. Neither the model of Freivalds et al. (1984) nor of Kromodihardjo and Mital (1986b) predicts imposed stresses well.

3.2.1.9.3 LIFTING CAPACITY DETERMINATIONS
The compressive strength limits can be substituted into the El-Bassoussi equations to obtain the lifting capacity in terms of the *BLE*. The *BLE* can then be resolved to deter-

mine the weight or size limits of the load. Table 3.4 shows the *BLE* for each of the critical values determined in Table 3.3. Figure 3.22 contains a set of weight by size curves for the *BLE* ranges in Table 3.5. The curves for the 36–50-year range are the only ones presented. Additionally, the values for the back and leg lifts are very close and for practical purposes are of negligible difference. For this reason only the values for the leg lift are presented.

Table 3.4. Prediction equations for the compressive forces on the L4/L5 and L5/S1 discs

Leg lifts

    Upper surface of S1 = 367 + 19·37819 BLE_*
    Lower surface of L5 = 371 + 19·44315 BLE
    Upper surface of L5 = 366 + 19·49252 BLE
    Lower surface of L4 = 366 + 19·5204 BLE

Back lifts

    Upper surface of S1 = 379 + 23·00299 BLE
    Lower surface of L5 = 390 + 23·04803 BLE
    Upper surface of L5 = 380 + 22·72843 BLE
    Lower surface of L4 = 376 + 22·77000 BLE

*BLE, the biomechanical lifting equivalent in kg-m. Compressive force in kg.

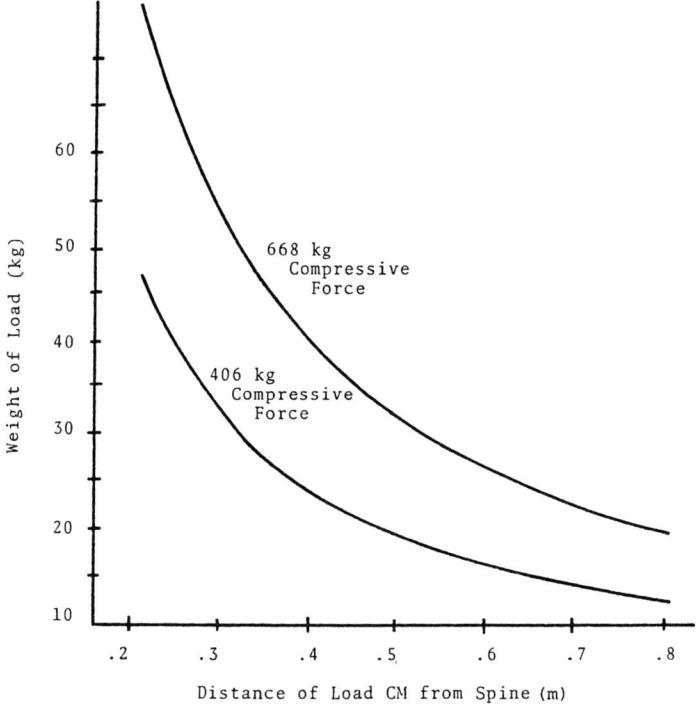

Figure 3.22. Variation of compressive force with distance of the load from the spine

3.3 The Physiological Approach

Physiological stress of interest in MMH is a stress acting upon the cardiovascular system which can be assessed through the measurement of physiological responses such as oxygen consumption, heart rate, blood pressure and lactic acid accumulation. These responses vary according to operator, task and environmental characteristics.

Table 3.5. Lifting capacity as a BLE for weight and age categories (from Ayoub et al., 1983)

Age	Male BLE (kg-m)	Female BLE (kg-m)
	Leg lifts	
21–35	28·820	
36–50	15·550	9·46
51–70	0·056	0·45
	Back lifts	
21–35	27·99	
36–50	14·83	9·02
51–70	−0·06	−0·05

A lifting task is considered to be a semi-dynamic activity because it is composed of both dynamic and isometric activities. This section deals with the physiological responses to lifting and is intended to arrive at a physiological design criterion for repetitive lifting tasks.

3.3.1 Classification of Activity

Physical activity can be classified into three categories: dynamic, isometric and semi-dynamic activities.

3.3.1.1 Dynamic Activity

A dynamic activity is an activity which requires a muscle group to vary its length when stimulated. Examples of this type of activity are walking, jogging, running, bicycling, arm cycling, lifting or lowering an object at a high frequency.

In a progressive exercise protocol, minute ventilation (VE) and the rate of oxygen uptake (VO_2) increase linearly until a level corresponding to approximately 40–60% of the subject's VO_2 max is reached. Above that level of work intensity, VE rises curvilinearly, reflecting an inordinate ventilatory response for the metabolic demand (Figure 3.23). The curvilinear increase in VE is in response to the increase in lactic acid production at high work intensities.

In a constant work-rate exercise protocol, the ventilatory response can be described by three phases (Figure 3.24).

Phase 1 is the immediate increase upon the beginning of the activity; phase 2 is the slow rise to the steady state level, and phase 3 is the steady state plateau (Fox and Mathews, 1981). The magnitude of the ventilatory response in phase 1 varies individually and with the level of work intensity. For mild intensity, phase 1 may constitute as

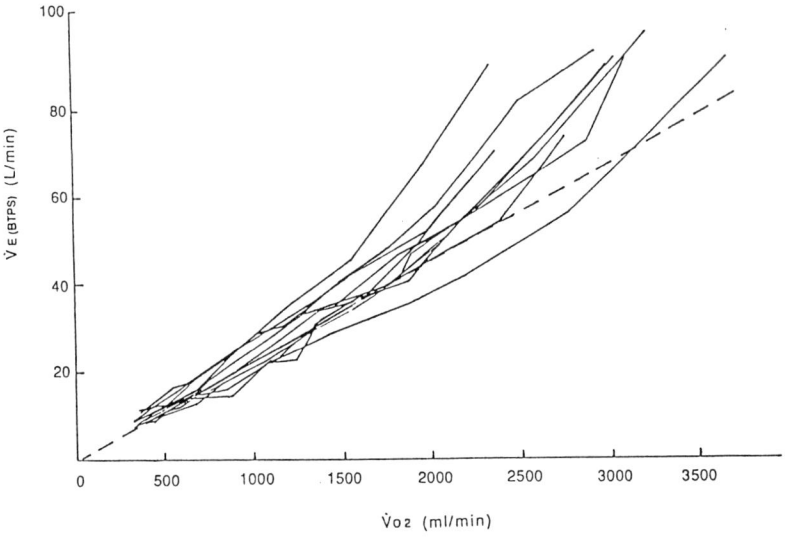

Figure 3.23.   Nonlinear relationship between V_E and V_{O_2}

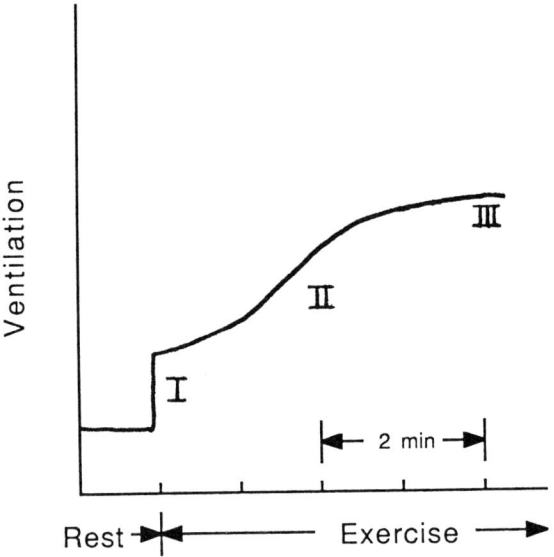

Figure 3.24.   Development of steady-state condition (Fox and Mathews, 1981)

much as 50% of the total ventilatory response in phase 3. For a high intensity workload, the initial response is just a small portion of the ventilatory response in phase 3.

Interestingly, O_2 consumption is linearly related to the amount of work done. Therefore, it is possible to measure the intensity of workload by measuring the VO_2. That is to say, VO_2 max can be achieved at maximum workload.

3.3.1.2 Isometric Activity

An isometric activity requires the muscle to contract with no change in length. Since the distance is zero in the isometric activity, no mechanical work is done. Holding or carrying boxes or suitcases are typical examples. This activity also demands energy and can be very fatiguing.

As is true in any muscular activity, muscle consumes energy when it contracts: metabolites are formed, oxygen, if available, is utilized, and carbon dioxide, water and heat are produced. Physiological responses due to isometric activity, however, are different from those found in dynamic activity (Lind, 1977). Therefore, maximum voluntary contraction (MVC) is introduced in an equivalent form to VO_2 max, which is found in dynamic activity. Fatigue occurs at tensions of about 20% MVC in isometric activity (Astrand and Rodahl, 1977), while fatigue in dynamic activity develops at the intensity of work which demands about 40% of VO_2 maximum and above.

3.3.1.3 Semi-dynamic Activity

A semi-dynamic activity is an activity which comprises both dynamic and isometric components. When the semi-dynamic activity is performed, the physiological responses of the isometric component are superimposed on those found in the steady state conditions of dynamic activity. In lifting tasks, the isometric component can be anticipated to be fairly large depending upon the origin and height of lift which require the exertion from different groups of muscle.

3.3.2 Physiological Response to Lifting

In contrast to the biomechanical approach, which is only applicable to the occasional lifting type (frequency of lift per minute is less than 4 lifts/min), the physiological approach is applied to repetitive lifting where the weight of the load is presumed to be within the *physical strength* of the workers. During this type of task, a person's endurance is primarily limited by the capacity of the oxygen transport system. When muscles become active, their increased metabolism demands an increase in the delivery of oxygen and nutrients to the tissue if the activity is to be continued.

In any physical work, irrespective of activity type, energy cost, heart rate, blood pressure and blood lactate are common physiological responses used to measure physiological stress. For a lifting task, energy cost and heart rate are probably the most widely used to determine the physiological limit. Petrofsky and Lind (1978a) reported the use of blood lactate analysis in finding the safe limit for lifting task. Blood pressure as a response to lifting task (two-hand lift) has not been reported in the literature up to the present time, probably because of the difficulties of the measurement technique.

3.3.2.1 Energy Cost

Karvonen (1974) reported the survey of work metabolic rates of occupational activities. Table 3.6 illustrates such values converted to kcal/min for a male of 77 kg with related tasks.

Ayoub et al. (1981a) studied the energy expenditure of various tasks performed by low seam coal miners in the United States. They reported average energy expenditure per shift (8 h minus a 30-min lunch break) of the most demanding tasks as shown in Table 3.7.

Table 3.6. Energy expenditure for various tasks (adapted from Karvonen, 1974)

Task description	Energy expenditure (kcal/min)
Sitting crane operator	3·37
Light welding	4·04
Masonry, painting, paper hanging	5·39
Walking	5·39
Carrying trays, dishes, etc.	5·66
Lifting and carrying objects (arm work)	
20–44 lb	6·06
45–64 lb	8·08
65–84 lb	10·11
85–100 lb	11·45
Shovelling, picking	10·78
Laying railroad track	9·43
Hammering, sawing	8·08

Table 3.7. Energy expenditure for mining tasks (adapted from Ayoub et al., 1981a)

Task description	Energy expenditure (kcal/shift)
Roof bolting	
Drilling	
Tramming machines	
Setting jacks	
Timbering	
Shovelling	
Resting, waiting, idle	
Travelling	2102·38
Miner Helping	
Bending bolts	
Moving cables	
Transporting supplies	
Rock dusting	
Timbering	
Shovelling	
Resting, waiting, idle	
Travelling	2788·78

It is obvious that the average expenditure per day or shift varies from task to task. This is due to the fact that energy cost required to perform the task is dependent upon the amount of muscle groups active during task performance. These muscles need oxygen and an energy source in the form of ATP (adenosinetriphosphate) so that they can contract efficiently. Furthermore, the difference in working postures from task to task cannot be overlooked.

Muller (1953) and Bink (1962) proposed the upper limit of the energy expenditure required for daily work should be 5–5·2 kcal/min. Bonjer (1962) measured energy expenditure to perform two tasks to validate Bink's (1962) model. The time to perform both tasks (milk delivery in four-storey apartment houses, and activities in cleansing departments) in the experiments required approximately 6–9 h. He proposed that if the ratio of average energy consumption required to perform task (M) to the upper limit (A) calculated from Bink's (1962) model was less than unity ($M/A < 1$), then a sufficient recovery during the periods was warranted irrespective of the peak demands of energy that could occur.

Michael et al. (1961) conducted cycle ergometry and treadmill walking experiments at difficult loads and speeds continuing for 8 h without interruption. They concluded that 35% of the maximum aerobic capacity (VO_2 max) was the limit of work that could be performed without undue fatigue.

Astrand (1967b) reported that 50% of VO_2 max (determined by bicycling) was probably not attainable by all subjects (building construction workers), and indeed, her subjects spontaneously chose between 25–55% of VO_2 max with a mean of 39%. In a study by Petrofsky and Lind (1978a), it was reported that lifting could be performed for up to 4 h if the workload was less than 50% of VO_2 (determined by lifting) for that individual for each specific weight of a box. They also concluded that lifting a box without fatigue was at a VO_2 of about 25% of VO_2 max (determined by

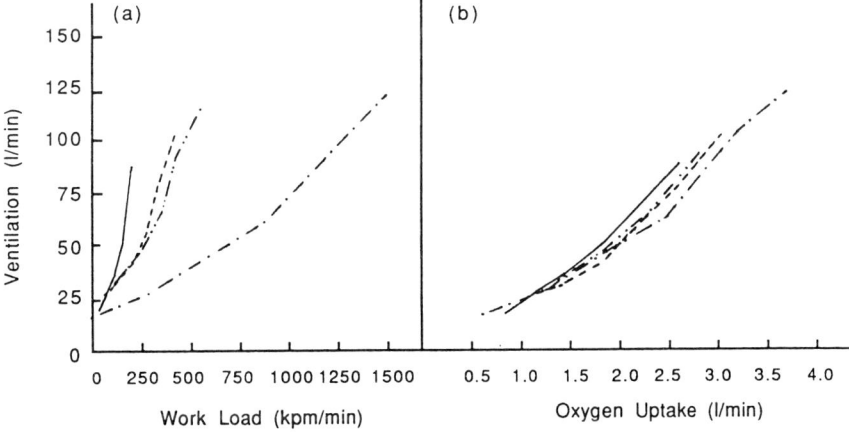

Figure 3.25. (a) Minute ventilation recorded at different workloads while lifting 6·82 kg (—), 22·73 kg (———) and 36·36 kg (—··—) boxes and during cycling (—·—). Each point on the figure illustrates the mean of two determinations on each of four subjects. (b) Minute ventilation plotted against VO_2 of the workloads. After Petrofsky and Lind (1978a)

lifting) for that individual for each specific weight of a box. They also concluded that lifting a box without fatigue was at a VO_2 of about 25% of VO_2 max (determined by a bicycle ergometer).

These differences of recommendations are not only affected by the age, bodyweight, training, genetic factors and sex of the workers, but by task type as well. Furthermore, a factor that influences the performance of a worker in any job is the VO_2 max for that job, particularly when the VO_2 max for that job is lower than for bicycling (Figure 3.25).

3.3.2.2 Heart Rate

Several investigators have also defined continuous work capacity in terms of heart rate criterion. The recommendations range between 99–130 beats/min. Brouha (1967) recommended that the mean heart rate should not exceed 115 beats/min which corresponded to the conclusion of Suggs and Splinter (1961). Snook and Irvine (1969) took a more conservative estimate on the mean heart rate criterion. They recommended that the mean heart rate should not exceed 112 beats/min for leg tasks and 99 beats/min for arm tasks. Lind (1977), however, suggested that: (1) industrial work had a greater isometric component than dynamic activity, (2) semi-dynamic activity resulted in a higher heart rate than dynamic activity, (3) HR/VO_2 for isometric activity was greater than HP/VO_2 for dynamic activity (HR, heart rate; HP, heart pressure), and (4) fatigue occurred in isometric activity faster than in dynamic activity. These facts lead to the conclusion that VO_2 and heart rate are task dependent.

3.3.2.3 Blood Pressure

Blood pressure is also task dependent. Astrand and Rodahl (1977) concluded that the blood pressure was significantly higher in arm exercise than in leg work. The typical response is an elevation in diastolic pressure. In isometric activity, the intrathoracic pressure is raised from 80–200 mm Hg or more causing a sharp increase in pressure, both systolic and diastolic. So far, there is no report on the blood pressure pattern in a lifting task, probably due to the difficulty of the measurement technique. However, Astrand et al. (1968) studied carpenters using a hammer to nail at different heights which could be considered as a semi-dynamic activity. They reported that heart rate and intra-arterial blood pressure when hammering into the ceiling were significantly higher than when hammering at bench level.

3.3.2.4 Blood Lactate Level

Petrofsky and Lind (1978b) employed blood lactate analysis to determine the safe limit for a lifting task. They used three male medical students as test subjects. All subjects had been trained in lifting, bicycling and isometric strength training for hand grip and back muscles for 12 weeks. Submaximal works were performed at 25, 40, 55 and 70% of the subject's aerobic lifting capacity for three different weights of boxes. The box weight was kept constant for each lifting task and only frequency was varied to reach the required VO_2 at each submaximal work load (Petrofsky and Lind, 1978a). The height of lift was from floor to 60 cm. EMG and blood lactate analysis were used to determine at what percentage of VO_2 max lactic acid began to accumulate in the venous return or when muscle fatigue occurred.

A 22·73 kg box was lifted for 4 h at about 50% of aerobic lifting capacity for each subject and the workload was increased the next day to check if the 4 h of lifting could be completed. They reported that no subject completed the operation if the workload was more than 60% of his aerobic lifting capacity, and concluded that lifting could be performed up to 4 h if the workload was less than 50% of the aerobic lifting capacity for that individual for each specific weight of a box. This limit was equivalent to 25% of his aerobic capacity determined by a bicycle ergometer. Figure 3.26 illustrates the patterns of lactic acid accumulation as the workload increased.

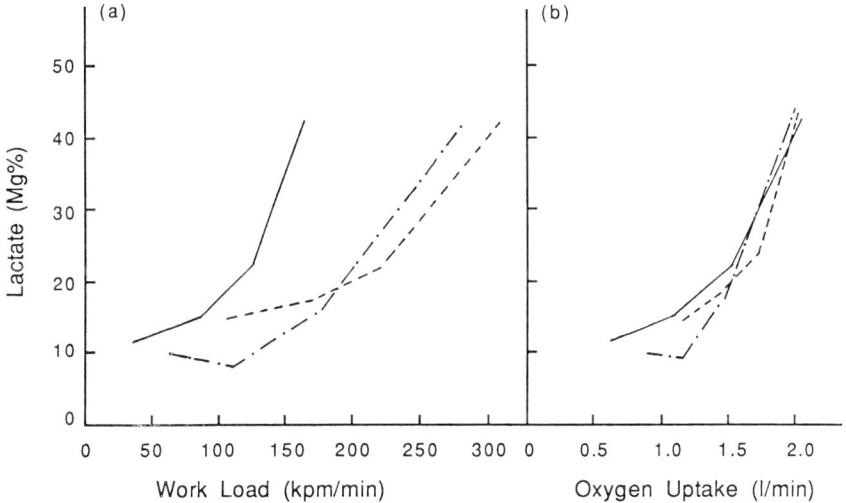

Figure 3.26. The mean arterial lactate concentrations measured between 15 and 30 seconds after the completion of a 1-hour bout of work lifting the 6·82 (—), 22·73 (·—·) and 36·36 (——) kg boxes. (a). Results plotted against the absolute workloads. (b). The same data plotted against percent VO_2 maximum. After Petrofsky and Lind (1978b)

3.3.3 Physiological Fatigue

For all types of physical activity, there are limits that cannot be exceeded, i.e., the work intensity that can be sustained, the amount of work that one can perform, and working time. If one attempts to perform a physical task beyond these limits, a decrease in working performance of that person will result. This physical phenomenon is termed physiological fatigue and is usually defined operationally as a decrease in the physical capacity of an individual to perform work (Capaln, 1971). This situation is often associated with an impairment of parts of the working element, i.e., the contractile mechanism. There are several factors that have been suspected to be the major causes of such impairment. Some of these are: (1) accumulation of lactic acid, (2) depletion of ATP and high energy phosphate stores, (3) depletion of muscle glycogen stores, and (4) lack of oxygen and inadequate blood flow to the muscle fibres.

3.3.3.1 Endurance Time

Endurance is defined as the ability to persist in an activity. Fatigue and endurance are related terms. Low endurance shows up as early fatigue. For a given physical task, an individual with higher physical work capacity is able to endure longer than an individual with lower physical capacity.

Endurance time limit is the time duration an individual is able to sustain the physical work until he is forced to give up due to fatigue. A number of research investigations have been made on determining the endurance time limits for given levels of work. In the case of industrial work, the interest is in the maximum level of physical work that can be sustained over a work day of 8 h or so. This level, in the opinion of the researchers, varies from about 20–50% of the maximal aerobic capacity. The difference is primarily due to the fact that the total amount of muscular mass involved in the work and the amount of rest pauses in between work bouts varies from case to case.

Based on his 40 years of experience, Simonson (1971) proposed the limits presented in Table 3.8 to be used as guidelines for this purpose. The limiting figures, for a given work duration from one minute to one year, are those underlined. The other figures are equivalent along any horizontal line. Thus, 4·6 kcal/min is equivalent to 275 kcal/h or 2200 kcal/8-h workday, or 13 200 kcal/48-h work-week, and so on.

Table 3.8. Maximum performance limits in terms of excess kilocalories (over resting rate) (from Simonson, 1971)

Year	Month	Week	Day	Hour	10 min	1 min
<u>616000</u>	57200	13200	2200	275	46	4·6
	<u>62920</u>	14500	2420	305	51	5·1
		<u>15480</u>	2640	330	55	5·5
			<u>3080</u>	385	64	6·4
				<u>523</u>	87	8·7
					<u>129</u>	12·9
						25·0

3.3.3.1.1 BINK'S LOGARITHMIC FORMULA

Bink (1962) developed a logarithmic relationship between working time (min) and energy required to work during that time (kcal/min) based on an average aerobic capacity (VO_2 max) of 15·1 kcal/min, an 8-h continuous work of 5·2 kcal/min, and a 24-h overall energy cost of 2·85 kcal/min. The acceptable level of energy expenditure (E) that an individual can perform for a certain time (min) is:

$$E = (\log 5700 - \log t) \times \text{aerobic capacity}/3·1$$

For an 8-h shift, the acceptable energy expenditure is 0·34 × aerobic capacity.

3.3.3.1.2 GRAPHICAL INTERPRETATION, WITH EXAMPLE

Figure 3.27 is a graphical presentation of the endurance time versus the energy cost when using the population aerobic capacity estimated from bicycling by Texas Tech University. It is recommended from the graph that the energy expenditure required to work an 8-h day be less than 5 and 3·35 kcal/min for male and female, respectively. Energy costs of 6·35 and 4·20 kcal/min to work for 4 h should be the upper limit for male and female, respectively.

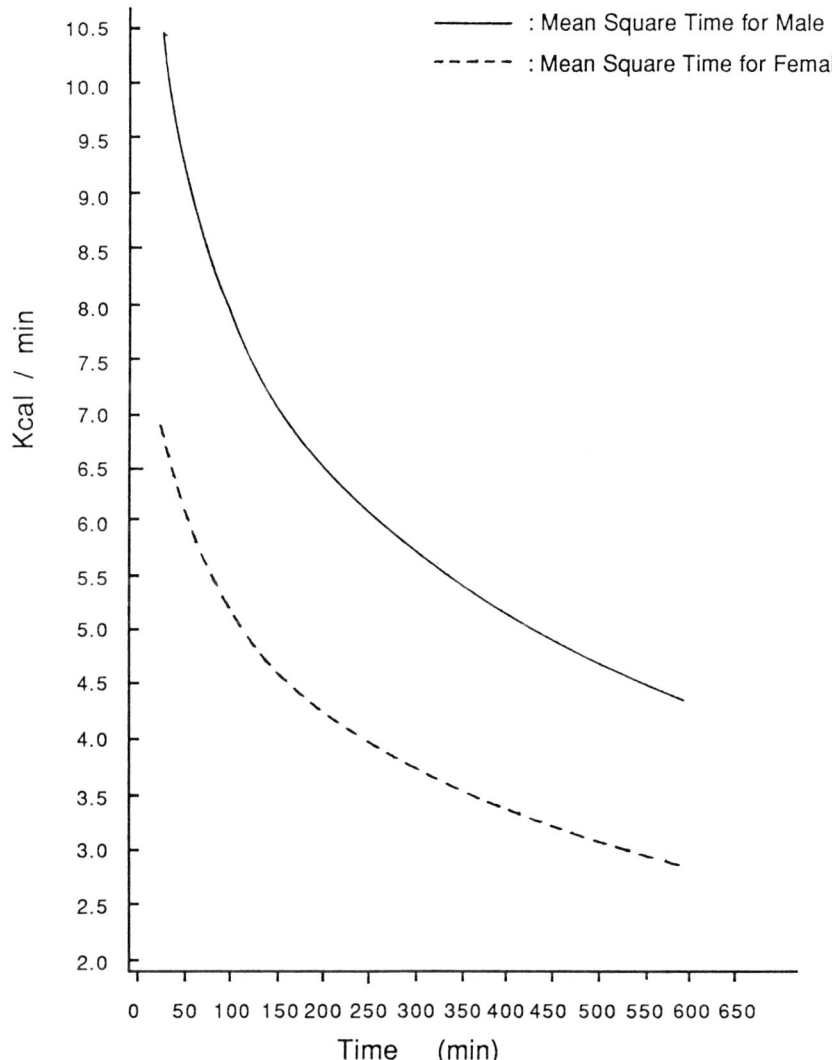

Figure 3.27. Endurance time (min.) vs. energy cost (kcal/min)

By using data from Bobo et al. (1983), endurance time for a roof bolter is 206 min based on the mean aerobic capacity and mean age for low seam coal miners of 13·64 kcal/min and 31·54 years, respectively. For miner helpers, however, the predicted endurance time is only 274 min. This is probably because most of the miner helpers were younger than the other group, but Bink's formula is based on the mean age of 35 years.

From this result, it is concluded that the predicted endurance time is more accurate if it is determined by age, group and sex.

Another source (Deivanayagam and Ayoub, 1979) states that endurance time has a logarithmic relationship which is:

$$\text{endurance time} = 10^{(0\cdot 26144 - E)/0\cdot 0712}$$

where E = gross energy requirement of the job (kcal/kg/min).

This formula can be interpreted such that the endurance time for a task that requires 100% max aerobic capacity is 4 min; and the endurance time for a task that requires 33% max aerobic capacity is 8 h. With this known, the energy requirement of a task, the endurance time, can be obtained analytically or numerically.

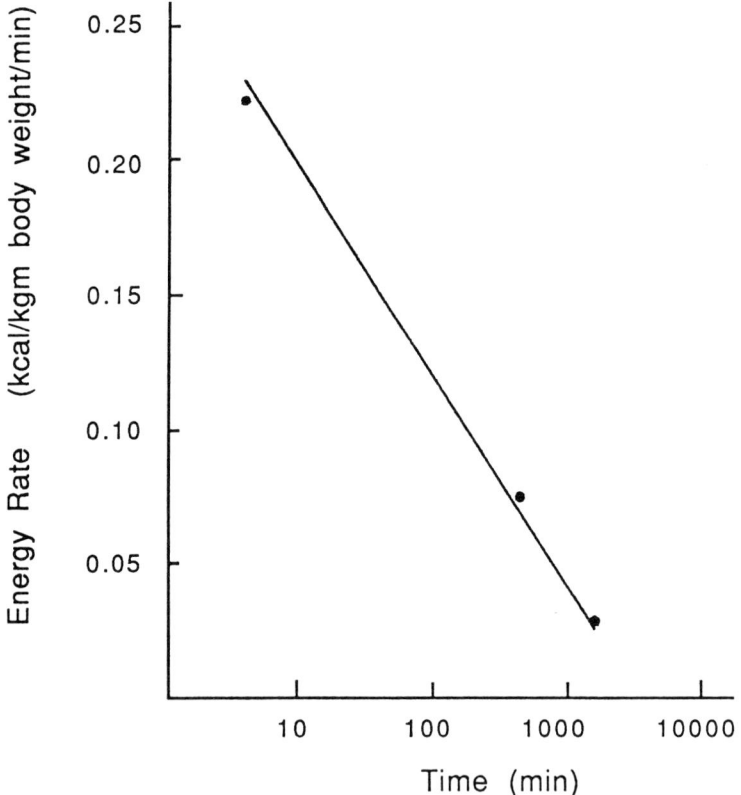

Figure 3.28. *Endurance time limit (ETL) model*

An example is used to illustrate the analytical method of calculating the endurance time, as shown in Figure 3.28.

Example
Physical Work Capacity (PWC) of operator = 3 l/min,
Oxygen consumption for task = 1·25 l/min,
Resting oxygen consumption = 0·25 l/min.

3.3.4 Physical Fitness for Lifting

Physical fitness, in general, means a level of capability to generate enough power that a person can utilize per day for his occupation, recreational activity and emergency needs. Physical fitness of an individual depends on health-related variables such as cardio-respiratory endurance, strength, flexibility and body composition (also see Chapter 2).

3.3.4.1 Aerobic Capacity

It has long been accepted that an individual who possesses a high aerobic capacity also has a high degree of physical fitness. In other words, he is capable of performing physical activity which demands high energy expenditure (Astrand and Rodahl, 1977; deVries, 1980; Fox and Mathews, 1981).

Aerobic capacity can be defined as the maximum level of metabolism of which a person is capable. An individual's aerobic capacity depends on the capacity to deliver oxygen to the working muscles, as well as procedure and equipment used to measure sex, age, body composition, genetic factors, training and level of the aerobic capacity at the start of the training (McNab et al., 1967; Kamon and Pandolf, 1972; Girandola and Katch, 1973; Astrand and Rodahl, 1977; Petrofsky and Lind, 1978a).

Methods and procedures to determine the aerobic capacity for an individual have been discussed extensively by Kamon and Ayoub (1976). A favourable piece of equipment to be used is a bicycle ergometer because it enjoys the advantages of objectively and accurately measured power output. It also lessens the problems of instrumentation, producing better ECG signals and providing a stable mouthpiece because of the lack of upper body movement.

Attempts have been made to estimate the population aerobic capacity (Hettinger et al., 1961; Cumming, 1967), but the results vary greatly. Texas Tech University employed three submaximal work load on a bicycle ergometer to predict the aerobic capacity by extrapolating heart rate to the maximum heart rate (220-age) for low seam coal miners, hard rock miners, university employees and college students (Table 3.9). The estimated VO_2 max is very close to those proposed by the NIOSH (1981) lifting guide. The estimated value of female aerobic capacity is about 67% of male aerobic capacity.

This section will use these values as a reference to establish the physiological design criterion.

Table 3.9. Estimated population aerobic capacity (working data of Texas Tech University)

Sex	n	Age year mean	(s.d.)	Weight lbs mean	(s.d.)	V_{O_2} max l/min mean	(s.d.)	kcal/min mean	(s.d.)
Male	409	28·9	(8·7)	175·2	(25·58)	2·84	(0·63)	14·19	(3·16)
Female	88	29·9	(6·6)	140·5	(25·34)	1·89	(0·42)	9·46	(2·09)

3.3.3.4.1.1 Aerobic Lifting Capacity

As previously mentioned, aerobic capacity for an individual is task-dependent. Petrofsky and Lind (1978a) recommend using aerobic lifting capacity as a measure of physical fitness for lifting. A standard procedure for testing aerobic lifting capacity has not yet been established. There are, however, two ways to increase workload to estimate the aerobic lifting capacity, i.e. (1) increasing frequency of lift, and (2) increasing weight of lift. Increasing height of lift seems to be another method for increasing the workload but it may not be suitable since the task or muscle groups used to perform the lifting task are altered from one level to another level.

Petrofsky and Lind (1978a) used the first approach, by increasing the frequency of lift and keeping the weight and height of lift constant for each lifting experiment, to determine the aerobic lifting capacity. Each subject was asked to lift a specific box until the maximum working capacity was established. They reported that using a lighter weight to lift produced less aerobic lifting capacity. They used four boxes weighing 0·91, 6·82, 22·73 and 36·36 kg (2, 15, 50 and 80 lbs) at different rates of lift. However, the result from this experiment may not be accurate, since varying the speed of the bicycle ergometer (in terms of rev/min) can result in variations in predicted V_{O_2} max up to 10% (Lind and McNicol, 1967; Hermansen and Saltin, 1969). Analogously, one may argue that varying the 'speed' of a lifting task in terms of frequency of lift, as was done in this study, may result in variations in the predicted V_{O_2} max.

Intaranont (1983) used the second approach by increasing the weight to lift and sub-maximal procedure to determine an individual's aerobic lifting capacity. Lifting a box from floor to knuckle height and from knuckle height to shoulder height with three frequencies of lift (6, 7·5, 9 lifts/min) was used in his study. The submaximal procedure required each subject to lift the box at three different workloads for 4 min at each load. Heart rate and V_{O_2} were monitored and recorded accordingly. Aerobic lifting capacity was calculated by extrapolating the regression line of V_{O_2} and heart rate at steady state to the predicted maximum heart rate (220-age), and projecting for V_{O_2} max in l/min as described by Kamon and Ayoub (1976), deVries (1980), and McArdle et al. (1981). Table 3.10 illustrates the summary of the results as compared to the V_{O_2} max obtained from bicycle and arm cycling.

It is apparent that aerobic lifting capacity is less than the aerobic capacity determined by bicycling which is in agreement with Petrofsky and Lind (1978a). Interestingly, the aerobic capacities determined from arm-work tasks are almost the same. It is also speculated that when lifting a box from floor to knuckle height at higher frequency, the aerobic lifting capacity will approach the aerobic capacity determined by bicycling

Table 3.10. Comparison of aerobic capacity of various tasks (Intaranont, 1983)

Task description	l/min	Aerobic capacity ml/kg/min	kcal/min
Bicycling	3·54	48·36	17·69
Arm Cycling	2·85	39·30	14·39
Lifting:			
From floor to knuckle			
6 lifts/min	2·94	40·06	14·71
7·5 lifts/min	2·89	39·54	14·47
9 lifts/min	3·12	42·73	15·63
From knuckle to shoulder			
6 lifts/min	2·96	40·74	14·80
7·5 lifts/min	2·92	40·04	14·62
9 lifts/min	2·85	39·14	14·27

since the pause time between successive lifts becomes shorter. Besides, more muscle groups will be active and, therefore, more oxygen is delivered.

3.3.4.2 Anaerobic Threshold

The body depends upon the utilization of oxygen to supply energy for muscular work. As long as the intensity of the work equals the capacity of the body to transport and utilize oxygen, no fatiguing metabolites (i.e., lactic acid, H-ion) are produced within the system. This is regarded as 'aerobic work'. If, however, the intensity of work exceeds the body's ability to supply oxygen, lactic acid begins to build up in the tissues and work, if continued, must utilize 'anaerobic' energy sources. The point at which lactic acid begins to accumulate in the blood stream has been termed the anaerobic threshold (Wasserman et al., 1973).

The anaerobic threshold (AT) has been identified as a new parameter that may be closely related to the percentage of the capacity that one can maintain for a prolonged period of time (deVries, 1980). Wasserman and Whipp (1975) introduced a non-invasive (bloodless) test to determine the AT during graded exercise on the bicycle ergometer. Davis et al. (1979) showed a 95% correlation between the values of the AT determined by an invasive and a non-invasive method.

Graded exercise protocol is required to identify the AT. For the non-invasive test which uses gas exchange indices, one can detect the AT at a point of (1) non-linear increase in VE, (2) non-linear increase in V_{CO_2}, (3) an increase in end-tidal oxygen with no corresponding decrease in end-tidal CO_2, (4) a sudden increase in respiratory quotient, (5) a sudden increase of the fraction of oxygen in the expired air (FEO_2), and/or (6) a systematic increase in the ventilatory ratio for oxygen (VE/VO_2) without an increase in the ventilatory ratio for CO_2 (VE/V_{CO_2}) (Davis et al., 1976). Caiozzo et al. (1982) reported that the use of VE/VO_2 for non-invasive detection of the AT gave higher correlation with the invasive test than any other detection criteria. Ready and Quinney (1982), however, used other criteria in their study when the VE/VO_2 was not apparent.

In the case of repetitive lifting, Petrofsky and Lind (1978b) were the first investigators to study the development of fatigue invasively. They performed blood lactate analysis to determine when lactic acid occurred. This point is known as the AT. They reported that fatigue occurred when the level of work exceeded 50% of the aerobic lifting capacity.

Intaranont (1983) used a graded exercise protocol for a lifting task by increasing the weight of lift every minute in order to determine the AT non-invasively. Three frequencies (6, 7·5, and 9 lifts/min) and two heights of lift (floor to knuckle height and knuckle height to shoulder height) were used in his study. He used a relationship between VE and VO_2, which is a straight line up to the AT, by employing a piecewise linear regression to determine the change of linear relationship as the principal criterion. He also used other criteria, a sudden increase of FEO_2 in particular, when the first criterion was not apparent. At the workload which demanded 90% of the AT, each subject was asked to lift a box for about 30 min to verify the principle of the AT. Table 3.11 shows the results from his experiment.

The results show that the aerobic threshold is also a function of the height of lift. Percentage of the AT to the aerobic lifting capacity for lifting from floor to knuckle seems to agree with that reported (50%) by Petrofsky and Lind (1978a), but the percentage of the AT to the bicycling aerobic capacity does not. This is always the case when using subjects who are not used to bicycling. However, for lifting from knuckle to shoulder, such percentages are significantly less than 50%. This is probably because the individual's muscular strength rather than his physiological endurance becomes the limiting factor as pointed out by Asfour (1980).

There is a similar conclusion drawn from the experiment by Garg (1983) when he attempted to employ the physiological approach to determine the lifting capacity for one-handed lift in the horizontal plane. He concluded that physiological responses (VO_2, blood pressure, heart rate), as well as grip strength and endurance time were not significantly changed as the work intensity changed. This is probably because this activity requires a small group of muscles to be active, therefore localized muscle fatigue occurs prior to any significant physiological responses.

Table 3.11. Comparison of aerobic lifting capacity and AT (Intaranont, 1983)

Task description	A	B	C	D
Lifting				
From floor to knuckle				
6 lifts/min	2·94	1·53	52·31	43·29
7·5 lifts/min	2·89	1·55	53·93	43·84
9 lifts/min	3·13	1·63	52·53	46·21
From knuckle to shoulder				
6 lifts/min	2·96	1·20	41·41	34·04
7·5 lifts/min	2·92	1·13	38·60	31·81
9 lifts/min	2·85	1·10	38·75	31·11

A = average aerobic lifting capacity, l/min.
B = average anerobic threshold, l/min.
C = percentage of AT to the corresponding value in A.
D = percentage of AT to the bicycling aerobic capacity.

3.3.5 Design Criterion

Energy expenditure (or O_2 uptake) and heart rate were used extensively as the reliable physiological responses to the workload since both variables are directly proportional to the workload at steady state conditions (Astrand and Rodahl, 1977; deVries, 1980; Fox and Mathews, 1981). However, energy expenditure is more popular as a measure of physiological stress due to its relationship to the workload as shown in the works of Durnin and Passmore (1967), Aquilano (1968), Hamilton and Chase (1969) and Ayoub et al. (1981a), for example. Additionally, Fernandez (1986) noted that an increased heart rate correlates with decreased O_2 consumption. The energy expenditure (kcal/min) is considered to be $5 * VO_2$ where VO_2 is expressed in l/min. Since more researchers have accepted the use of energy expenditure (as well as VO_2) as the measure of physiological stress than heart rate, the authors will also use energy expenditure as the major design parameter.

3.3.5.1 Criterion Limit

There are two major areas of disagreement in the determination of lifting capacity:

1. At what percentage of such capacity should an individual work?
2. What type of aerobic capacity should be used (as determined by bicycling or lifting activity)?

To circumvent these areas of controversy, this section will attempt to generate the lifting capacity from various mathematical models and compile necessary information in a graphical form, so that a job design can be accomplished with greater flexibility and ease. The safe lifting limit will be determined for a certain percentage of the population's aerobic capacity estimated by Texas Tech University within a certain range of lifting frequencies and between certain heights of lift.

3.3.5.2 Factors Affecting the Design Criterion

To develop a predictive model of the design criterion for lifting tasks, a number of factors must be considered. They are basically classified into three types: worker, task and environmental factors (see Chapter 2 for details).

3.3.5.3 Models Using the Physiological Criterion

Four predictive models are presented, each of them having some limitations. Each model has advantages over another under certain conditions.

3.3.5.3.1 Frederick's Model

Frederick (1959) developed a predictive model to estimate the energy expenditure resulting from manual lifting and constructed a graph from his experiment showing the relationship between weights to lift and energy consumption in gram-calories per foot-pound at different lifting ranges (Figure 2.15, Chapter 2). The formula is:

$$E = F \times a \times W \times C/1000$$

where F = the frequency of lift, lifts/h; a = vertical lifting ranges, ft; W = weight to lift, lbs; C = energy consumption, g-cal/ftlb.

For example, suppose a lifting task requires an operator to lift a box (15 × 15 × 10 in) weighing 40 lbs from floor to 20 in above the floor with a frequency of 8 lifts/min, the energy expenditure can be determined as follows:

From Figure 2.15 C is 6·75 g-cal/ft-lb. W is 40 lbs, a is 20/12 ft and F is 8 × 60 lifts/min. Substituting these values into the formula, therefore,

$$E = 8 \times 60 \times (20/12) \times 40 \times 6\cdot75/1000$$
$$= 216 \text{ kcal/h or } 3\cdot6 \text{ kcal/min.}$$

He also recommended that the energy expenditure should not be more than 3·33 kcal/min (200 kcal/h) for an average man to work all day.

3.3.5.3.2 GARG'S MODEL

Garg et al. (1978) developed regression equations to predict energy expenditure for MMH operations. They assumed that a complex task could be divided into several subtasks. If the energy consumption for each subtask was known, the summation of all energy expenditures required for each subtask would be equal to the energy expenditure required for the main task. The following equations are for lifting with three different postures (standing, stoop and squat).

Arm lift (standing position assumed):

$$E = 0\cdot024BW$$
$$+ (0\cdot062BW(H_2 - 0\cdot81) + (3\cdot19L - 0\cdot52SL) \times (H_2 - H_1))$$
$$\times F/100,$$

for $0\cdot81 < H_1 < H_2$

Stoop lift (standing position assumed):

$$E = 0\cdot024BW$$
$$+ (0\cdot325BW(0\cdot81 - H_1) + (1\cdot41L + 0\cdot76SL) \times (H_2 - H_1))$$
$$\times F/100,$$

for $H_1 < H_2 < 0\cdot21$.

Squat Lift (standing position assumed):

$$E = 0\cdot024BW$$
$$+ (0\cdot514BW(0\cdot81 - H_1) + (2\cdot19L + 0\cdot62SL) \times (H_2 - H_1))$$
$$\times F/100$$

for $H_1 < H_2 < 0\cdot81$, where E = energy expenditure, kcal/min; BW = body-weight, kg; H_1 = the starting point of lift, m; H_2 = the ending point of lift, m; S = gender, male is 1, female is 0; F = frequency of lift, lifts/min; L = weight lifted (kg).

If the previous example is used, $H_1 = 0$, $H_2 = 0\cdot508$, $F = 8$, $L = 18\cdot14$ (40 lbs), and assuming a male (S = 1) weighing 77 kg and stoop lift is chosen, then substituting into Garg's model,

$$E = 0\cdot024 \times 77$$
$$+ (0\cdot325 \times 77 \times (0\cdot81 - 0) + (1\cdot41 \times 18\cdot14 + 0\cdot76 \times 1 \times 18\cdot14))$$
$$\times (0\cdot508 - 0) \times 8/100 = 5\cdot069 \text{ kcal/min}.$$

3.3.5.3.3 ASFOUR'S MODEL

Asfour (1980) used stepwise multiple linear regression techniques to develop energy cost prediction models for lifting and lowering tasks.

The following model is for lifting (lowering) from floor up to 30 in above the floor.

$$VO_2 = 545\cdot7538 - 106\cdot4477TA$$
$$+ FLL\ (35002\cdot65 - 350\cdot58L)/1000000$$
$$+ 17\cdot47FLHWBANG/1000000$$
$$+ 16435\cdot22BWFF/1000000.$$

The following model is for lifting (lowering) from table height (30 in) up to 50 in above the floor:

$$VO_2 = 371\cdot5055 - 51\cdot9573TA$$
$$+ BWFF(31856\cdot54 - 2332\cdot8F)/1000000$$
$$+ 12684\cdot91FLL/1000000$$
$$+ 12\cdot31FHLWBLBANG/1000000$$

where VO_2 = oxygen consumption, ml/min; TA = task type, TA = 1 for lifting, TA = 2 for lowering; BW = body-weight, lbs; F = frequency of lift, lifts/min; L = weight to lift, lbs; H = height of lift, in; WB = box width, in; LB = box length, in; ANG = angle of twist, ANG = 1 for 0°, ANG = 2 for 90°.

Asfour's model can also be applied to the previous example as follows:

$$TA = 1, F = 8, L = 40, H = 20, WB = 15, LB = 15, ANG = 1 \text{ and } BW = 169\cdot76.$$

Substitute these values into the first equation:

$$VO_2 = 545\cdot7538 - 106\cdot4477 \times 1$$
$$+ 8 \times 40 \times 40 \times (35002\cdot65 - 350\cdot58 \times 40)/1000000$$
$$+ 17\cdot47 \times 8 \times 40 \times 20 \times 15 \times 15 \times 1/1000000$$
$$+ 16435\cdot22 \times 169\cdot76 \times 8 \times 8/1000000$$
$$= 578\cdot704 \text{ ml/min or } 2\cdot894 \text{ kcal/min}.$$

3.3.5.3.4 INTARANONT'S MODEL

Intaranont (1983) used stepwise multiple linear regression techniques to develop models of the anaerobic threshold (l/min) and capacity (lbs).

Lifting from floor to knuckle height (R − square = 0·8692):

$$AT = (471892\cdot555 + 1\cdot439WTFF - 3461\cdot837\ PB$$
$$- 11\cdot744WTWT - 3771\cdot16WT \exp R$$
$$+ 24\cdot964\ LBWLBW) \times 10 \exp(-5)$$

where AT = the anaerobic threshold, l/min; WT = body-weight of a subject, lbs; F = frequency of lift, lifts/min; LBW = lean body-weight of a subject, lbs; $R = LBW/WT$; $PB = PWCB \times 1000 \times 2 \cdot 2046/LBW$, ml/kg (LBW) – min; $PWCB = PWC$ determined by bicycling, l/min.

Lifting from floor to knuckle height (R – square = $0 \cdot 8714$):

$$L90 = (1044206 \cdot 994 - 764422 \cdot 134\, F + 229233 \cdot 277\, AK \\ + 86454 \cdot 21\, PBW) \times 10 \exp(-5)$$

where $L90$ = lifting capacity at 90% of the AT, lbs; F = frequency of lift, lifts/min; $AK = 0 \cdot 9 \times AT \times 1000 \times 2 \cdot 2046/WT$, ml/kg (WT) – min; AT = the anaerobic threshold (predicted or actual), l/min; WT = body-weight of a subject, lbs; $PWB = PWCB \times 1000 \times 2 \cdot 2046/WT$, ml/kg (WT) – min; $PWCB = PWC$ determined by bicycling, l/min.

Lifting from knuckle to shoulder height (R – square = $0 \cdot 8944$):

$$L90 = (3018662 \cdot 771 - 616833 \cdot 995\, F + 330678 \cdot 86\, AKB \\ + 10152 \cdot 833\, LBW) \times 10 \exp(-5)$$

where $L90$ = lifting capacity at 90% of the AT, lbs; F = frequency of lift, lifts/min; $AKB = 0 \cdot 9 \times ATB \times 1000 \times 2 \cdot 2046/WT$, ml/kg (WT) – min; ATB = the anaerobic threshold (predicted or actual) for arm lift, l/min; WT = body-weight of a subject, lbs; LBW = lean body-weight of a subject, lbs.

3.3.6 Comparisons of the Models with Limitations

Up to the present time, there are three mathematical models for predicting the energy cost of lifting as described in the above section. Unfortunately, each one has certain limitations. Since Frederik's model is limited to certain ranges of lift, only the models by Garg et al. (1978) and Asfour (1980) will be used to estimate lifting capacity.

Garg's models can be applied to both genders and both stoop and squat lifts, but the assumption of additivity creates some doubtful outcomes. Asfour's models involve many independent variables making the model complex and can be applied to both lowering and lifting tasks in the sagittal plane and with twisting motion. These, to a certain extent, will reduce the reliability of the results if one attempts to use the model to predict energy cost for lifting in the sagittal plane alone.

Lifting capacity is generated from the energy cost prediction models developed by Garg et al. (1978) and Asfour (1980) for certain heights and techniques of lift. But only lifting a box (frequencies of lift range from 3–12 lifts/min) in the sagittal plane is presented. Male and female with weights of 77 and 62 kg, respectively, are used. Box length and box width are assumed to be $38 \cdot 1 \times 38 \cdot 1$ cm (15×15 in). A criterion limit is at 33% of the population aerobic capacity estimated ($4 \cdot 73$ l/min for males and $3 \cdot 15$ l/min for females). Figure 3·29 illustrates the comparison between lifting capacity generated from both models for lifting from knuckle to shoulder heights. The models differ in their results. This is probably due to the above mentioned deficiencies and the difference in experimental methodology. Figure 3.30 (a) and (b) show the lifting capacity generated by Garg's models for various types of lifts, for males and females, respectively.

Design Criteria 115

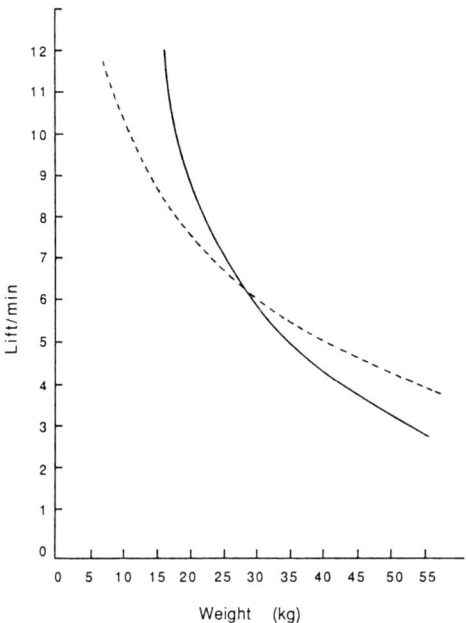

Figure 3.29. Comparison of lifting capacity between the Asfour (- - -) and Garg (—) models for male lifting from 0·81 to 1·27 m

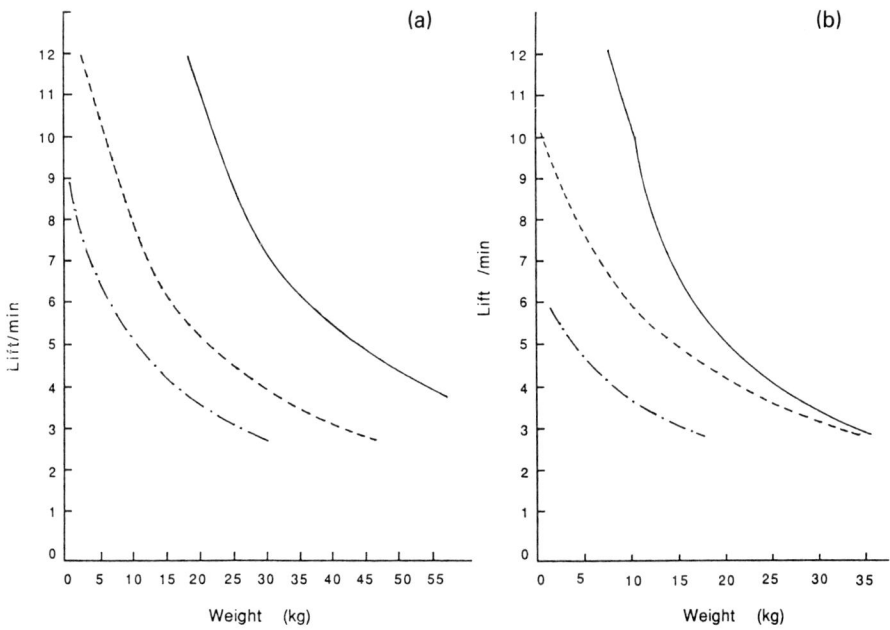

Figure 3.30. Comparison of lifting capacities (a) for males and (b) for females, using Garg models for different types of lift. (—) lifting from 0·81 to 1·27 m; (- - -) stoop lift from floor to 0·76 m; (—·—) squat lift from floor to 0·76 m

Physiological stress is defined in terms of energy cost of the activity. Biomechanical stress is defined as the compressive force acting on the spine due to the lift. It is obvious that biomechanical stress is always present during lifting tasks even with a light lifting weight because body and/or body segments are lifted and lowered. Therefore, it can be concluded that lifting capacity determined by evaluating physiological stress alone is not sufficient. The magnitude of the biomechanical stress varies according to frequency of lift, weight to lift, and posture required to lift. Lifting a box from table to shoulder height involves small muscle mass, therefore, physiological stress may not be significant as opposed to biomechanical stress.

With regard to the first area of disagreement, Mital (1983c) reports a decrease of energy expenditure over the work shift duration. Subjects chose an acceptable load weight for an 8-h shift. Adjustments of load weight were permitted throughout the experiment. Males lifted 65% of the estimated value, while females lifted 84% of the estimated value. When the work shift was increased to 12 h, males lifted 70% of the estimated value while females lifted 77% of the estimated value. Energy expenditure rates decreased over the 12 h shift while heart rate remained fairly constant. Mital (1983c) suggests that subjects were aware of their physiological strain and adjusted the load weight to maintain constant circulatory burden.

3.4 The Psychophysical Approach

3.4.1 Rationale for Use of the Psychophysical Design Criterion

In the previous sections two approaches for the determination of lifting capacity have been introduced, the biomechanical approach and the physiological approach. In this section a third approach, the psychophysical design criterion, will be introduced. It is the assertion of this book that the psychophysical design criterion is one of the most appropriate design criterion for the establishment of MMH capacities.

The use of psychophysics in lifting tasks requires the subject to adjust the weight of load or in some instances, frequency of handling, according to his or her perception of effort such that the MMH task does not result in over-exertion or excessive fatigue. As such, the psychophysical design criteria is a measure of perceived stress.

The psychophysical design criterion is based on the assumption that both biomechanical and physiological stresses are present in any MMH task, and that these stresses are integrated or combined under the measure of perceived stress. The validity of this assumption was recently demonstrated empirically by Karwowski (1982) and Karwowski and Ayoub (1984b).

Karwowski hypothesized that a combination of the acceptability of biomechanical and physiological stresses imposed during manual lifting leads to an overall measure of the lifting task acceptability, expressed by the acceptability of the psychophysical stress. To test this hypothesis, the fuzzy sets theory proposed by Zadeh (1965) was utilized. Fuzzy sets theory was developed to deal with classes in which there are grades of membership intermediate to full membership and non-membership.

The primary concept in fuzzy sets theory is the development of a set in which membership values can vary over a continuum from 0 (non-membership) to 1 (full membership). Membership functions were developed by Karwowski for the biomechanical, physiological and psychophysical stress, with membership values ranging from 0 (i.e., totally unacceptable stress) to 1 (i.e., totally acceptable stress). The stresses, given a value of 0 to 1, were selected based on past research in the areas of acceptable and unacceptable biomechanical, physiological and psychophysical stresses. For example, the membership function for the acceptability of the psychophysical stress was based on the lifting capacity norms presented in Figure 3.31. A measure of combined stress was then determined by taking the algebraic product of the acceptability measures of the biomechanical and physiological stress. Following this, mathematical procedures stemming from fuzzy sets theory were used to determine the relationship between the maximum acceptable weight of lift from the psychophysical and combined standpoints. Based on these mathematical procedures, it was concluded that the maximum acceptable weight of lift based on the psychophysical design criterion appeared to be the result of the integration of the biomechanical and physiological stresses imposed by the lifting task.

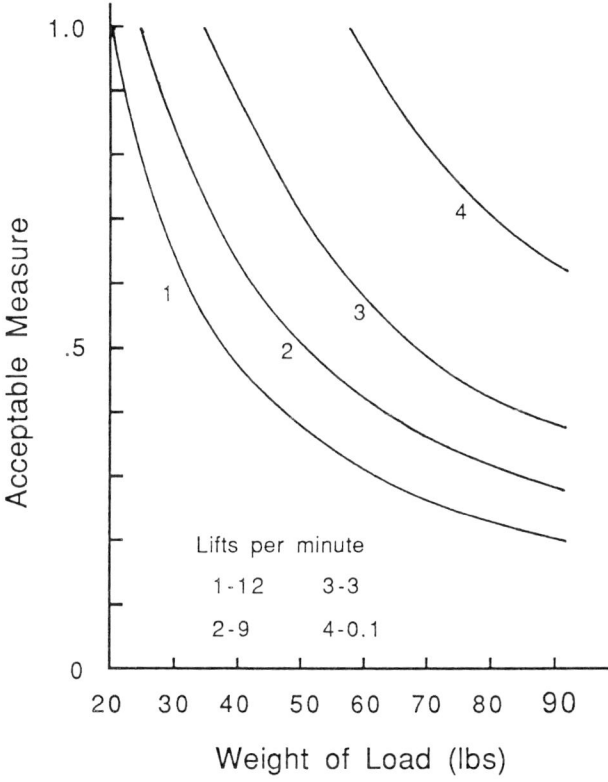

Figure 3.31. Membership function of psychophysical stress

Hafez (1984) conducted a psychophysical lifting study to assess the lifting capabilities of individuals in hot environments. A fuzzy sets based model was used to test the hypothesis that a combination of the acceptability of the biomechanical, physiological and thermal stresses leads to an overall measure of the lifting task acceptability. Like Karwowski (1982), Hafez determined the combination of acceptability measures of the biomechanical and physiological stresses that could predict the overall acceptability of the lifting task based on the psychophysical design criterion. However, use of the fuzzy sets model indicated that thermal stress could not be expressed independently when combining the stresses due to an apparent confounding with the physiological stress.

3.4.2 Principles of the Psychophysical Design Criterion

Psychophysics is a branch of psychology concerned with the relationship between sensations and their corresponding physical stimuli. Stevens (1960) indicated that the strength of a sensation (S) is directly related to the intensity of its physical stimulus (I) by means of a power function:

$$S = kI^n$$

where k is a constant which is a function of the particular units of measurement that are used, and n is the slope of the line that represents the power function when plotted in log–log coordinates. Exponents determined experimentally for different types of stimuli were found to be 1·6 for the perception of muscular effort and force, and 1·45 for lifting weights.

The use of psychophysics in the study of MMH tasks requires the subject to adjust one of the task variables (weight of load or frequency of lift) according to his or her perception of physical strain. All other variables such as container size and range of lift are controlled. Subjects are instructed to adjust their workload to the maximum amount they can sustain without strain or discomfort, and without becoming unusually tired, weakened or overheated, or out of breath. The final workload decided on by the subject is referred to as the maximum acceptable weight of load (MAWL).

The psychophysical methodology has been successfully used by several researchers for determining load lifting capacities of individuals or groups of individuals. An advantage of the psychophysical design criterion over other methods is that it may be applied to all types of industrial tasks and MMH activities without the need for maximum exertions. One of its disadvantages, however, is that the special controls necessary for psychophysical measurements frequently restrict its use to laboratory investigations rather than field studies.

The type of subject used in a psychophysical study is very important. If the study is investigating industrial tasks, then industrial workers should be used as subjects. If a student population is used, efforts should be made to match the characteristics of the student population with that of the appropriate industrial population. As a minimum, the students should be matched in terms of height and weight with the industrial population. It is also advisable to have the same subjects returning for several days or weeks in order to overcome the motivational effects created by the experimental atmosphere which will frequently cause the subjects to overestimate their capabilities. In this same vein, a training period prior to actual psychophysical data collection is suggested. This allows the subject practice in determining psychophysical MAWLs, and eliminates confounded results due to increases in strength and/or stamina as a result of performing the psychophysical tasks. The effects of motivational factors and training are discussed in greater detail later (chapter 8).

Instructions given to the subjects must be specific and carefully worded to insure complete understanding. Generally, they are instructed to imagine that they are on piece-work, getting paid for the amount of weight that they can lift or for the number of lifts that they can make, but working a normal 8-h shift that allows them to go home without feeling excessively tired. In other words, they are asked to work as hard as they can without becoming unusually tired, overheated, or out of breath.

Subjects are encouraged to make changes in the object weight by starting them with a very light or a very heavy weight. Building a false bottom into the object being handled is also desirable in order to reduce visual cues.

Tasks under study should be run at least twice, each time with a different initial weight randomly selected. If the results of the subject's first test are within 15% of the second test, as suggested by Snook (1978a, b), the average of the two results is recorded. Otherwise, the results should be discarded and the test rerun at another time.

In a psychophysical study the subject is given control of one of the task variables, usually the weight of the object being handled. All other variables such as frequency of lift, load dimensions and height, are controlled by the experimenter.

3.4.3 Factors Affecting the Psychophysical Design Criterion

Manual lifting is affected by three main categories of variables: (1) individual or worker variables, (2) task variables, and (3) environmental variables. A summary of these variables is presented in Chapter 2.

Individual or worker variables include physical, physiological, psychological and experiential factors. Physical factors include age, sex, anthropometric and strength characteristics. See Chapter 2 for the influence of these factors on MMH capability.

Table 3.12 shows the equations used to fit experimental data. These equations indicate that differences exist in the psychophysical MAWL as a function of the range of lift. These differences are a function of two principles. First, mechanical work is a function of the distance moved. Second, different muscle groups are involved in different ranges of lift. As Table 3.12 indicates, the psychophysical MAWL was highest in the case of lifting from floor to knuckle height. This finding is in agreement with the results of other research (Ayoub *et al.*, 1978; Snook, 1978a, b; Asfour, 1980). Analogously, Snook (1978a, b) reported that the maximum acceptable weight in a psychophysical carrying task decreased overall as carrying distance was increased from 2·1–4·3–8·5 m.

The effect of box size in the sagittal plane on psychophysical lifting capacity has received most of the researchers' attention; this is due to the fact that increasing box size in the sagittal plane increases the box centre of gravity relative to the subject. Asfour (1980) varied box size in both the sagittal plane (box size = 15–26 in) and frontal plane (15–26 in). A significant ($P < 0.01$) decrease in psychophysical MAWL was reported as box size was increased in the two planes. Bakken (1983) reported no significant difference in psychophysical MAWL as box size in the sagittal plane was increased from 14–20 in. Ayoub et al. (1978) reported a linear relation between the psychophysical MAWL and box size in the sagittal plane, the MAWL being inversely proportional to the box size. Table 3.13 (from Ayoub et al., 1983b) presents adjustment factors for lifting capacity based on changes in box size in the sagittal plane.

Table 3.12. Prediction equations for lifting capacity based on frequency of lift for males

Range of lift (male capacity)	Frequency of lift (lifts/min)
	$0.1 < FY^* 361.0$
Floor to knuckle	$57.2† (FY)_{**} (-0.184697)$
Floor to shoulder	$51.2 (FY)_{**} (-0.184697)$
Floor to reach	$49.1 (FY)_{**} (-0.184697)$
Knuckle to shoulder	$52.8 (FY)_{**} (-0.138650)$
Knuckle to reach	$50.0 (FY)_{**} (-0.138650)$
Shoulder to reach	$48.4 (FY)_{**} (-0.138650)$
	$1.0 < FY < 12.0$
Floor to knuckle	$-2.0 (FY-1)$
Floor to shoulder	$-2.0 (FY-1)$
Floor to reach	$-2.0 (FY-1)$
Knuckle to shoulder	$-2.0 (FY-1)$
Knuckle to reach	$-2.0 (FY-1)$
Shoulder to reach	$-2.0 (FY-1)$

*FY = frequency of lift (lifts/min).
†57.2 = mean capacity for lift based on data from Ayoub et al. (1978) and Snook (1978a) for the various ranges of lift for the 50th percentage and 1.0 lift/min.
$**$ = exponentiation (e.g., FY to the power of -0.184697).

Table 3.13. Adjustment of lifting capacity based on box size for males (from Ayoub et al., 1983b)

Range of lift (male capacity)	Box size (in. in sagittal plane)
	$12" < BX† < 18"$
Floor to knuckle	$CAP^* + 1.65 (18 - BX)$
Floor to shoulder	$CAP + 1.65 (18 - BX)$
Floor to reach	$CAP + 1.65 (18 - BX)$
Knuckle to shoulder	$CAP + 1.10 (18 - BX)$
Knuckle to reach	$CAP + 1.10 (18 - BX)$
Shoulder to reach	$CAP + 1.10 (18 - BX)$
	$BX = 18"$
Floor to knuckle	$CAP + 0.8 (18 - BX)$
Floor to shoulder	$CAP + 0.8 (18 - BX)$
Floor to reach	$CAP + 0.8 (18 - BX)$
Knuckle to shoulder	$CAP + 0.8 (18 - BX)$
Knuckle to reach	$CAP + 0.8 (18 - BX)$
Shoulder to reach	$CAP + 0.8 (18 - BX)$

*CAP, capacity of lift in lbs.
†BX, box size (inches) in the sagittal plane.

Chapter 4
Lifting and Lowering Activity Data Bases

4.1 Strength versus Capacity

A clear distinction between strength and capacity must be made prior to presenting data norms for MMH activities. According to Snook and Irvine (1969), strength implies what a person can do in a singular attempt, whereas capacity implies what a person can do, repeatedly for an extended period of time without overexertion or excessive fatigue. The researchers, however, are vague in defining what constitutes 'infrequent' materials handling. It is not clear, for instance, after what time interval the difference in strength exertions would become insignificant.

Most experts agree that if at least 5 min elapse before the next attempt, the recovery from static fatigue would be complete. However, most individuals do not perceive it this way. According to Snook (1978a, b) and Ciriello and Snook (1983), significant differences exist in maximum acceptable weights and forces for MMH activities when tasks are performed every 5 min, every 30 min or one in 8 h. This contradicts recommendations from Poulsen (1970) and Poulsen and Jorgensen (1971) that the magnitude of allowable weight for single lifting at intervals of 30 or 60 min can be a constant (70% of maximum isometric back strength).

There have been relatively few attempts to verify the hypothesis that the maximum isometric strength equals maximum dynamic exertion for occasionally performed MMH activities. While Garg et al. (1980) reported lack of consistency in the ability of isometric lifting strength to predict dynamic lifting strength for occasional lift, Pateman (1981) reported only a 6% difference in the measured average maximal dynamic lift of young males and the values predicted by Poulsen's formula (Poulsen, 1970; Poulsen and Jorgensen, 1971).

Several other studies (Mital and Ayoub, 1980; Pytel and Kamon, 1981; Kamon et al., 1982; Mital et al., 1986c), however, have also been unable to predict dynamic lifting capabilities of individuals for either frequent or occasional handling just by using an individual's isometric strengths. While isokinetic strengths have been found to be better predictors of maximum dynamic exertion for occasional handling, other operator variables must be included in the formula to increase prediction reliability (Pytel and Kamon, 1981; Kamon et al., 1982; Mital et al., 1986c).

The above discussion clearly indicates that although strength plays the pivotal role in the determination of MMH capabilities of an individual, when such tasks are performed occasionally or infrequently, it alone cannot be used to estimate capability. Additionally, capability appears to be substantially lower than isometric or isotonic strength maxima.

The strength data reviewed in this and the next chapter are sometimes based on isometric strength maxima and sometimes on actual, but infrequent, dynamic exertion. Strength oriented tasks, which are tasks performed with at least a 15 min time span between them, and tasks performed more frequently than every 15 min are included in this section. Caution should be taken when using data bases which strictly use isometric strength as the basis since, in all likelihood, the actual capability for infrequently performed MMH tasks, even though called strength, would be substantially lower (approximately 30% lower based on the works of Poulsen, 1970; Poulsen and Jorgensen, 1971, and Pateman, 1981). Martin and Chaffin (1972) have also suggested somewhat similar limits. The data bases for tasks which are performed more frequently than once every 15 min are classified as capacity data and are addressed in Chapters 4 and 5.

4.2 Strength Data for Two-Handed Lifting and Lowering Activities

The MMH strength capability data are mainly based on the psychophysical and the biomechanical approaches. In this section, studies dealing with strength capabilities of individuals for lifting and lowering activities are reviewed and data norms are presented in descriptive and, where appropriate, tabular form. Methodologies and procedures of research are reported since they can affect recommendations derived from experimental results.

During manual lifting of materials, individuals move objects from one location to another while working against gravity. Individuals work with gravity during lowering activities. Since individuals have to work against gravity lifting objects, lifting activities are considerably more stressful than lowering activities.

Emanuel et al. (1956) instructed 19 young students to lift 'the greatest weight possible without a feeling of possible injury'. Subjects then varied the weight of the object lifted by adding or subtracting 4·54 kg bags of lead shot. The age of subjects ranged from 17–18 years. They were required to lift with a straight, though not necessarily vertical, back. The subjects were classified as untrained in lifting and lifted an F-86H ammunition case: 27·3 cm high, 67·77 cm wide and 17·17 cm deep with two handles at the top. The maximum acceptable weight for a single lift was observed to decrease with the increase in height of lift (104·64 kg for 0–0·3 m height of lift to 25·82 kg for 0–1·52 m height).

Similar to Emanuel et al.'s study, Switzer (1962) instructed 75 male college students to find reasonable weights that could be lifted without excessive strain or discomfort for a single lift. Subjects varied the weight lifted by adding or subtracting 2·26, 4·53 and 9·06 kg bags of lead shot. The results of this study indicated a decrease in weight lifted as the height of lift increased (56·35 kg for 0–0·46 m height to 23·9 kg for 0–1·58 m height). While the weight for the 0–1·58 m height is comparable with that lifted by Emanuel et al.'s subjects, large differences exist in weights lifted for the lower height. It appears that instructions given to the subjects by Emanuel et al. may have caused these differences.

It should also be noted that the reported results are based on only 33 men who represented the first to fifteenth percentile stature range. The subjects lifted a sheet metal box 15·24 × 30·48 × 30·48 cm, with 11·43 cm handles.

In another experiment, Switzer (1962) studied the weight lifting capability of 75 men. Standardized and controlled lifting procedures were used to simulate lifting tasks. Three heights of lift were used: floor to 0·46 m high, floor to 1·06 m high and floor to 1·58 m high. Lifting was performed in the sagittal plane using the right hand. The subjects were divided into three groups: a short stature group (1–15 percentile Air Force stature group), a medium stature group (45–60 percentile), and a tall stature group (85–99 percentile). The results indicated that the amount of weight lifted decreased with the increase in the height of lift. However, the amount of weight lifted increased with the height of subjects for all three heights-of-lift levels. This again points out that heavier weights are lifted in the lower lifting height regions, possibly due to greater muscle strength capability, than in the higher height regions. It is interesting to note that shorter individuals are at a disadvantage in manual lifting; this contradicts normal expectation (see Chapter 2, Discussion on Stature).

In 1962, the International Labour Organization (ILO), using a consensus of medical experts, issued limits for occasional lifting (Table 4.1). These issued limits are important because this was the first attempt to issue limits recognizing gender and age differences.

Poulsen in 1970, and Poulsen and Jorgensen in 1971, proposed weight limits for occasional lifts (once or twice per hour) based on 70% of the maximum isometric back strength (Table 4.2). These limits also recognize the importance of gender and age, but overall, are substantially greater than the ILO limits (Table 4.1). Pateman (1981) verified the procedure which led to the determination of the limits given in Table 4.2.

Table 4.1. International Labour Office suggested limits for occasional weight lifting (ILO, 1962)

Age (years)	Men	Women
14–16	14·57*	9·78
16–18	18·45	11·72
18–20	22·63	13·66
20–35	24·46	14·57
35–50	20·59	12·74
50+	15·59	9·78

*Kilogram force or Kp.

Table 4.2. Allowable weight limit (kg) for single lift based on 70% of isometric back strength (Poulsen, 1970; Poulsen and Jorgensen, 1971)

Age (years)	Male	Female
15	46–61*	24–35
25	56–75	27–44
45	58–77	27–38
55	54–74	22–31

*Based on stature: males, 1·6–1·85 m females, 1·5–1·75 m.

Table 4.3 Maximum acceptable weight of lift (kg) for males and females as a function of box-length (cm), vertical distance of lift (cm), and lifting height level (Snook, 1978a, b)

Box length	Vertical distance	Height level	Sex	Once/30 min $\bar{x}^*$	S†	Once/480 min $\bar{x}$	S
75	76	Floor to knuckle	Male	33	7·30	36	7·30
			Female	21	3·65	23	3·65
		Knuckle to shoulder	Male	28	6·68	30	6·68
			Female	17	2·43	18	2·43
		Shoulder to reach	Male	26	6·08	28	6·08
			Female	14	1·82	15	1·82
	51	Floor to knuckle	Male	34	7·30	37	7·90
			Female	22	3·65	24	3·65
		Knuckle to shoulder	Male	32	6·68	34	7·30
			Female	19	2·43	20	3·04
		Shoulder to reach	Male	30	6·08	31	7·30
			Female	16	1·82	17	1·82
	25	Floor to knuckle	Male	39	7·90	42	9·12
			Female	25	4·25	27	4·25
		Knuckle to shoulder	Male	37	8·51	40	8·51
			Female	22	3·65	23	3·65
		Shoulder to reach	Male	35	7·90	37	7·90
			Female	19	2·43	20	2·43
49	76	Floor to knuckle	Male	39	8·51	42	9·73
			Female	25	3·65	27	4·25
		Knuckle to shoulder	Male	28	6·68	30	6·68
			Female	17	2·43	18	2·43
		Shoulder to reach	Male	26	6·08	28	6·08
			Female	14	1·82	15	1·82
	51	Floor to knuckle	Male	40	9·12	43	10·33
			Female	26	4·25	28	4·25
		Knuckle to shoulder	Male	32	6·68	34	7·30
			Female	19	2·43	20	3·04
		Shoulder to reach	Male	30	6·08	31	7·30
			Female	16	1·82	17	1·82
	25	Floor to knuckle	Male	45	10·33	49	10·94
			Female	29	4·86	32	4·86
		Knuckle to shoulder	Male	37	8·51	40	8·51
			Female	22	3·65	23	3·65
		Shoulder to reach	Male	35	7·90	37	7·90
			Female	19	2·43	20	2·43
36	76	Floor to knuckle	Male	43	9·73	46	10·94
			Female	28	4·25	30	4·25
		Knuckle to shoulder	Male	30	7·30	32	7·90
			Female	18	2·43	19	3·04
		Shoulder to reach	Male	28	6·68	30	6·68
			Female	15	1·82	16	1·82
	51	Floor to knuckle	Male	44	10·33	48	10·94
			Female	29	4·25	31	4·86
		Knuckle to shoulder	Male	34	7·90	36	8·51
			Female	20	3·04	21	3·65
		Shoulder to reach	Male	32	7·30	34	7·90
			Female	17	2·43	18	2·43
	25	Floor to knuckle	Male	50	11·55	54	12·76
			Female	32	5·47	35	5·47
		Knuckle to shoulder	Male	40	9·73	43	9·73
			Female	24	3·65	25	4·25
		Shoulder to reach	Male	37	9·12	40	9·12
			Female	21	1·82	22	2·43

*$\bar{x}$ = mean; †S = standard deviation.

Table 4.4 Maximum acceptable weight of lowering (kg) for males and females as a function of the variables of Table 4.3 (Snook, 1978a)

				Frequency			
				Once/30 min		Once/480 min	
Box length	Vertical distance	Height level	Sex	$\bar{x}$*	S†	$\bar{x}$	S
75	76	Floor to knuckle	Male	38	7·90	41	9·12
			Female	22	3·04	24	3·04
		Knuckle to shoulder	Male	29	6·68	30	7·29
			Female	18	3·04	19	3·04
		Shoulder to reach	Male	23	5·47	25	5·47
			Female	16	1·82	17	1·82
	51	Floor to knuckle	Male	40	7·90	43	9·12
			Female	23	3·04	25	3·04
		Knuckle to shoulder	Male	32	6·68	34	7·90
			Female	20	3·04	21	3·04
		Shoulder to reach	Male	26	6·08	28	6·08
			Female	17	1·82	18	2·43
	25	Floor to knuckle	Male	45	9·12	48	10·33
			Female	26	3·04	28	3·65
		Knuckle to shoulder	Male	38	9·12	40	9·73
			Female	23	3·65	24	3·65
		Shoulder to reach	Male	31	7·29	33	7·29
			Female	20	2·43	21	2·43
49	76	Floor to knuckle	Male	45	9·73	48	10·94
			Female	26	3·04	28	3·65
		Knuckle to shoulder	Male	29	6·68	30	7·29
			Female	18	3·04	19	3·04
		Shoulder to reach	Male	23	5·47	25	5·47
			Female	16	1·82	17	1·82
	51	Floor to knuckle	Male	46	10·94	50	11·55
			Female	27	3·04	29	3·65
		Knuckle to shoulder	Male	32	7·29	34	7·90
			Female	19	3·04	21	3·04
		Shoulder to reach	Male	26	6·08	28	6·08
			Female	17	1·82	18	2·43
	25	Floor to knuckle	Male	52	12·16	56	13·37
			Female	30	4·25	33	4·25
		Knuckle to shoulder	Male	38	9·12	40	9·73
			Female	23	3·65	24	3·65
		Shoulder to reach	Male	31	7·29	33	7·29
			Female	20	2·43	21	2·43
36	76	Floor to knuckle	Male	49	11·55	53	12·16
			Female	28	3·65	31	3·65
		Knuckle to shoulder	Male	31	7·29	33	7·29
			Female	20	2·43	21	3·04
		Shoulder to reach	Male	25	6·08	27	6·08
			Female	17	1·82	18	2·43
	51	Floor to knuckle	Male	51	11·55	55	12·76
			Female	30	3·65	32	4·25
		Knuckle to shoulder	Male	35	7·90	37	8·51
			Female	21	3·04	22	3·65
7		Shoulder to reach	Male	28	6·68	30	6·68
			Female	18	2·43	19	2·43
	25	Floor to knuckle	Male	58	13·37	62	14·59
			Female	33	4·25	36	4·86
		Knuckle to shoulder	Male	41	9·73	44	9·73
			Female	25	3·65	26	4·25
		Shoulder to reach	Male	33	7·90	35	8·51
			Female	21	3·04	22	3·04

*$\bar{x}$ = mean; †S = standard deviation.

Snook (1978a, b) briefly reviewed previous psychophysical studies performed at the Liberty Mutual Insurance Company (Snook, 1965, 1971; Snook and Irvine, 1966, 1967, 1968, 1969; Snook et al., 1970; Snook and Ciriello, 1974a, b; Ciriello and Snook, 1978). With this information, he was able to detail some meaningful recommendations for the design of MMH tasks. The experimental details of some of these studies are given in section 4.3.1. Tables 4.3 and 4.4 present the strength data, based on the psychophysical approach, for industrial males and females for lifting and lowering activities, respectively.

The data in Tables 4.3 and 4.4 were based on the assumption that female responses to object size and vertical distance of lift are proportionately similar to male responses, to object size and vertical distance of lift and the responses to frequencies (between once per minute and once per 480 min) are not significantly different from those found in a small pilot study (Snook, 1978a, b). These assumptions were tested by Ciriello and Snook (1983) on 12 female and 10 male industrial workers. The experimental details are given in section 4.3. Lifting and lowering tasks were performed in two height regions: floor to knuckle height (low) and knuckle to shoulder height (centre). The results are shown in Table 4.5. The 8-h interval frequency produced considerably higher weight than the 30-min interval frequency, indicating that the assumption made earlier (Snook, 1978a, b) was not valid and data in Tables 4.3 and 4.4 must be adjusted.

Table 4.5 Maximum acceptable weights (kg) for males and females performing lifting and lowering (Ciriello and Snook, 1983)

Task	Height level	Sex	Frequency			
			Once/30 min		Once/480 min	
			$\bar{x}$*	S†	$\bar{x}$	S
Lifting	Floor to knuckle	Male	52	13·8	61	16·3
		Female	19	2·6	25	4·8
	Knuckle to shoulder	Male	41	7·2	44	8·1
		Female	16	2·7	18	3·1
Lowering	Floor to knuckle	Male	55	18·3	70	14·0
		Female	20	3·0	27	7·3
	Knuckle to shoulder	Male	42	9·0	52	10·5
		Female	17	3·9	17	4·7

*$\bar{x}$ = mean; †S = standard deviation, Box length = 36 cm; vertical distance = 51 cm.

Garg and Saxena (1980) used six college students (21–47 years of age, 62·7–86·4 kg body-weight, and 165–185 cm heights) to lift six different tote boxes and three different mail bags. All six boxes were 25 cm high. The dimensions (width × length) were: 38 × 51, 51 × 38, 51 × 51, 51 × 64, 64 × 51 and 64 × 64 cm. Boxes were lifted with and without handles. Also, three mail bags were used which are the same as those used by the United States Postal Service. Their dimensions (width × diameter) were: 46 × 51, 89 × 56 and 91 × 71 cm.

Using the psychophysical methodology, each subject determined the maximum weight he was willing to lift every 15 min from the floor to a 76-cm high table. The procedure was replicated for each experimental condition. Tables 4.6 and 4.7 show the maximum weights acceptable to males, who were inexperienced in manual lifting, for lifting six tote boxes, with and without handles, and three mail bags, respectively. The weights acceptable for lifting tote boxes are considerably lower than the weights acceptable for lifting a comparable size box, given in Tables 4.4 and 4.5, but for 30-min interval frequency.

Table 4.6 Maximum acceptable weights of lift (kg) for occasional lifting of tote boxes from 0 to 76 cm (Garg and Saxena, 1980)

Box type	Box dimensions (cm)		$\bar{x}$*	S†	Range
	Width	Length			
With handles	38	51	30·3	8·5	17·4–40·4
	51	38	30·4	8·2	16·1–41·8
	51	51	27·8	7·5	15·8–37·9
	51	64	27·3	6·9	16·0–36·2
	64	51	27·2	6·7	15·4–36·2
	64	64	26·4	7·8	13·7–34·3
Without handles	38	51	27·0	7·0	17·9–39·5
	51	38	26·9	6·4	17·5–38·2
	51	51	26·7	5·7	17·1–36·1
	51	64	25·8	7·0	14·3–35·3
	64	51	26·1	6·7	14·1–37·3
	64	64	24·5	8·3	11·4–32·5

*$\bar{x}$ = mean; †S = standard deviation.

Table 4.7 Maximum acceptable weights of lift (kg) for occasional lifting of mail bags from 0 to 76 cm (Garg and Saxena, 1980)

Mail bag Size (cm)		$\bar{x}$*	S†	Range
Diameter	Width			
51	46	25·9	9·9	14·3–41·6
56	89	27·2	8·5	18·2–39·8
71	91	29·3	7·4	20·2–39·1

*$\bar{x}$ = mean; †S = standard deviation.

Whitney (1958) studied the position of the foot and its effect on the lifting force. Eight male subjects were instructed to exert a steady, maximum lifting force for 3 s on a horizontal bar (3·2 cm diameter) placed in a frontal plane. The distance between the two hands was maintained at 40 cm. Three levels of grasp height were investigated: 12·5, 25 and 50 cm above ground level. Three distances of foot placement were considered: 30, 40 and 50 cm. Subjects adopted one of two types of lifting action: a derrick action or a knee action. The results indicated the lifting force decreased as the foot distance increased. This decrease was more significant than the decrease in the lifting force due to the increase in grasp height. The maximum isometric force that could be exerted on a horizontal bar varied from 14·95–119·6 kg for the different lifting conditions. The results thus demonstrate that foot placement is at least as critical, if not more so, as the height of force exertion.

Martin and Chaffin (1972), using computer simulation of subject's maximum strength data, predicted lifting strengths of males and females. Forty-one men (19–52 years of age) and 48 women (18–48 years of age), all industrial workers, participated in the study. The results indicated that the lifting strength is not only influenced by gender, but also the horizontal and vertical locations of the load. Table 4.8 gives the lifting capability of men and women for occasional lifting based on isometric strength simulation. The data are applicable to lifting in the sagittal plane only.

Table 4.8 Lifting capabilities (kg) of industrial men and women based on isometric strength simulation (Martin and Chaffin, 1972)

Horizontal distance (m)	Vertical distance (m)	Male		Female	
		50%	95%	50%	95%
0·3	0·3	51*	45	23	7
0·3	0·9	44	39	19	11
0·3	1·5	47	29	11	5
0·6	0·3	24	9	9	0
0·6	0·9	28	15	6	1
0·6	1·5	21	11	5	0
0·9	0·3	5	0	0	0
0·9	0·9	10	1	1	0
0·9	1·5	7	0	0	0

*Values converted from graphical representation.

Chaffin (1974) studied the lifting strength of industrial workers in standard job positions. A total of 103 jobs held by 411 men and women were evaluated for physical stress. Modal value for females (11·77 kg) was 65% of the modal value for men (18·12 kg). Women had a lifting strength mean of 13·59 kg, with a standard deviation of 7·25 kg, while the mean for men was 23·55 kg, with a standard deviation of 10·42 kg.

In a further study, Chaffin et al. (1977a) determined isometric arm, leg and torso lifting strengths (see Chapter 8 for definitions) of 443 males and 108 females. Table 4.9 shows the distribution of these strengths for males and females. The values for females were approximately 50% of the values for males.

Table 4.9 Isometric lifting strength (kg) distribution for males and females (Chaffin et al., 1977a)

Strength (kg)	Male (M)		Female (F)		Ratio (%)
	$\bar{x}$*	S†	$\bar{x}$	S	(F/M)
Arm lifting	38·95	12·98	20·38	8·00	52
Leg lifting	96·16	34·73	42·58	20·15	44
Torso lifting	55·57	24·88	27·19	14·07	49

*$\bar{x}$ = mean; †S = standard deviation.

Warwick et al. (1980) measured isometric lifting strength of 29 adult males (21–73 years of age, 60.9–105 kg body-weight, and 165–194 cm height). Force was exerted at shoulder height (142 cm above the floor) and knee height (60 cm above the floor) on each of two handles, placed 43 cm apart, by one or both hands. Feet placement positions were such that the subject either faced the handles (anterior), or the handles were on his right, in the frontal plane (right), or on his left, in the frontal plane (left). The other two positions (left posterior and right posterior) were obtained by rotating the body clockwise and counter-clockwise such that the frontal plane of the subject's body formed $-45°$ angle from the 'left' or 'right' position, respectively. Table 4.10 gives the average magnitude of the lifting forces exerted by both hands in various foot positions.

Table 4.10 Average lifting forces (kg) exerted by males at shoulder and knee heights using both hands (Warwick et al., 1980)

Foot position	Shoulder height	Knee height
Anterior	39·35	28·03
Right	24·36	13·66
Right Posterior	24·36	—
Left	—	12·94

The Materials Handling Research Unit of the University of Surrey (Davis and Stubbs, 1980) has prepared a guide which puts limits on forces that can be exerted by males, with two hands, for occasional lifting when in standing, squatting, sitting or kneeling positions. The force limits take into account the age of the worker and position of his arms. The occasional lifting is defined as performing the activity not more than once a minute. The force limits are developed such that intra-abdominal pressure of 90 mm Hg is reached in a worker whose height and weight coincide with the fifth percentile limits of the British population. Thus, if the working forces do not exceed the prescribed limits, almost any male can apply them without undue risk of back injury. Table 4.11 gives the value of forces for two-handed tasks, with hands

Table 4.11 Lifting forces* (kg) for two-handed lifting with arms in front of the body in the horizontal plane of the shoulders (Davis and Stubbs, 1980)

Position	Acromial – grip distance (cm)	Age (years)		
		up to 40	41–50	51–60
Standing/squatting	65	10	9	8
	40	20	18	16
	20	30	27	23
	10	40	37	31
	At the shoulder	50	46	39
Sitting	70	11	11	10
	50	15	15	13
	25	25	25	22
	15	30	30	27
Kneeling on one knee	60	14	14	10
	55	20	20	14
	40	30	30	21
	30	40	40	28
	15	50	50	36

*Values converted from graphical representation.

directly in front, in the horizontal plane, in standing (or squatting), sitting and kneeling positions.

As evident from the review of studies in this section, several different kinds of data bases are available and most differ from each other. Differences in experimental methodologies and experimental conditions investigated make it inappropriate to either combine data from different studies or to recommend any particular set of data. Weight recommendations for manual lifting based on isometric strengths also differ considerably from weight recommendations based on actual lifting. It would be wise, therefore, to adopt recommendations which best approximate the actual lifting conditions.

4.3 Capacity Data for Two-Handed Lifting and Lowering Activities

MMH studies dealing with individual capabilities for repetitive performance (time interval between successive lifts less than 15 min) are reviewed in this section.

Snook and Irvine (1968, 1969) used a psychophysical approach to determine the maximum acceptable frequency of lift for different combinations of height of lift and box weight. Three heights of lift were used; floor level to knuckle height, knuckle to shoulder height, and shoulder to reach height. Two weights were used, 15·85 and 22·65 kg. Eight male workers from local industry performed as subjects. Their age ranged from 26–37 years (average age 31 years). Each subject performed a daily schedule of push-ups, pull-ups, and deep knee bend exercises in their homes during the week prior to their scheduled participation in the experiment. During the first 4 days of the experiment, subjects were instructed in the correct lifting procedures, were given 18 practice trials at adjusting the weight of the lifted objects, and were gradually conditioned to the task by performing repeated lifting for varying periods of time. Eighteen 5-min sessions and three 30-min sessions of repeated lifting were performed by each subject before any measurements were recorded for analysis. The subjects were asked to adjust the frequency of lift according to their sensations or feeling of fatigue. They were specifically instructed to imagine that they were on piece-work, getting paid for the number of lifts each could make, but working a normal 8-h shift that allowed each subject to go home without becoming tired, weakened, overheated or out of breadth. The experiment was replicated three times for each subject. The heart rate of the subjects was monitored and recorded. The results indicated that there were no significant heart rate differences between the two loads — 15·85 and 22·65 kg — while there were significant heart rate differences among the three heights of lift. Subjects maintained a mean heart rate of 112 beats/min for floor to knuckle lift, while they maintained a mean heart rate of 99 beats/min for the other two heights of lift. Oxygen consumption was estimated to be 1·09 l/min for leg lifts and 0·97 l/min for arm lifts.

The final frequencies of lift selected by each worker were converted into meter-kilogram per minute to provide a comparison between the 15·85 kg and 22·65 kg weights of lift. A significant difference was found for these workloads among the three heights and two weights. The maximum workloads acceptable to male industrial workers were highest for the knuckle to shoulder height (62·68 kg-m/min for the 22·65 kg weight, and 57 kg-m/min for the 15·85 kg weight); followed by the shoulder to reach height (51·64 kg-m/min for the 22·65 kg weight, and 47·22 kg-m/min for the 15·85 kg weight); and then the floor to knuckle height (43·77 kg-m/min for the 22·65 kg weight, and 33·69 kg-m/min for the 15·85 kg weight).

Snook (1971) and Snook *et al.* (1970) conducted another study to determine the maximum weights and workloads acceptable to male industrial workers. The MMH activities involved lifting, lowering, pushing, pulling, walking and carrying tasks. Twenty-eight male subjects from local industries participated in the experiment. They were all in good health. Three height ranges were used for lifting and lowering tasks: floor level to knuckle height, knuckle height to shoulder height, and shoulder height to arm reach height (Figure 4.1).

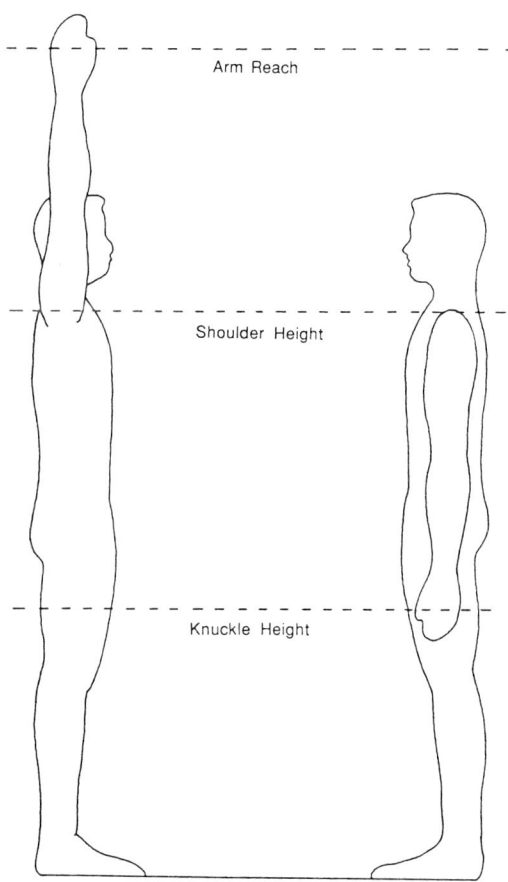

Figure 4.1. Classification of lifting heights (Snook, 1978b).

Force was the dependent variable for all tasks, except during walking. During lifting and lowering, the force was adjusted by varying the weight of the object being handled (tote box: 34·3 × 48·26 × 13·97 cm, with two 17·78 × 4.13 cm handles). All tasks except walking were performed at high, medium and low rates of work. The tote box used was similar to the one used earlier (Snook and Irvine, 1968, 1969). Subjects were instructed to adjust their workload to the maximum amount they could perform without strain, discomfort, weakness or getting tired. Each task took about 40 min. The acceptable weight was higher for the lowering tasks as compared to the lifting tasks, with the exception of shoulder to reach height level. All acceptable workloads, however, were consistently higher for lowering tasks. Subjects had a higher lifting and lowering workload for the knuckle to shoulder height than for the floor to knuckle height level or shoulder to reach height level. This pattern was similar to the one reported by Snook and Irvine (1969).

Snook and Ciriello (1974a) performed a similar study on female workers to find the maximum weight and workloads acceptable to them. Thirty-one women served as subjects. Sixteen of these were housewives with a mean age of 35·6 years. The remaining 15 were workers from local industry with a mean age of 38·5 years. The experimental set-up was identical to the one described in the previous studies (Snook, 1971; Snook et al., 1970). The results of this study were very similar to the study by Snook et al. (1970). The lifting and lowering capacity of industrial women was higher than that of the housewives for all height levels. Industrial women also accepted a higher workload. The acceptable workload was highest for the knuckle to shoulder height level, followed by the shoulder to reach, and then the floor to knuckle height level.

Occasionally, individuals have to work in hot and humid conditions. Such conditions impose additional circulatory burdens on the individuals. Snook and Ciriello (1974b) studied the effects of heat stress on lifting, pushing and carrying tasks. Two different environmental conditions were used: 21·5°C dry bulb and 45% relative humidity (17·2°C WBGT); and 30·3°C dry bulb and 65% relative humidity (27·2°C (WBGT). The workload was reduced by 20% for lifting tasks, while the heart rate increased by about 9–10 beats/min, while the increase in rectal temperature was about 0·2–0·3°C.

Snook (1978a, b) and Ciriello and Snook (1978) conducted another study in which the height of lift, frequency of lift, and size and weight of the object were studied. The psychophysical methodology was used, and 61 different lifting, pushing and pulling tasks were performed. Fifteen male subjects participated in the study. Three lifting heights (25, 51 and 76 cm), three object lengths (36, 49 and 75 cm — in the sagittal plane), and two object widths (57 and 89 cm — in the frontal plane) were investigated. The subjects lifted weight every 5, 9 and 14 s; every 1, 2, 5 and 30 min; and once in 8 h. The results indicated no significant weight differences in object width. However, the differences in the weight handled for different length objects were statistically significant but only for the floor to knuckle height at low frequencies. The effects of lifting distances and sex on the maximum acceptable weight of lift were also significant.

The data were integrated with data generated in previous studies to yield capacity of industrial males and females. The following assumptions were made: (1) female's responses to object size, vertical distance of lift, and height of push/pull are proportionately similar to male's responses, and (2) the responses of males and females to intermittent task frequencies between 1 min and 8 h are not significantly different from those found in a pilot study. Tables 4.12 and 4.13 show the capacity data of males and females for repetitive lifting and lowering tasks.

Table 4.12a Maximum acceptable weight of lift (kg) for males (Snook, 1978a, b)

Box length (cm)	Vertical distance (cm)	Height level*	One lift every (second)											
			5		9		14		60		120		300	
			$\bar{x}$†	S‡	$\bar{x}$	S	$\bar{x}$	S	$\bar{x}$	S	$\bar{x}$	S	$\bar{x}$	S
75	76	F-K	13	3·0	17	4·2	19	4·2	22	4·9	24	4·9	30	6·7
		K-S	14	3·0	19	3·6	21	4·9	21	4·9	23	4·9	27	5·5
		S-R	11	1·8	15	3·6	18	3·6	20	4·2	21	4·9	25	5·5
	51	F-K	14	3·0	18	3·6	20	4·2	23	4·9	25	4·9	31	6·7
		K-S	15	3·0	20	4·2	23	4·9	24	5·5	25	6·1	30	6·7
		S-R	12	2·4	17	3·0	19	4·2	22	5·5	23	5·5	28	6·1
	25	F-K	16	3·6	21	4·9	23	5·5	26	5·5	28	6·1	35	7·9
		K-S	18	3·6	24	4·9	27	5·5	28	6·7	30	6·7	35	7·9
		S-R	14	3·0	20	4·2	22	5·5	26	6·1	28	6·1	33	7·3
49	76	F-K	15	3·0	19	4·2	21	5·5	26	6·1	28	6·1	35	8·5
		K-S	14	3·0	19	3·6	21	4·9	21	4·9	23	4·9	27	5·5
		S-R	11	1·8	15	3·6	18	3·6	20	4·2	21	4·9	25	5·5
	51	F-K	15	3·6	20	4·2	22	5·5	27	6·1	29	6·7	37	8·5
		K-S	15	3·6	20	4·2	23	4·9	24	5·5	25	6·1	30	6·7
		S-R	12	2·4	17	3·0	19	4·2	22	5·5	23	5·5	28	6·1
	25	F-K	18	4·2	23	5·5	26	6·1	30	7·3	33	7·3	41	9·7
		K-S	18	3·6	24	4·9	27	5·5	28	6·7	30	6·7	35	7·9
		S-R	14	3·0	20	4·2	22	5·5	26	6·1	28	6·1	33	7·3
36	76	F-K	17	4·2	22	5·5	25	6·1	29	6·7	31	6·7	39	9·1
		K-S	14	3·0	20	3·6	23	4·2	23	5·5	24	6·1	29	6·7
		S-R	11	3·0	16	3·6	19	4·2	21	5·5	22	5·5	27	6·1
	51	F-K	17	4·2	23	5·5	26	6·1	30	6·7	32	7·3	40	9·7
		K-S	16	3·0	21	4·2	24	4·9	26	6·1	27	6·7	32	7·3
		S-R	13	3·0	18	4·2	20	4·9	24	5·5	25	6·1	30	6·7
	25	F-K	20	5·5	27	6·1	30	7·3	34	7·3	36	8·5	46	10·3
		K-S	19	3·6	25	4·9	28	6·1	31	6·7	32	7·3	38	8·5
		S-R	15	3·6	22	4·9	24	5·5	28	6·7	30	6·7	35	8·5

*F = floor, K = knuckle, S = shoulder, R = reach; † $\bar{x}$ = mean; ‡S = standard deviation.

Ciriello and Snook (1983) verified the assumptions made in the earlier studies (Snook, 1978a, b). They performed two series of experiments. Twelve female and ten male industrial workers performed 51 variations of lifting, lowering, pushing, pulling and carrying tasks. Task frequency varied from once every 5 s to once every 8 h. The vertical distance of lifting and lowering was 51 cm. Two height levels were studied: floor level to knuckle height (low) and knuckle height to shoulder height (centre). Box width and box length were 57 cm and 36 cm, respectively.

In the second series of experiments 12 female industrial workers performed 53 variations of lifting, pushing and pulling. Two box widths of 57 cm and 89 cm were investigated. For each width, three box lengths were studied: 36, 49 and 75 cm. All boxes were 14·4 cm high. Lifts were performed once every 5, 14 and 60 s for the floor

Table 4.12b Maximum acceptable weight of lift (kg) for females* (Snook, 1978a, b)

Box length (cm)	Vertical distance (cm)	Height level	One lift every (second)									
			5		9		14		60		120		300	
			$\bar{x}$†	S‡	$\bar{x}$	S	$\bar{x}$	S	$\bar{x}$	S	$\bar{x}$	S	$\bar{x}$	S
75	76	F-K	9	1·8	12	1·8	13	2·4	14	2·4	15	2·4	20	3·0
		K-S	9	1·8	12	1·2	12	1·2	13	1·8	13	2·4	16	2·4
		S-R	6	0·6	9	1·2	10	1·2	11	1·2	11	1·8	13	1·8
	51	F-K	9	1·8	12	1·8	14	1·8	15	2·4	16	2·4	20	3·0
		K-S	10	1·2	13	1·2	13	1·2	14	2·4	15	2·4	18	2·4
		S-R	6	0·6	11	0·6	12	1·2	12	1·8	13	1·2	15	1·8
	25	F-K	11	1·8	14	2·4	16	2·4	17	2·4	18	3·0	23	3·6
		K-S	12	1·2	15	1·8	15	1·8	17	2·4	18	2·4	21	3·0
		S-R	8	0·6	13	1·2	14	1·2	14	1·8	15	1·8	18	1·8
49	76	F-K	10	1·8	13	2·4	15	2·4	17	2·4	18	3·0	23	3·6
		K-S	9	1·2	12	1·2	12	1·2	13	1·8	13	2·4	16	2·4
		S-R	6	0·6	9	1·2	10	1·2	11	1·2	11	1·8	13	1·8
	51	F-K	10	1·8	13	2·4	15	2·4	17	3·0	19	2·4	24	3·6
		K-S	10	1·2	13	1·2	13	1·2	14	2·4	15	2·4	18	2·4
		S-R	6	1·2	11	0·6	12	1·2	12	1·8	13	1·2	15	1·8
	25	F-K	12	1·8	16	2·4	18	3·0	20	3·0	21	3·0	27	4·2
		K-S	12	1·2	15	1·8	15	1·8	17	2·4	18	2·4	21	3·0
		S-R	8	0·6	13	1·2	14	1·2	14	1·8	15	1·8	18	1·8
36	76	F-K	11	2·4	15	2·4	17	2·4	19	2·4	20	3·0	25	4·2
		K-S	10	1·2	13	1·2	13	1·2	14	1·8	14	2·4	17	2·4
		S-R	6	0·6	10	1·2	11	1·2	12	1·2	12	1·8	15	1·2
	51	F-K	12	1·8	16	2·4	17	3·0	19	3·0	21	3·0	26	4·2
		K-S	10	1·2	14	1·2	14	1·2	15	2·4	16	2·4	19	3·0
		S-R	7	0·6	11	1·2	13	1·2	13	1·8	14	1·2	16	1·8
	25	F-K	14	2·4	18	3·0	20	3·6	22	3·0	23	3·6	30	4·2
		K-S	12	1·8	16	1·8	16	1·8	18	3·0	19	3·0	22	3·6
		S-R	8	1·2	14	1·2	15	1·8	16	1·2	16	1·8	19	3·0

*See Table 4.12a for legend.

level to knuckle height (low), knee height to elbow height (mid), and knuckle height to shoulder height (centre). Each lift was performed at a vertical distance of 51 cm. The 57 × 36 cm box was also lifted at 25 and 76 cm vertical distance.

Using the psychophysical approach, subjects determined the maximum acceptable weight of lift, or force. Subject's heart rates and oxygen consumption were also recorded. Test sessions were conducted between 9 am and 1 pm (4 h). During each session, five 40-min tasks were performed. A 10-min rest break was provided between tasks.

The results indicated that the weights for the 5-min task frequencies were overestimated in Tables 4.12 and 4.13 by 10–15%. Both males and females selected weights that produced similar cardiovascular strain. The oxygen consumption for very high frequencies, however, was greater than the widely accepted physiological fatigue criterion.

The second series of experiments indicated that the pattern of female responses was similar to that of males. Box width also had an influence on the maximum acceptable weight of lift. In general, the results were similar to the results obtained in previous studies (Snook, 1978a, b; Ciriello and Snook, 1978).

Table 4.13a Maximum acceptable weight of lowering (kg) for males* (Snook, 1978a, b)

Box length (cm)	Vertical distance (cm)	Height level	One lower every (second)										
			5		9		14		60		120		300	
			$\bar{x}$†	S‡	$\bar{x}$	S	$\bar{x}$	S	$\bar{x}$	S	$\bar{x}$	S	$\bar{x}$	S
75	76	F-K	14	1·8	18	3·6	21	4·2	26	4·9	27	6·1	35	7·3
		K-S	17	3·0	18	4·2	20	4·9	22	4·9	23	4·9	25	5·5
		S-R	12	2·4	14	3·6	16	3·6	18	3·6	19	3·6	22	4·9
	51	F-K	14	3·6	18	4·2	21	4·9	27	5·5	28	6·1	36	7·9
		K-S	18	4·2	19	4·9	22	4·9	24	6·1	26	5·5	29	6·1
		S-R	13	3·0	15	3·6	17	4·2	20	4·2	21	4·9	25	5·5
	25	F-K	17	3·6	21	5·5	25	5·5	30	6·7	32	6·7	41	8·5
		K-S	21	4·9	23	5·5	26	5·5	29	6·7	30	7·3	34	6·7
		S-R	15	3·6	18	4·2	20	4·9	23	5·5	25	5·5	29	6·7
49	76	F-K	16	3·6	21	4·9	24	5·5	30	6·7	32	7·3	41	9·1
		K-S	17	3·6	18	4·2	20	4·9	22	4·9	23	4·9	27	6·1
		S-R	12	2·4	14	3·6	16	3·6	18	3·6	19	4·2	22	4·9
	51	F-K	16	4·2	21	5·5	25	5·5	31	7·3	33	7·9	42	9·7
		K-S	18	4·2	19	4·9	22	4·9	24	6·1	26	5·5	30	7·3
		S-R	13	3·0	15	3·6	17	4·2	20	4·2	21	4·9	25	5·5
	25	F-K	19	4·9	25	6·1	29	6·7	35	7·9	38	8·5	48	10·9
		K-S	21	4·9	23	5·5	26	5·5	29	6·7	30	7·3	36	7·9
		S-R	15	3·6	18	4·2	20	4·9	23	5·5	25	5·5	29	6·7
36	76	F-K	17	4·2	23	5·5	26	6·7	33	7·3	35	8·5	45	10·3
		K-S	18	4·2	20	4·2	22	5·5	23	5·5	25	5·5	29	6·7
		S-R	13	3·0	15	4·2	17	4·2	19	4·2	20	4·9	24	5·5
	51	F-K	18	4·2	24	5·5	27	6·7	34	7·9	37	9·1	47	10·3
		K-S	19	4·9	21	4·9	23	6·1	26	6·1	28	6·1	33	7·9
		S-R	14	3·0	16	4·2	18	4·9	21	4·9	22	5·5	26	6·7
	25	F-K	21	5·5	28	6·7	32	7·9	39	8·5	41	9·7	53	12·1
		K-S	23	5·5	25	5·5	28	6·7	31	7·3	33	7·3	39	8·5
		S-R	16	4·2	19	4·9	22	4·9	25	6·1	26	6·7	31	7·3

*See Table 4.12a for legend.

Ayoub et al. (1978) conducted a study to determine and model the lifting capacity of male and female industrial workers, using 73 male and 73 female subjects. Six different height levels and four frequencies were employed. The height levels were: floor to knuckle, floor to shoulder, floor to reach, knuckle to shoulder, knuckle to reach, and shoulder to reach. The frequencies studied were 2, 4, 6 and 8 lifts/min. The experimental procedure was similar to the one described by Snook et al. (1970). Subjects were asked to lift a tote box, which had a dimension in the sagittal plane of 30·48, 45·72 or 60·96 cm. Isometric strength and anthropometric measurements were recorded for each subject and were later used to develop lifting capacity prediction models. Table 4.14 gives the lifting capability data of males and females for repetitive lifting tasks.

The results indicated that the lifting capability decreased almost linearly with the increase in the box size and frequency. The lifting capacity also decreased with the increase in height of lift and with the point of origin of the lifting task. For the same vertical height of lift, lifting capability decreased when the point of origin of lift changed from knuckle to shoulder. Age had no effect on lifting capability. Males lifted substantially more weight than females for all six height levels.

Legg and Myles (1981) employed 10 soldiers (mean age 24·1 years, mean weight

Table 4.13b Maximum acceptable weight of lowering (kg) for females* (Snook, 1978a, b)

Box length (cm)	Vertical distance (cm)	Height level	One lower every (second)											
			5		9		14		60		120		300	
			$\bar{x}$†	S‡	$\bar{x}$	S	$\bar{x}$	S	$\bar{x}$	S	$\bar{x}$	S	$\bar{x}$	S
75	76	F-K	8	1·2	12	1·8	13	2·4	15	1·8	16	1·8	20	2·4
		K-S	13	1·8	13	1·8	13	1·8	14	1·8	14	2·4	17	2·4
		S-R	9	1·8	11	1·8	11	1·8	12	1·2	12	1·8	15	1·8
	51	F-K	8	1·2	13	1·8	14	1·8	15	2·4	16	2·4	21	2·4
		K-S	14	2·4	14	2·4	14	2·4	15	1·8	15	2·4	18	3·0
		S-R	11	1·2	12	1·8	12	1·8	13	1·8	13	1·8	16	1·8
	25	F-K	9	1·2	15	1·8	16	2·4	17	2·4	19	2·4	24	3·0
		K-S	17	3·0	17	3·0	17	3·0	17	3·0	18	3·0	22	3·0
		S-R	13	1·8	15	1·8	15	1·8	15	1·8	16	1·8	18	2·4
49	76	F-K	9	1·2	14	1·8	15	2·4	17	2·4	19	1·8	24	2·4
		K-S	13	1·8	13	1·8	13	1·8	14	1·8	14	2·4	17	2·4
		S-R	9	1·8	11	1·8	11	1·8	12	1·2	12	1·8	15	1·8
	51	F-K	10	1·2	14	2·4	16	1·8	18	2·4	19	2·4	24	3·6
		K-S	14	2·4	14	2·4	14	2·4	15	1·8	15	2·4	18	3·0
		S-R	11	1·2	12	1·8	12	1·8	13	1·2	13	1·8	16	1·8
	25	F-K	11	1·2	17	2·4	18	3·0	20	3·0	22	2·4	28	3·0
		K-S	17	3·0	17	3·0	17	3·0	17	3·0	18	3·0	22	3·0
		S-R	13	1·8	15	1·8	15	1·8	15	1·8	16	1·8	18	2·4
36	76	F-K	10	1·8	16	2·4	17	3·0	19	2·4	20	3·0	26	3·0
		K-S	13	2·4	13	2·4	13	2·4	15	2·4	16	1·8	18	3·0
		S-R	10	1·2	12	1·2	12	1·2	13	1·8	13	1·8	16	1·8
	51	F-K	11	1·2	16	2·4	18	2·4	20	2·4	21	3·0	27	3·6
		K-S	15	2·4	15	2·4	15	2·4	16	2·4	17	2·4	20	3·0
		S-R	11	1·8	13	1·8	13	1·8	14	1·8	14	1·8	17	2·4
	25	F-K	12	1·8	19	3·0	21	3·0	22	3·0	24	3·0	30	4·2
		K-S	18	3·0	18	3·0	18	3·0	19	2·4	20	3·0	23	3·6
		S-R	13	2·4	15	2·4	15	2·4	16	2·4	17	1·8	20	2·4

*See Table 4.12a for legend.

69·3 kg, mean stature 172·7 cm) to determine the maximum acceptable workloads. Subjects lifted and lowered a pallet (50 × 26 × 11 cm) at the rate of once every 24 s. The mean load selected by soldiers, for an 8-h day, was 17·5 kg which was substantially lower than those reported by Snook (1978a, b). The average heart rate was 92 beats/min while the mean oxygen uptake was 21% of their maximum oxygen uptake.

In 1981, the National Institute for Occupational Safety and Health (NIOSH) published a work practice guide for manual lifting. The guide considers epidemiology of musculoskeletal injury, biomechanical stresses, and physiological and psychophysical limits. Its application is limited to two-handed, symmetrical, and smooth lifting directly in front of the body (no twisting or turning) using handles. No postural restrictions are considered, and lifting is expected to be performed in favourable ambient conditions.

The NIOSH *Work Practice Guide* requires six different factors to determine the lifting weight for continuous, frequent or infrequent tasks: (1) object weight (L), (2) horizontal distance (H) between the ankles and the hands at the origin of lift, (3) vertical distance (V) between the hands and the floor at the origin of lift, (4) vertical travel distance (D) of hands from origin to destination of object being lifted,

Table 4.14 Maximum acceptable weight of lift (kg) for males and females (Ayoub et al., 1978)

Height level*	Sex†	Box (cm)	Frequency (lifts/min)											
			2			4			6			8		
			30–48	45–72	60–96	30–48	45–72	60–96	30–48	45–72	60–96	30–48	45–72	60–96
F–K	M		27·6 (9·4)‡	27·9 (8·6)	24·9 (5·7)	24·4 (5·8)	27·6 (9·5)	25·4 (7·5)	28·6 (8·0)	25·0 (5·7)	21·0 (5·2)	27·8 (7·8)	23·2 (6·3)	20·6 (8·0)
	F		15·0 (1·5)	13·4 (3·2)	12·6 (2·4)	15·9 (2·5)	14·7 (4·9)	11·8 (1·6)	13·8 (3·5)	12·5 (3·1)	12·8 (2·0)	13·2 (3·2)	14·7 (2·0)	13·6 (2·7)
F–S	M		22·9 (6·1)	24·1 (3·2)	21·0 (5·1)	20·5 (5·5)	22·0 (5·7)	22·0 (4·3)	22·5 (6·1)	21·6 (4·8)	22·7 (3·7)	19·0 (6·2)	21·2 (6·7)	20·3 (6·3)
	F		13·2 (3·3)	14·3 (2·8)	13·1 (2·8)	12·9 (1·5)	12·9 (3·3)	10·3 (1·9)	14·2 (3·3)	12·0 (4·7)	11·1 (2·4)	12·6 (1·9)	11·7 (2·1)	11·3 (3·6)
F–R	M		21·4 (5·0)	24·5 (6·2)	18·5 (5·3)	21·6 (5·8)	16·5 (5·8)	21·4 (3·7)	21·0 (3·1)	19·7 (3·9)	18·3 (4·0)	17·9 (3·7)	14·8 (3·7)	16·6 (4·3)
	F		12·4 (2·2)	12·6 (2·0)	12·0 (2·9)	12·3 (2·6)	11·0 (2·2)	11·5 (3·3)	12·0 (2·8)	10·2 (1·9)	10·3 (2·3)	10·8 (1·7)	10·9 (1·9)	10·3 (1·9)
K–S	M		24·3 (5·4)	25·9 (7·1)	25·7 (7·3)	24·1 (6·3)	24·9 (8·8)	22·6 (7·5)	23·5 (5·8)	24·5 (6·4)	22·0 (8·7)	24·6 (5·4)	20·1 (5·3)	21·1 (7·3)
	F		14·7 (3·4)	14·0 (3·8)	12·6 (2·6)	12·3 (1·5)	12·4 (2·2)	11·3 (3·2)	13·6 (2·2)	12·8 (2·2)	13·7 (2·9)	11·1 (3·6)	11·4 (3·0)	11·0 (2·4)
K–R	M		21·3 (6·6)	23·9 (3·9)	23·7 (6·0)	21·2 (4·0)	23·4 (3·8)	18·2 (3·3)	18·8 (4·8)	14·5 (3·2)	25·5 (2·0)	16·5 (3·5)	18·0 (5·2)	18·1 (3·4)
	F		12·7 (3·5)	11·7 (1·6)	11·2 (2·2)	11·0 (2·7)	11·1 (1·4)	13·0 (2·7)	11·7 (2·1)	12·2 (1·9)	11·5 (1·2)	11·0 (3·1)	11·0 (1·9)	10·7 (2·2)
S–R	M		22·1 (2·3)	18·7 (4·3)	18·3 (5·1)	19·5 (6·5)	19·9 (3·4)	17·2 (4·3)	16·4 (2·8)	17·4 (7·8)	19·3 (4·8)	15·1 (3·8)	15·1 (4·1)	18·6 (4·1)
	F		11·9 (2·2)	10·4 (1·9)	11·1 (2·1)	10·4 (1·7)	10·3 (1·5)	10·0 (1·8)	10·8 (1·5)	10·3 (1·6)	10·5 (0·9)	12·9 (2·2)	10·5 (2·1)	9·1 (1·4)

*F = floor, K = knuckle, S = shoulder, R = reach; †M = male, F = female.
‡Numbers in parentheses are standard deviations.

(5) frequency of lifting (F/min) averaged over period of lifting, and (6) duration of the lifting task (1 h, or less or 8 h). Figure 4.2 shows the H and V measurements.

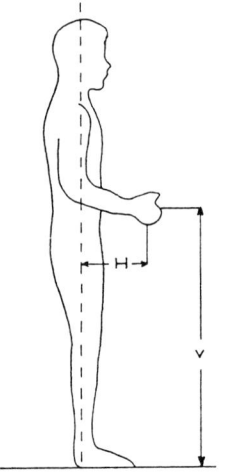

Figure 4.2. Horizontal distance (H) of hands from ankles and vertical distance (V) of hands from floor (NIOSH, 1981)

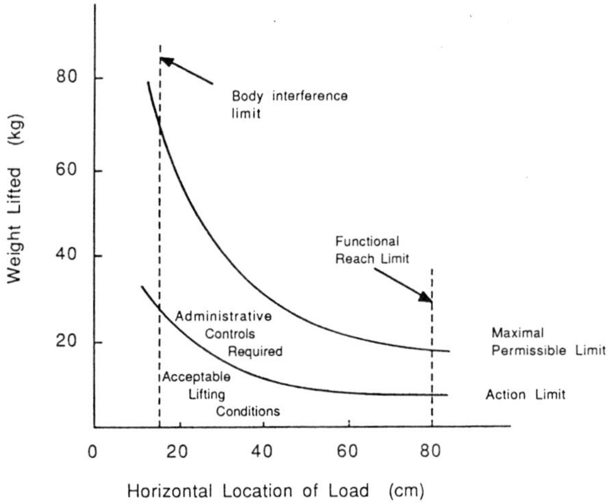

Figure 4.3. Maximum weight vs. horizontal location for infrequent lifts from floor to knuckle height (NIOSH, 1981)

Using the above factors, action limit (AL) and maximal permissible limit (MPL = 3AL) for a job can be determined. Figure 4.3 shows AL and MPL for different horizontal location of loads lifted from floor (V = 15 cm) to knuckle height (D = 60 cm) on an infrequent basis ($F < 0·2$).

The action limit weight for a job may be determined by using the following equation:

$$\text{AL} = 392 \frac{(15)}{H} (1 - (0·004|V - 75|)) (0·7 + \frac{7·5}{D}) \frac{(1-F)}{F_{\max}}$$

The various factors (H, V, D and F) may be determined from Figure 4.4. The value of $F_{\max}$ can be determined from Table 4.15.

Mital and Manivasagan (1983a) investigated the effects of lifting frequency (2, 4 and 6 lifts/min), load centre of gravity (c.g.) location (0, 12·7 and 25.4 cm from the midsagittal plane in the frontal plane), three different materials (water, dry sand and lead shot, with densities of 1, 2·02 and 10·68 g/cm^3, respectively), and offsets of c.g. in the direction of preferred (stronger hand) and non-preferred (weaker hand) on the maximum acceptable weight of lift. Ten college males (21–23 years of age) participated in the experiment. The results indicated that lifting capability is strongly influenced by the material density, frequency of lift, location of c.g. and the direction of c.g. offset. Table 4.16 gives the average weight accepted by subjects for lifting.

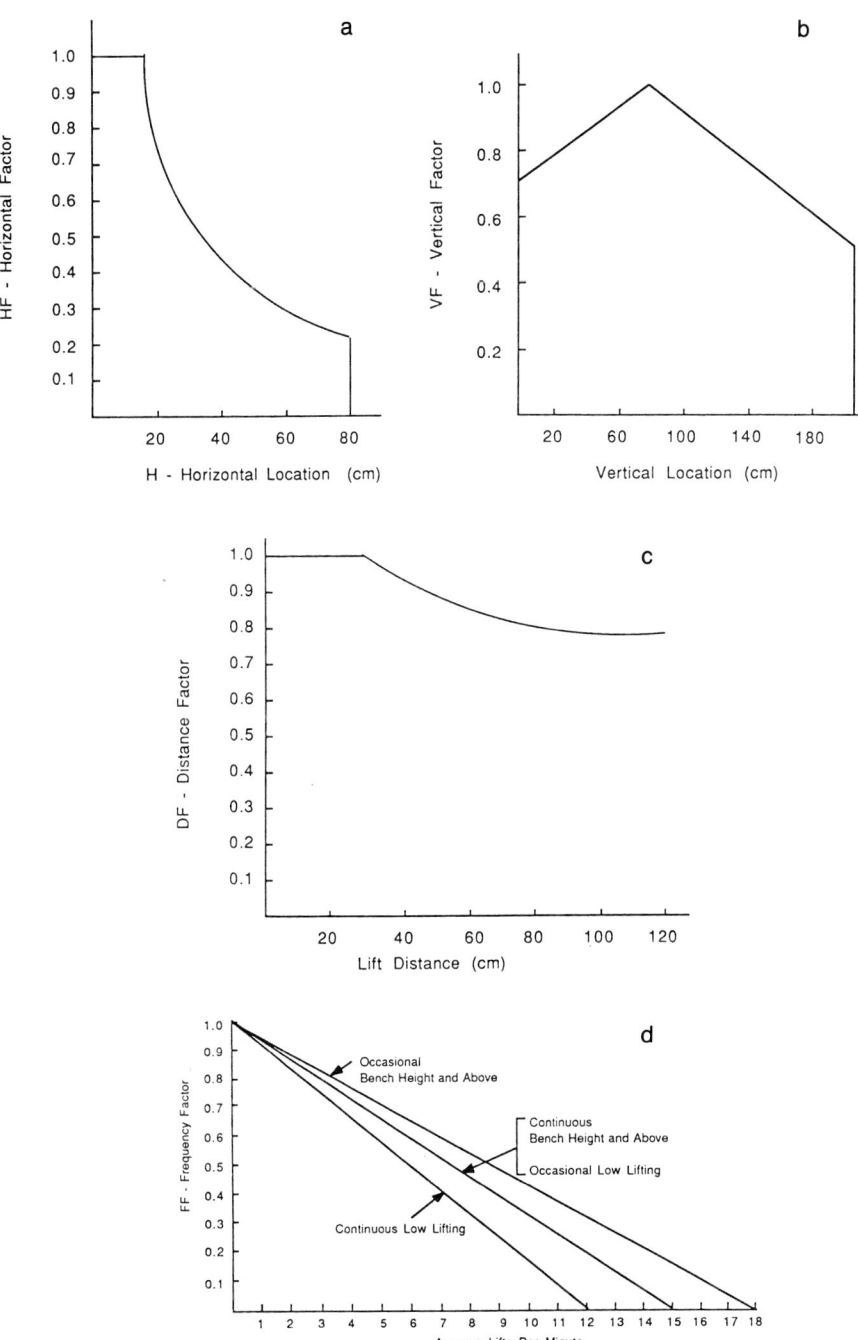

Figure 4.4(a). Nomograms for (a) horizontal factor; (b) vertical factor; (c) vertical distance factor; (d) frequency of lift factor. All from NIOSH (1981)

Table 4.15. *Maximum frequency of lift (per minute), F_{max} (NIOSH, 1981)*

Lifting duration (hours)	$V > 75$ cm (standing)	$V \leq 57$ cm (stooped)
1	18	15
8	15	12

Table 4.16. *Mean maximum acceptable weights of lift (kg) for various task factors (Mital and Manivasagan, 1983a)*

Task factor	Levels			Respective means[*]		
Material	Water	Sand	Lead Shot	19·11	19·58	20·46
Frequency (lifts/min)	2	4	6	21·00	19·41	18·74
c.g. location from midsagittal plane (cm)	0	12·7	25·4	20·44	19·81	18·90

[*]Means underscored by the same line are not significantly different at the 5% level.

On average 2·8% more weight was lifted when the c.g. offset was in the direction of the stronger hand.

Karwowski and Ayoub (1984a), using nine male students (mean age 20·66 years, mean weight 80·37 kg and average height 178·45 cm), determined the influence of frequency (0·1, 3, 9 and 12 lifts/min) on the maximum acceptable weight of lift. On average 44·8, 29·9, 19·3 and 15·1 kg weight was lifted at frequencies of 0·1, 3, 9 and 12 lifts/min. The oxygen uptake and heart rate also increased significantly with frequency. The only exception was heart rate at 9 and 12 lifts/min. At very high frequencies subjects accepted weights which violated the physiological fatigue criteria. This finding is consistent with those of Ciriello and Snook (1983).

Asfour *et al.* (1984b, 1985) and Khalil *et al.* (1985) used 8–10 male students (average age 21·9 years and average height 178·8 cm) in two experiments to determine lifting capability for low and high frequencies. In the first experiment 6·82, 13·64, 20·45, 27·27, 34·09 and 40·91 kg load was lifted at 1, 3, 5, 7 and 9 lifts/min, and the oxygen consumption and heart rate of the individual were recorded. In the second experiment, subjects determined the maximum acceptable weight of lift for different frequencies. Using the free-style lifting technique, subjects lifted weight in a 38·1 × 38·1 × 25·4 cm box across three different height levels: floor to 76·2 cm, floor to 127 cm, and 76·2 to 127 cm. Table 4.17 shows the recommendations based on the psychophysical and the physiological approaches.

The lifting capability data given in Table 4.17 are substantially lower than the recommendations made by Snook (1978a, b) and Ayoub *el al.* (1978). This once again demonstrates the lifting capabilities of inexperienced individuals are substantially lower than the lifting capabilities of experienced individuals.

The Materials Handling Research Unit of The University of Surrey (Davis and Stubbs, 1980) recommends that the lifting strength values given in Table 4.11 should be reduced by 30% in case the task frequency is greater than once per minute.

Mital *et al.* (1982) integrated data from several previous studies to arrive at com-

Table 4.17. *Lifting capability of males (in kg) based on the psychophysical and the physiological approaches (Asfour et al., 1985)*

Height (cm)	Lifting capability*	Frequency (lifts/min)				
		1	3	5	7	9
0–76·2	$\bar{x}$	31·7	22·8	21·6	13·4	10·7
	S	8·9	5·9	5·4	4·1	3·1
0–127	$\bar{x}$	24·5	18·2	16·3	12·5	9·1
	S	7·2	5·0	4·4	3·9	2·5
76·2–127	$\bar{x}$	25·4	18·9	18·0	10·0	4·4
	S	7·3	4·7	4·3	2·1	0·9

*$\bar{x}$ = mean, S = standard deviation.

binations of load (weight) and frequency that met the physiological fatigue criterion of 5 kcal/min for males. Table 4.18 lists the various recommended combinations for the floor to knuckle and knuckle to shoulder height levels. The physiological approach was the basis in all studies used to develop these recommendations.

Table 4.18. *Recommended combinations of load and frequency for an energy expenditure rate of 5 kcal/min (Mital et al., 1982)*

Height	Load (kg)	Frequency (lifts/min)
Floor to knuckle*	4·54	12·0
	11·36	7·0
	15·91	6·0
	20·45	4·5
	27·27	4·0
Knuckle to shoulder†	4·54	25·0
	11·36	12·5
	15·91	9·4
	20·45	8·4
	25·00	6·8

*Floor to 0·76–0·84 m; †0·51 m.

Mital and Fard (1986) using 18 males (mean age 22·6 years, mean weight 78·6 kg and mean height 176·6 cm) investigated the influence of lifting frequency, height of lift, load asymmetry, box size and symmetrical and asymmetrical lifting on the maximum acceptable weight of lift and the resulting physiological costs (heart rate and oxygen uptake). Table 4.19 shows the average values of all the three responses for various task factors.

The results indicated that the maximum acceptable weight of lift was strongly influenced by all the task factors investigated. As the physical stresses increase, less weight is accepted. With the exception of the effect due to lifting frequency, heart rate and oxygen uptake were not affected. This observation is similar to the one made by Ciriello and Snook (1983) — subjects reduce the weight to maintain the same circulatory strain. In this case, subjects also maintained the same metabolic energy expenditure rate for all task factors; lifting frequency being the only exception.

Table 4.19. Average maximum acceptable weight of lift (kg), heart rate (bpm), and oxygen uptake (l/min) for various task factors (Mital and Fard, 1986)

Factor	Weight	Heart rate	Oxygen uptake
Lifting			
Symmetrical	17·44 (4·06)*	103·15 (11·74)	0·67 (0·29)
Asymmetrical	15·97 (3·57)	103·55 (12·32)	0·65 (0·29)
Box size (cm) (length × width)			
45·72 × 30·48 (symmetrical)	16·79 (3·41)	103·18 (12·33)	0·67 (0·28)
60·96 × 30·48 (symmetrical)	15·94 (3·22)	105·18 (12·72)	0·69 (0·31)
30·48 × 35·56 (symmetrical)	17·79 (4·03)	102·23 (11·43)	0·64 (0·28)
30·48 × 35·56 (asymmetrical − 10·16 cm GC offset)	17·08 (4·77)	103·40 (12·33)	0·64 (0·28)
30·48 × 35·56 (asymmetrical − 20·32 cm CG offset)	15·93 (3·54)	102·76 (11·28)	0·65 (0·29)
Height of lift			
Floor to 81·3 cm	17·45 (4·13)	104·79 (12·20)	0·71 (0·30)
81·3–152·4 cm	15·96 (3·48)	101·91 (11·69)	0·61 (0·27)
Frequency (lifts/min)			
1	18·06 (4·18)	89·55 (10·59)	0·43 (0·20)
4	16·79 (3·73)	100·53 (10·73)	0·69 (0·24)
8	15·26 (3·19)	109·97 (11·89)	0·85 (0·26)

*Standard deviations are given in parentheses.

Mital (1984a, b) used 37 male and 37 female industrial workers (18–61 years of age, 154·1–186·6 cm height, and 44·1–121·71 kg weight) to determine maximum acceptable weights of lift for regular (8-h) and extended (12-h) work shifts. All subjects recruited for participation in the study were experienced in lifting goods manually for 12 h each working day. Four lifting frequencies (1, 4, 8 and 12 lifts/min), three height levels (floor to knuckle, knuckle to shoulder and shoulder to reach), and three box sizes (30·48, 45·72 and 60·96 cm long in the sagittal plane) were investigated. A balanced incomplete factorial design was used and each subject performed 9 of the 36 possible treatment combinations (4 frequencies × 3 height levels × 3 boxes). Subjects were stratified across ages by dividing them into three age groups (group 1 — up to 29 years of age; group 2 — 30–39 years of age; group 3 — above 39 years of age). There were 12 male and 12 female subjects in group 1 and group 3; group 2 had 13 males and 13 females. This stratification permitted the subject population to be representative of the industrial population across a wide age range. The room temperature was maintained at 21–22°C; the relative humidity was 45–55%. All boxes were fitted with handles (12·7 × 7·6 cm) located 12·7 cm above the base of the box and were 15·24 cm high and 30·48 cm wide. Lifting was performed using a free-style technique.

A modified psychophysical approach was used by the subjects to determine the maximum acceptable weight of lift and the experimental procedure was the same for each subject. The subjects started with either a light or heavy load (lead shot and lead pieces) in the box and were then allowed to make adjustments by adding or removing weight from the box. They were asked to assume an 8-h work shift (including meal and rest breaks) and make as many adjustments as necessary to arrive at the maximum weight that they felt they could comfortably lift, without feeling tired or exhausted. This process took approximately 20–25 min to complete after which this acceptable weight of lift was measured. The oxygen uptake and heart rate of the subject, at this maximum acceptable weight of lift for an 8-h period, were also measured. The subject was then asked to assume an extra 4-h work period (4 h of overtime having just completed an 8-h shift; no rest allowances were provided in between) and resume the lifting activity. During this phase, the subject was permitted to make further adjustments to the weight in the box. This phase also lasted approximately 20–25 minutes, at the end of which, the maximum acceptable weight of lift, the oxygen uptake and the heart rate were again measured.

This new weight was lifted for a full 12 h. There were two reasons for using this procedure: (i) to obtain additional data for 8-h shifts and (ii) frequently workers are asked to put in an extra 4 h of work at the end of the 8-h shift. It was considered that if the new weight was lower than the original estimate, it would be due primarily to the extended shift duration. The 40–45 min work period was considered sufficient to allow individuals to determine the weight they could lift for 12 h even if it included 4 h of overtime about which they had no prior warning.

The above procedure was repeated for all planned treatment combinations for each subject. Treatments were performed in a random order. No rest allowance was provided between the treatments except the time necessary to change the equipment setting for the next treatment combination.

The psychophysical approach used in this study and many past studies was validated on industrial subjects in an independent study (Mital, 1983c). It revealed that individuals cannot continue to accept lifting the weight estimated during a trial period of 20–45 min throughout the work shifts. The actual weight lifted decreased with time and the rates of decrease, assuming the initial psychophysical estimate for 8 h to be 100%, were approximately 3·4%/h for males and 2%/h for females. The heart rate decreased by about 1·9%/h for males and 0·8%/h for females.

As the maximum acceptable weight of lift decreased, so did the metabolic cost of lifting it. The decline in metabolic energy expenditure rate was approximately 2·6%/h for males and 1·9%/h for females (the actual declines for all three responses were non-linear and steeper than indicated by these linear estimates). There were two reasons for using linear estimates: (i) simplicity and (ii) the maximum acceptable weight of lift for 8 or 12 h, arrived at after these adjustments, is the weight that is acceptable after 8 or 12 h of lifting higher weights.

It is very likely that if subjects started with a weight lower than that estimated after 25 or 45 min, the weight after 8 or 12 h would have been higher due to slower onset of fatigue. While it is not known what optimum weight can be lifted for 8 or 12 h without adjustments, linear estimates of weight decline would provide maximum acceptable weights of lift closer to the optimum weights.

All data generated by Mital (1984a, b,) were adjusted by the multipliers given above. Such adjustments are essential if the psychophysical methodology is used to determine actual lifting capability of individuals.

Table 4.20a. Maximum weight of lift acceptable to male industrial workers for 12 h and their physiological responses at that weight (Mital, 1984a)

Lift height	Box size (cm)		1			4			8			12		
			30·48	45·72	60·96	30·48	45·72	60·96	30·48	45·72	60·96	30·48	45·72	60·96
Floor to knuckle	Weight (kg)	$\bar{x}$*	13·33	12·02	13·05	12·20	10·56	10·66	11·89	9·39	8·59	8·19	7·54	7·34
		S	3·63	2·92	2·70	3·87	2·77	3·12	4·66	1·79	1·71	2·93	2·53	1·84
	Heart rate	$\bar{x}$	81·44	72·82	74·19	88·03	88·69	92·64	102·91	101·24	100·65	100·88	109·90	107·36
	(beats/min)	S	13·29	10·34	13·24	15·27	8·38	18·64	18·60	14·03	18·55	14·87	17·10	11·55
	Oxygen uptake	$\bar{x}$	0·40	0·34	0·40	0·69	0·73	0·68	0·85	0·86	0·91	0·99	0·87	0·88
	(l/min)	S	0·21	0·15	0·20	0·19	0·25	0·20	0·23	0·20	0·16	0·23	0·24	0·19
Knuckle to shoulder	Weight	$\bar{x}$	12·44	12·82	11·88	10·51	11·67	10·69	9·90	9·48	8·40	8·69	9·10	7·93
		S	3·28	3·75	3·26	2·35	2·19	3·37	2·04	2·34	2·25	2·18	2·69	1·95
	Heart rate	$\bar{x}$	83·33	80·79	78·12	95·88	94·61	89·72	98·51	91·11	91·00	97·52	106·54	98·10
		S	14·13	12·02	11·60	14·87	13·44	15·68	13·75	9·79	16·78	14·91	12·96	12·40
	Oxygen uptake	$\bar{x}$	0·27	0·38	0·39	0·51	0·57	0·59	0·63	0·72	0·65	0·75	0·75	0·58
		S	0·13	0·20	0·11	0·18	0·19	0·23	0·19	0·25	0·16	0·19	0·21	0·12
Shoulder to reach	Weight	$\bar{x}$	10·11	9·58	10·86	9·69	10·05	9·36	8·95	9·23	9·80	8·85	7·70	7·87
		S	2·75	1·76	3·10	3·40	3·57	1·80	2·44	1·63	3·59	2·24	1·72	1·78
	Heart rate	$\bar{x}$	69·50	76·09	72·18	81·63	84·99	83·30	90·39	88·82	96·74	94·77	85·67	89·55
		S	11·27	8·45	15·15	12·31	13·37	16·64	11·00	10·09	16·81	20·61	10·98	13·58
	Oxygen uptake	$\bar{x}$	0·30	0·30	0·31	0·46	0·52	0·52	0·54	0·53	0·70	0·68	0·62	0·68
		S	0·15	0·14	0·16	0·23	0·17	0·18	0·18	0·20	0·28	0·18	0·20	0·21

*$\bar{x}$ = mean; S = standard deviation.

Table 4.20b. Maximum weight of lift acceptable to female industrial workers for 12 h and their physiological responses at that weight (Mital, 1984a)

Lift height	Box size (cm)			1			4			8			12		
				30·48	45·72	60·96	30·48	45·72	60·96	30·48	45·72	60·96	30·48	45·72	60·96
Floor to knuckle	Weight (kg)	x̄*		11·90	11·71	10·18	10·43	9·46	10·17	8·87	7·55	7·53	7·42	8·37	8·17
		S		2·27	2·38	2·06	2·73	2·37	2·67	2·44	2·10	1·00	2·16	2·64	2·47
	Heart rate (beats/min)	x̄		89·08	92·46	87·18	103·15	99·60	97·54	108·61	111·09	98·16	116·44	121·78	113·51
		S		7·22	10·11	9·92	15·02	17·13	14·07	19·08	8·01	10·18	17·67	14·86	12·43
	Oxygen uptake (l/min)	x̄		0·36	0·30	0·36	0·52	0·44	0·55	0·64	0·54	0·50	0·57	0·66	0·56
		S		0·17	0·15	0·17	0·13	0·20	0·27	0·37	0·13	0·16	0·17	0·22	0·20
Knuckle to shoulder	Weight	x̄		10·06	10·58	9·41	9·07	9·32	8·91	8·15	7·86	7·93	6·93	7·91	7·50
		S		1·97	2·47	2·18	1·64	2·17	2·29	1·58	1·30	2·11	0·89	1·67	2·00
	Heart rate	x̄		90·89	89·07	88·27	98·57	98·33	103·52	107·57	100·41	102·08	114·70	120·51	102·73
		S		12·11	9·45	13·25	9·76	10·95	12·98	12·64	8·67	14·65	12·43	13·79	16·54
	Oxygen uptake	x̄		0·25	0·26	0·16	0·34	0·34	0·37	0·43	0·45	0·45	0·45	0·59	0·47
		S		0·13	0·15	0·05	0·18	0·21	0·24	0·13	0·07	0·22	0·15	0·17	0·15
Shoulder to reach	Weight	x̄		9·38	7·92	7·66	8·57	7·98	8·03	7·93	7·93	7·57	7·33	6·46	7·15
		S		1·89	1·34	0·96	1·94	1·39	1·68	1·03	1·66	1·50	0·93	0·91	1·04
	Heart rate	x̄		86·80	86·65	83·07	97·58	93·58	84·12	106·20	94·96	105·83	102·44	110·27	109·67
		S		8·91	9·10	6·66	11·87	14·03	13·04	12·85	11·57	14·58	15·11	10·13	18·52
	Oxygen uptake	x̄		0·15	0·24	0·18	0·35	0·28	0·23	0·49	0·30	0·47	0·31	0·39	0·46
		S		0·09	0·18	0·09	0·12	0·17	0·08	0·27	0·20	0·23	0·10	0·18	0·19

*x̄ = mean; S = standard deviation.

Table 4.21a. Maximum weight of lift acceptable to male industrial workers for 8 h and their physiological responses at that weight (Mital, 1984b)

Lift height	Box size (cm)		Frequency of lift (lifts/min)											
			1			4			8			12		
			30·48	45·72	60·96	30·48	45·72	60·96	30·48	45·72	60·96	30·48	45·72	60·96
Floor to knuckle	Weight (kg)	$\bar{x}$*	16·83	15·12	16·46	15·42	14·28	13·60	14·99	12·46	11·11	10·57	10·12	9·62
		S	4·68	3·48	3·47	4·52	3·92	3·85	5·34	2·30	1·69	3·50	2·55	2·10
	Heart rate (beats/min)	$\bar{x}$	90·72	82·36	81·04	96·68	97·88	101·75	114·85	111·15	109·60	111·74	118·03	114·87
		S	14·33	12·35	13·83	16·76	7·86	20·80	20·16	14·93	22·31	19·52	20·06	13·88
	Oxygen uptake (l/min)	$\bar{x}$	0·47	0·46	0·49	0·83	0·79	0·81	1·03	1·03	1·08	1·20	0·96	1·08
		S	0·22	0·19	0·22	0·20	0·21	0·23	0·26	0·24	0·19	0·23	0·20	0·31
Knuckle to shoulder	Weight	$\bar{x}$	15·46	16·27	14·86	13·56	14·47	13·51	12·96	11·89	10·63	11·31	11·88	10·35
		S	3·96	4·17	3·81	3·15	2·53	4·15	2·70	2·68	2·48	2·73	3·03	2·56
	Heart rate	$\bar{x}$	92·32	91·96	84·76	105·72	102·62	100·48	109·80	99·86	101·73	107·37	119·20	109·32
		S	15·97	10·83	12·29	18·72	13·33	17·54	13·22	12·72	20·36	18·76	14·83	13·53
	Oxygen uptake	$\bar{x}$	0·34	0·46	0·44	0·62	0·70	0·72	0·77	0·83	0·75	0·92	0·92	0·72
		S	0·15	0·22	0·11	0·21	0·23	0·26	0·21	0·25	0·09	0·26	0·28	0·13
Shoulder to reach	Weight	$\bar{x}$	13·05	12·19	13·62	12·02	12·81	11·94	11·28	11·56	12·26	11·44	9·63	10·34
		S	2·94	2·49	3·74	4·13	4·06	2·18	3·02	1·97	4·24	2·74	1·97	2·47
	Heart rate	$\bar{x}$	76·76	82·67	78·26	90·07	93·97	92·80	100·63	97·31	105·18	103·61	94·12	100·72
		S	13·08	8·30	18·76	12·27	12·53	18·36	13·15	10·90	18·85	20·67	20·67	16·46
	Oxygen uptake	$\bar{x}$	0·37	0·39	0·37	0·57	0·64	0·60	0·70	0·64	0·83	0·79	0·77	0·82
		S	0·18	0·16	0·14	0·25	0·21	0·19	0·19	0·21	0·30	0·20	0·21	0·24

*$\bar{x}$ = mean; S = standard deviation.

Table 4.21b. Maximum weight of lift acceptable to female industrial workers for 8 h and their physiological responses at that weight (Mital, 1984b)

Lift height	Box size (cm)		1			4			8			12		
			30·48	45·72	60·96	30·48	45·72	60·96	30·48	45·72	60·96	30·48	45·72	60·96
Floor to knuckle	Weight (kg)	x̄*	13·33	12·94	11·40	11·84	10·52	11·51	10·47	9·05	8·64	8·41	9·54	9·83
		S	2·39	2·63	2·19	3·47	2·64	2·74	3·76	2·07	1·22	2·30	2·88	2·72
	Heart rate (beats/min)	x̄	92·23	95·19	91·97	107·94	104·19	99·75	115·06	116·55	101·76	123·04	123·75	118·36
		S	7·28	10·30	13·13	14·59	17·38	15·12	22·74	8·87	10·19	17·60	18·09	12·43
	Oxygen uptake (l/min)	x̄	0·40	0·35	0·39	0·56	0·51	0·60	0·71	0·63	0·59	0·60	0·79	0·62
		S	0·19	0·16	0·17	0·11	0·25	0·28	0·44	0·16	0·21	0·15	0·21	0·16
Knuckle to shoulder	Weight	x̄	11·37	11·69	10·45	10·30	10·38	10·34	9·52	8·90	8·93	8·07	8·34	8·40
		S	2·01	2·73	2·38	1·73	2·38	3·00	1·85	1·28	2·23	1·16	2·04	2·31
	Heart rate	x̄	94·07	91·34	91·59	102·44	101·57	105·71	111·90	100·80	105·97	118·98	123·62	106·77
		S	12·47	9·37	14·51	9·93	11·15	11·00	12·83	10·35	15·61	14·00	14·47	17·50
	Oxygen uptake	x̄	0·28	0·28	0·22	0·41	0·39	0·44	0·54	0·51	0·53	0·53	0·67	0·53
		S	0·14	0·17	0·09	0·22	0·21	0·29	0·18	0·09	0·24	0·22	0·21	0·21
Shoulder to reach	Weight	x̄	10·53	8·94	8·68	9·65	9·32	8·97	8·95	9·10	9·02	8·31	7·68	8·21
		S	2·05	1·61	0·95	2·09	1·35	1·82	1·15	2·14	2·39	1·03	1·16	1·26
	Heart rate	x̄	90·16	89·75	86·54	100·19	96·95	86·70	109·34	100·12	109·35	106·39	115·14	114·89
		S	8·96	10·21	7·33	12·13	14·82	12·78	11·36	16·28	18·28	17·19	11·19	21·55
	Oxygen uptake	x̄	0·18	0·27	0·22	0·40	0·35	0·28	0·54	0·38	0·55	0·37	0·47	0·53
		S	0·11	0·19	0·10	0·15	0·17	0·10	0·27	0·23	0·24	0·14	0·21	0·21

(Frequency of lift (lifts/min))

*x̄ = mean; S = standard deviation.

Table 4.22. Decrement (%) in maximum acceptable weight of lift with frequency, box size and height of lift (Mital, 1984a,b)

Task factor	12-h shift		8-h shift	
	Males	Females	Male	Females
Box size (cm)				
30·48	100*	100	100	100
45·72	94	98	94	98
60·96	92	95	92	95
Frequency (lifts/min)				
1	100	100	100	100
4	91	92	92	93
8	81	80	81	83
12	69	76	71	78
Height of lift				
Floor to knuckle	100	100	100	100
Knuckle to shoulder	97	92	96	92
Shoulder to reach	89	84	87	84

*Numbers rounded to the nearest integer.

Table 4.20 gives the lifting capability of males and females for 12-h shifts. The lifting capability norms for 8-h shifts are given in Table 4.21 for males and females. Table 4.22 shows the decrement in maximum acceptable weight of lift with frequency, box size and height of lift for 12-h and 8-h shifts. The overall means of responses for independent variables for 12-h and 8-h shifts are given in Table 4.23.

The results of the study indicated that, in general, males were willing to lift more weight than females. The maximum weight of lift acceptable to males and females, for 12-h shifts, resulted in metabolic energy expenditure rates which were 23 and 24% of their aerobic capacity, respectively. For the 8-h shift, the corresponding values were 29 and 28%.

The maximum weights, acceptable for 12 h of lifting, elicited an average heart rate of 90 and 101 beats/min for males and females, respectively. For 8 h, the respective heart rates were 99 and 104 beats/min. As shown in Table 4.22, the maximum acceptable weight of lift decreased with box size and frequency. Maximum weight was lifted in the floor to knuckle height region and least in the shoulder to reach height region.

As mentioned in Chapter 2, age of the workers, males or females, had no effect on the lifting capability.

Since the experimental conditions investigated in many investigations are similar, it is possible to integrate data from these studies to arrive at some comprehensive recommendations. Details of data integration and comprehensive recommendations are provided in section 4.6.

4.4 Strength Data for One-Handed Lifting and Lowering Activities

Unlike two-handed MMH tasks, studies dealing with one-handed MMH tasks are relatively few. The authors have been able to identify only eight studies which have

Table 4.23. Overall means and standard deviations of responses for 12-h and 8-h shifts (Mital, 1984a,b)

		12-h shift				8-hr shift			
		Males		Females		Males		Females	
Variable	Response*	$\bar{x}$‡	S	$\bar{x}$	S	$\bar{x}$	S	$\bar{x}$	S
Box size (cm)									
30·48	WT	10·57	3·44	8·78	2·21	13·44	4·14	10·02	2·55
	HR	90·07	16·85	102·22	15·45	99·60	18·89	106·36	16·43
	OX	0·59	0·27	0·40	0·21	0·72	0·32	0·46	0·24
45·72	WT	9·91	2·87	8·59	2·31	12·70	3·44	9·78	2·46
	HR	90·27	15·77	101·74	16·00	99·42	16·70	105·09	17·01
	OX	0·60	0·27	0·40	0·21	0·72	0·29	0·47	0·24
60·96	WT	9·69	3·01	8·35	2·08	12·34	3·61	9·53	2·35
	HR	89·48	17·78	97·97	16·02	98·40	20·09	101·61	17·15
	OX	0·61	0·25	0·40	0·22	0·73	0·29	0·46	0·24
Frequency (lifts/min)									
1	WT	11·80	3·18	9·87	2·36	14·89	3·84	11·04	2·55
	HR	76·60	12·55	88·16	9·68	84·64	14·08	91·43	10·39
	OX	0·34	0·16	0·25	0·15	0·42	0·18	0·29	0·16
4	WT	10·76	3·12	9·09	2·18	13·70	3·80	10·32	2·48
	HR	88·75	14·66	97·33	13·82	97·86	15·29	100·61	14·00
	OX	0·60	0·21	0·38	0·20	0·71	0·23	0·44	0·22
8	WT	9·50	2·72	7·94	1·64	12·10	3·11	9·20	2·05
	HR	95·62	14·95	104·20	13·11	105·47	16·98	108·20	14·80
	OX	0·71	0·24	0·47	0·21	0·85	0·26	0·55	0·24
12	WT	8·14	2·21	7·48	1·79	10·59	2·65	8·65	2·03
	HR	98·92	15·89	112·56	15·55	108·77	17·78	116·85	16·73
	OX	0·76	0·23	0·50	0·20	0·91	0·26	0·57	0·22
Height of lift†									
F–K	WT	10·55	3·58	9·30	2·67	13·55	4·28	10·62	2·95
	HR	93·06	18·27	103·34	16·72	102·17	20·00	107·56	17·82
	OX	0·71	0·28	0·50	0·22	0·85	0·32	0·56	0·24
K–S	WT	10·28	3·01	8·60	2·06	13·07	3·58	9·79	2·28
	HR	92·03	15·17	101·86	14·98	102·02	17·21	105·13	15·78
	OX	0·57	0·23	0·38	0·19	0·68	0·27	0·46	0·22
S–R	WT	9·34	2·60	7·83	1·50	11·84	3·13	8·95	1·71
	HR	84·47	15·37	96·76	15·36	93·01	16·89	100·46	16·79
	OX	0·51	0·23	0·32	0·19	0·62	0.26	0·38	0·21

*WT = weight (kg), HR = heat rate (beats/min), OX = Oxygen uptake (l/min).
†F = floor, K = knuckle, S = shoulder, R = reach.
‡$\bar{x}$ = mean, S = standard deviation.

addressed the problems of one-handed materials handling tasks (Davis and Stubbs, 1980; Warwick et al., 1980; McConville and Hertzberg, 1968; Garg and Saxena, 1982; Garg, 1983; Mital and Asfour, 1983; Mital and Manivasagan, 1983b; Mital, 1985b). These studies are reviewed in the following sections.

McConville and Hertzberg (1968) used 30 males to determine the maximum weight that could be lifted by preferred (stronger) hand from the floor level to a table approximately 76 cm high. The goal was to develop upper limits on the weight of industrial or military equipment which must be lifted by one hand. The subject population selected was representative of the US Air Force population in the body weight and height. A special box was designed that permitted width variation from 15·24–81·28 cm. The results of the study indicated that the maximum weight that 95% of the population would be able to lift, but not necessarily carry, could be expressed by a linear equation $Y = 60\text{-}X$, where Y is the weight in pounds and X is the width in inches. According to this equation, a weight of approximately 27·24 kg could not be lifted by the majority of the population using just the preferred hand. Of course, as the box width increases, the lifting capability for infrequent tasks would go down linearly.

Warwick *et al.* (1980) used 29 healthy adult males (age 21–73 years, height 165–194 cm, and weight 60·9–105 kg). Force (isometric strength) exerted by subjects, using either left or right hand, on a handle located at either 142 cm height (shoulder height) or at 60 cm height (knee height) was measured. The feet of the subject were always placed 30 cm apart, in one of three positions (Figure 4.5); right, right posterior and left. The forces were exerted in six directions: vertically upward (lift); vertically downward (press); push forward; pull backwards; push left or right parallel to the frontal plane when the body is facing the work piece. Table 4.24 gives the average forces exerted.

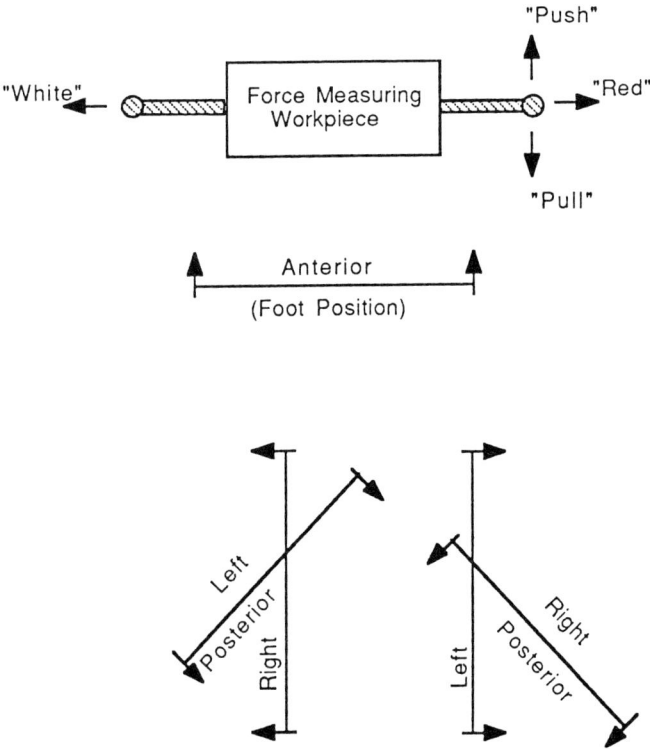

Figure 4.5 Placement of the subjects' feet relative to the workpiece (not to scale) viewed from above. Arrows show how subjects faced and where medial edges of the feet were placed. "Red" and "White" directions are the same as "Push Right" and "Push Left" directions (Warwick et al., 1980)

Table 4.24. Mean magnitudes of the forces exerted (kg) at shoulder and knee heights (Warwick et al., 1980)

Hand	Height	Foot position	Lift up	Press down
Left	Shoulder	Right	10·5	14·5
		Right posterior	7·8	12·1
	Knee	Left	16·1	24·1
		Right	11·5	22·1
Right	Shoulder	Right	13·8	16·3
		Right posterior	13·7	16·4
	Knee	Left	10·9	10·9
		Right	10·3	27·7

The Materials Handling Research Unit of the University of Surrey (Davis and Stubbs, 1980) has recommended forces that can be exerted occasionally for one-handed lifting tasks when standing, squatting, sitting or kneeling. These forces are for males and take into consideration his age and location of the grip. Table 4.25 gives the force values for one-handed lifts in various postures for males of different age groups. The hand (preferred) in all cases is directly in front. The same forces may be applied if the arm is in a plane in line with the shoulders or it is in a plane in between the front and the side positions.

The recommendations given in Table 4.25 are most comprehensive, are available in the literature and should be used when lift force are to be exerted by only one hand (the preferred hand). If the non-preferred (weaker) hand is to be used, the values should be reduced by approximately 10%.

Table 4.25. Lifting forces* (kg) for one-handed lifting with arm in the front of the body in the horizontal plane of the shoulder (Davis and Stubbs, 1980)

Position	Acromial-grip distance (cm)	Age (years)		
		Up to 40	41–50	51–60
Standing/squatting	65	10	10	9
	60	12	12	11
	50	15	15	13
	35	20	20	18
	20	25	25	22
	5	30	30	27
Sitting	65	10	9	8
	50	15	14	12
	40	20	18	16
	30	25	23	20
	15	30	27	24
	5	35	32	28
Kneeling	65	12	10	9
	60	15	12	11
	55	17	13	12
	30	20	16	15
	25	25	20	18
	15	30	23	22

*Values converted from graphical representation.

4.5 Capacity Data for One-Handed Lifting and Lowering Activities

One-handed lifting tasks have been studied by the Materials Handling Research Unit of the University of Surrey (Davis and Stubbs, 1980; Garg and Saxena 1981, 1982; Garg 1983 and Mital and Asfour 1983). While the last three studies deal with the determination of maximum acceptable frequency of lift and physiological fatigue limits, permissible force limits have been provided by the University of Surrey researchers. According to them, if a task is to be performed more frequently than once a minute, the force values given in Table 4.25 should be reduced by 30%. These recommendations are available only for male workers.

Garg and Saxena (1981, 1982) and Garg (1983) conducted a study to determine the maximum frequencies acceptable to female workers for one-handed lifts in the horizontal plane. Ten female college students (21–34 years of age, 50·9–78·3 kg bodyweight, 160–179·7 cm height, 74·1–99·1 cm functional arm reach, 1·8–2·55 l/min aerobic capacity on bicycle ergometer, and 0·77–1·35 l/min aerobic capacity on arm ergometer) continuously lifted three different loads (2·3, 4·5 and 5·7 kg for 38·1 cm

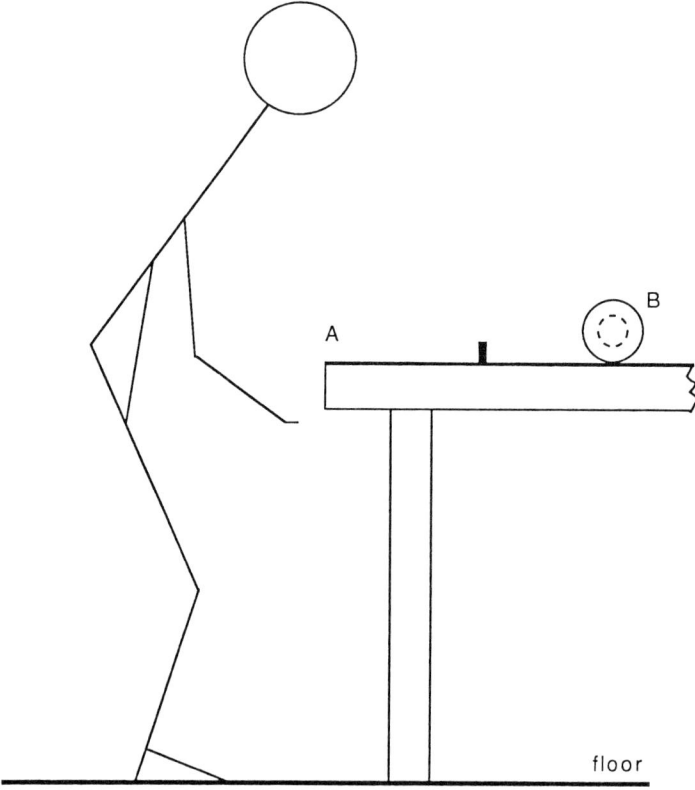

Figure 4.6 Schematic diagram of the lifting task (Mital, 1985). From Human Factors, Vol. 25 © 1983 by The Human Factors Society, Inc., and reproduced by permission

reach and 1·1, 2·3 and 4·5 kg for 63·5 cm reach) to two different reach distances (38·1 and 63·5 cm). All lifting was performed in the standing posture in front of a 91·4 cm high work surface. Each lifting cycle required subjects to reach for the load (position A to B, Figure 4.6), pick up the load, move the load away from the subject, release the load, return to the starting point without the load (position B to A).

A barrier, approximately 2 cm high, was placed in the path of lifting to assure that the load was indeed lifted, rather than dragged across the surface.

Subjects were trained for 8 weeks prior to data collection. Three training sessions (2 h each) per week, with one day rest between successive training sessions, were conducted for each subject. Using the psychophysical approach, subjects determined the maximum acceptable frequency of lift for 4 min of lifting before and after training. To determine maximum frequency, the metronome was initially set at a very high frequency. This initial rate was increased or decreased in steps of 2 lifts/min to determine the rate of lifting (frequency) that subjects could maintain for a 4-min period. During the training period, subjects were started at 50% of the maximum 4-min lifting frequency. The rate was slowly increased over several weeks. Subjects were trained for different weights. After the completion of 8 weeks of training, the maximum acceptable frequencies of lift for a 4-min period were determined again. Heart rate and oxygen uptake at the maximum 4-min frequencies were also recorded. Table 4.26 gives the frequencies for different load-reach combinations for a 4-min period.

Table 4.26. *Maximum 4-min lifting frequency, oxygen uptake and heart rate for one-handed lifts in the horizontal plane by female workers (Garg and Saxena, 1982; Garg, 1983)*

Lifting distance (cm)	Load (kg)	Frequency (lifts/min)			Oxygen uptake (lit/min)			Heart rate (beats/min)		
		$\bar{x}^*$	S†	Range	$\bar{x}$	S	Range	$\bar{x}$	S	Range
38·1	2·3	33·6	2·1	30–36	0·54	0·07	0·43–0·66	109	11·4	95–131
	4·5	31·8	2·2	30–36	0·66	0·11	0·49–0·86	123	13·9	94–144
	5·7	29·8	2·2	28–34	0·76	0·12	0·50–0·93	132	18·7	95–154
63·5	1·1	31·8	1·5	30–34	0·58	0·07	0·51–0·70	113	8·9	95–122
	2·3	29·6	1·6	28–32	0·70	0·08	0·51–0·82	123	16·2	92–139
	4·5	26·8	1·0	26–28	0·82	0·12	0·60–0·97	133	14·9	110–155

Subsequent to training, subjects, using the pyschophysical methodology, determined the maximum frequency of lift acceptable to them for an 8-h workday.
*$\bar{x}$ = mean; †S = standard deviation.

Subsequent to training, subjects, using the psychophysical methodology, determined the maximum frequency of lift acceptable to them for an 8-h workday. Subject lifted various loads for two reach distances. For each combination of load–reach distance (a total of six), subjects lifted for 1 h. Their heart rate, ratings of perceived exertion and the maximum acceptable frequency of lift were recorded every 15 min. Table 4.27 shows the final values of responses measured. Oxygen uptake values are not shown since they showed little change (increase) with increase in work intensity (Garg, 1983).

The results of the study indicated that both weight and lifting distance have a significant effect on the maximum frequency acceptable to subjects for one-handed lifting

Table 4.27. *Maximum lifting frequency, heart rate, and ratings of perceived exertion (RPE) for one-handed lifts in the horizontal plane by female workers for 8 hours (Garg and Saxena, 1981, 1982)*

Lifting distance (cm)	Load (kg)	Frequency (lifts/min)		Heart rate (beats/min)		RPE	
		$\bar{x}^*$	S†	$\bar{x}$	S	$\bar{x}$	S
38·1	2·3	18·8	4·1	98	16	11	2
	4·5	15·5	2·6	100	10	12	1
	5·7	13·7	2·6	105	13	13	2
63·5	1·1	18·0	5·0	97	9	11	1
	2·3	15·5	4·0	102	8	12	1
	4·5	12·7	2·7	104	8	13	2

*$\bar{x}$ = mean; †S = standard deviation.

in the horizontal (sagittal) plane. The average percentage of the maximum acceptable frequency was 51% (47–56%) of the maximum 4-min frequency. Even though subjects selected workloads that resulted in a mean heart rate of 101 beats/min (98–105 beats/min), the heart rate never stabilized over the 1-h lifting duration. Continuous increases in heart rate were observed (Figure 4.7).

The subjects rated the perceived exertion at the maximum acceptable frequency of lift ranging from 'fairly light' to 'somewhat hard'. The relationship between heart rate and ratings of perceived exertion (RPE) (heart rate = 10∗RPE) did not hold true for one-handed lifting tasks.

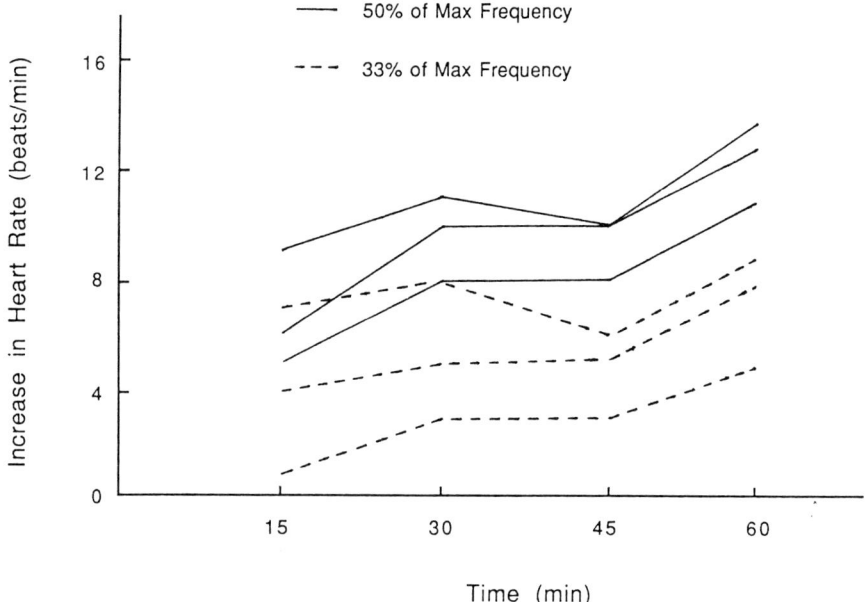

Figure 4.7 *Work pulse rate (work heartrate: standing – resting) with time for one-handed lifting (reach distance = 38·1 cm) (Garg and Saxena, 1981, 1982)*

Mital and Asfour (1983) extended the work of Garg (1983) and Garg and Saxena (1981, 1982) since the specific recommendations made are applicable only if the working time does not exceed 1 h. This was because the heart rates failed to stabilize. In order to establish permissible one-hand lift limits for the duration of the work shift, it is essential that heart rate for a given workload stabilize; otherwise increasing muscle fatigue will force the workers to stop. Data on males was also needed. The goal of the study by Mital and Asfour was to determine the maximum frequency acceptable to males lifting specified weights by preferred hand for a work period of 2 h. A 2-h continuous work period is a reasonable study time period since in most work situations a break (coffee or lunch) usually follows. This permits workers to rest and recover from fatigue. Other objectives of the study were to determine the effects of training and posture on maximum acceptable frequencies. The need for training workers for one-handed lifting tasks itself was also of interest. Asfour et al. (1984a) have shown that the maximum weight acceptable to males for two-handed lifting tasks increases with training. Garg (1983) and Garg and Saxena (1982) have also reported an increase with training in the maximum frequency acceptable to females for one-handed lifting for a 4-min period. Confirmation of positive training effects in the case of males performing one-handed lifting tasks was, therefore, of interest. For comparison with Garg (1983) and Garg and Saxena's (1981, 1982) subjects, data on a limited number of females were also collected.

Seven male (20–27 years age, 58·6–83·1 kg body-weight, 162·6–185·4 cm height, 58·4–61·6 cm mean seat height, 53·3–70·5 cm sitting functional reach, and 40·6–63·5 cm standing functional reach) and three female (21–26 years of age, 41·8–56·7 kg body-weight, 149·9–163·2 cm height, 61–64·8 cm mean seat height,

Table 4.28. *Maximum acceptable frequency of lift for 4-min period before and after training (Mital and Asfour, 1983)**

			Frequency (lifts/min)			
			Before training		After training	
Posture	Reach (cm)	Load (kg)	Males	Females	Males	Females
Sitting	38·1	2·27	28·57	20·33	28·71	22·67
		4·54	20·43	17·00	23·43	18·33
		6·81	12·86	—	14·14	—
	63·5	2·27	26·85	17·33	27·28	19·00
		4·54	17·28	10·33	19·14	12·67
		6·81	10·14	—	11·86	—
Standing	38·1	2·27	28·44	25·00	29·14	26·67
				(29·80)		(33·60)
		4·54	21·86	16·00	24·00	18·33
				(27·80)		(31·80)
		6·81	15·71	—	16·43	—
	63·5	2·27	27·00	19·00	28·43	20·67
				(26·00)		(29·60)
		4·54	17·28	12·00	20·00	14·67
				(21·60)		(26·80)
		6·81	11·86	—	13·28	—

*Numbers in parentheses are for Garg and Saxena's (1981, 1982) subjects.

Table 4.29. Means and standard deviations* of response variables for various combinations of posture, reach and load (Mital and Asfour, 1983)

			Males						Females					
Posture	Reach	Load	HR	RPEB	RPES	RPEA	Frequency	% MAX	HR	RPEB	RPES	RPEA	Frequency	% MAX
Sitting	38·1	2·27	81·98 (5·63)	8·66 (2·44)	10·52 (2·58)	10·41 (2·38)	15·32 (6·89)	53·36	76·04 (7·90)	8·37 (1·47)	9·45 (2·47)	10·67 (2·10)	10·84 (1·68)	47·81
		4·54	80·53 (10·98)	8·98 (2·89)	10·66 (2·85)	10·53 (2·91)	10·18 (3·48)	43·45	75·67 (7·55)	10·25 (2·61)	11·58 (2·22)	12·75 (2·62)	7·83 (0·56)	42·72
		6·81	90·14 (7·16)	9·57 (2·90)	11·30 (3·11)	10·86 (2·40)	7·00 (1·22)	49·50	—	—	—	—	—	—
	63·5	2·27	82·25 (5·34)	8·36 (2·40)	10·25 (2·26)	10·50 (2·54)	14·52 (5·86)	53·22	79·25 (9·89)	10·71 (1·27)	11·25 (1·87)	10·54 (1·50)	8·17 (0·92)	43·00
		4·54	83·04 (3·24)	9·16 (2·33)	11·32 (2·65)	11·16 (2·02)	11·20 (3·02)	58·52	73·71 (10·65)	10·37 (2·06)	10·91 (2·16)	11·67 (2·18)	5·42 (0·58)	42·78
		6·81	84·98 (5·87)	9·71 (2·97)	11·16 (2·97)	11·59 (2·56)	6·96 (1·40)	58·68	—	—	—	—	—	—
Standing	38·1	2·27	84·16 (6·17)	9·23 (2·41)	11·18 (2·83)	10·96 (2·67)	15·78 (6·21)	54·15	85·16 (8·82)	8·71 (1·94)	8·71 (2·37)	10·25 (1·67)	10·46 (0·93) 18·80‡	39·22
		4·54	84·16 (8·17)	9·59 (2·42)	10·84 (2·73)	10·82 (2·55)	11·16 (2·79)	46·50	89·50 (15·71)	11·21 (2·0)	11·42 (3·15)	11·83 (2·58)	6·58 (1·35) 15·50‡	35·89
		6·81	88·39 (10·05)	9·50 (3·01)	11·71 (2·90)	11·53 (2·79)	7·53 (1·40)	45·83	—	—	—	—	—	—
	63·5	2·27	81·45 (3·25)	8·28 (2·32)	9·95 (2·13)	10·82 (2·74)	13·93 (5·60)	49·00	89·12 (12·84)	9·83 (1·95)	10·42 (3·77)	11·71 (1·85)	9·08 (1·56) 15·50‡	43·94
		4·54	84·25 (8·92)	9·86 (2·82)	11·68 (2·66)	11·32 (2·38)	10·98 (3·36)	54·90	87·50 (6·98)	9·37 (2·04)	10·29 (3·14)	11·33 (2·14)	6·87 (1·59) 12·70‡	46·83
		6·81	86·93 (7·12)	9·87 (2·88)	11·87 (2·79)	11·82 (2·55)	6·95 (1·27)	52·33	—	—	—	—	—	—

*Standard deviations are given in parentheses.
HR = heart rate, RPEB = RPE back, RPES = RPE shoulder, RPEA = RPE arm, % MAX = acceptable frequency as a percentage of 4-min frequency.
‡ Average values for Garg and Saxena's (1981, 1982) subjects.

49·5–58·4 cm sitting functional reach, and 35·6–48·9 cm standing functional reach) college students participated in the experiment. Each male subject performed 12 combinations of posture (sitting and standing), reach distance (38·1 and 63·5 cm), and load (2·27, 4·54 and 6·81 kg). The third load level was heavier than the three loads used by Garg (1983) and Garg and Saxena (1981, 1982), since males in general can lift heavier loads. For female subjects, only the two lighter loads were used.

The task and the procedure were identical to that used earlier (Garg and Saxena, 1981, 1982; Garg, 1983). All subjects were trained, but only for a 6-week period. This period was considered adequate (Asfour et al., 1984a; Garg and Saxena, 1981, 1982). To determine the effects of training, the maximum acceptable frequency of lift for a 4-min period was determined for each posture–load–reach combination before and after training. Subjects were trained at 50% of the 4-min frequency for 4·5 weeks. The training frequency was increased to 75% of the 4-min frequency during the last 1·5 weeks. The training procedure was similar to that described earlier by Garg (1983) and Garg and Saxena (1981, 1982). Table 4.28 shows the maximum acceptable frequency of lift for a 4-min period before and after training.

Subsequent to training, subjects, using the psychophysical approach, determined the maximum acceptable frequency of lift for each posture–load–reach combination. Lifting was performed for a full 2 h, without a break. Heart rate, and ratings of the perceived exertion of the arm (RPEA), back (RPEB), and shoulder (RPES) were

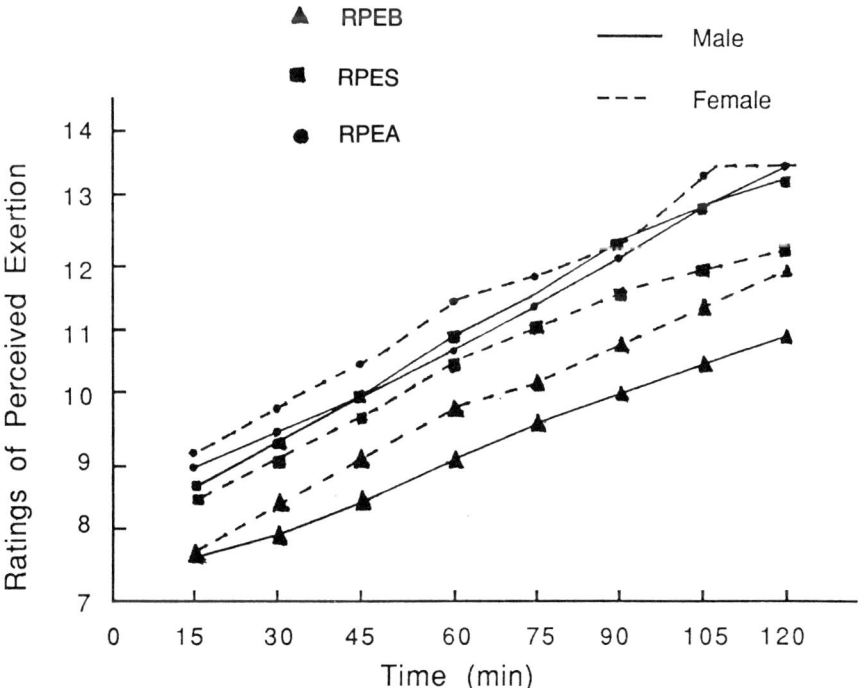

Figure 4.8 Variations of RPE with time (Mital and Asfour, 1983). From Human Factors, Vol. 25 © 1983 by The Human Factors Society, Inc., *and reproduced by permission*

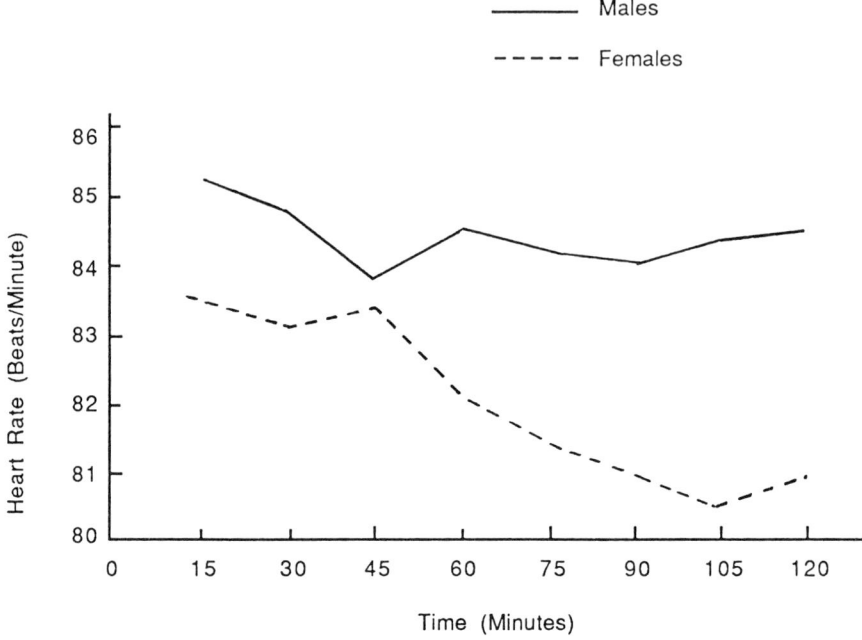

Figure 4.9 Variations of heart-rate with time (Mital and Asfour, 1983). From Human Factors, Vol. 25 © 1983 by The Human Factors Society, Inc., and reproduced by permission

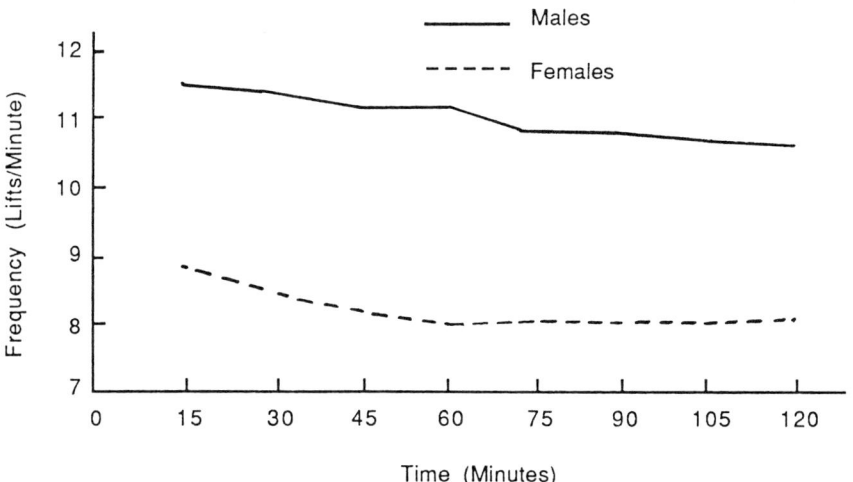

Figure 4.10 Variations of the maximum acceptable frequency of lift with time (Mital and Asfour, 1983). From Human Factors, Vol. 25 © 1983 by The Human Factors Society, Inc., and reproduced by permission

recorded along with the maximum acceptable frequency of lift every 15 min. Table 4.29 shows the responses for various posture, reach and load combinations.

The results indicated that the average maximum acceptable frequency of lift for one-handed lifting tasks for males was 51% (43·4–58·7%) of the 4-min frequency. The acceptable frequency of lift decreased with reach and load. The effect of posture was not significant. The sitting posture was, however, less demanding physiologically (lower heart rates). The ratings of perceived exertion ranged from 'very light' to 'somewhat hard' and did increase with time (Figure 4.8).

Heart rate did stabilize (Figure 4.9), but showed no systematic change with time, indicating that a decrease in heart rate due to the reduction in acceptable frequency (Figure 4.10), was compensated for by increased fatigue. The overall heart rate was approximately 85 beats/min.

The levels of acceptable frequency (average of 42·78% of 4-min frequency) were significantly lower than for Garg and Saxena's subjects. The increase in the 4-min frequency due to training was 7·23% for males (0–29% range) and 8·57% (4–11·9% range) for females. The training gains for females were also significantly lower than those for Garg and Saxena's subjects (mean gain 16%, range 9·7–25·4%). Thus it appears that training effects are neither consistent nor always significant.

4.6 Work-rate Recommendations for Two-Handed Lifting Tasks

Snook (1978a, b), Ayoub *et al.* (1978), and Mital (1984b) have collected lifting capability data on industrial workers for repetitive lifting tasks. These data are compared and integrated in this section to develop a comprehensive MAWL database for males and females for 8-h shifts.

In all three studies, lifting frequency, box size and height of lift were the independent variables and the psychophysical approach was used to determine the MAWL. Figures 4.11–4.13 show the effect of different task factors on the maximum acceptable weight of lift as determined in the three studies. To make comparisons, the maximum acceptable weight of lift was expressed as a percentage. Weight lifted either in the 30·48 cm box or at a frequency of 1 lift/min or from the floor to knuckle height level was considered as the 100% level. Acceptable weight of lift at other levels, of respective task factors, were expressed as percentages of the weight lifted at these datum levels.

4.6.1 Box Size Effect

The effect of box size on the maximum acceptable weight of lift is plotted in Figure 4.11, and tabulated in Table 4.30. In general, there was closer agreement between Mital's (1984b) data and Ayoub *et al.*'s (1978) data for males. Differences with Snook's (1978a, b) data for males were large at the intermediate level. For females, however, the three studies showed different trends. The difference in decrement in maximum acceptable weight of lift ranged from 3–9%. Since all three studies show similar trends (decrease in the maximum acceptable weight of lift with box size) and all subjects were experienced industrial workers, the differences were attributed to the variation in population capabilities.

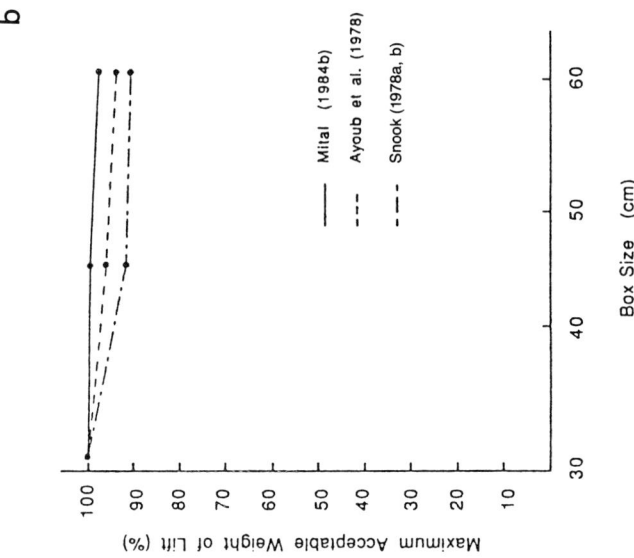
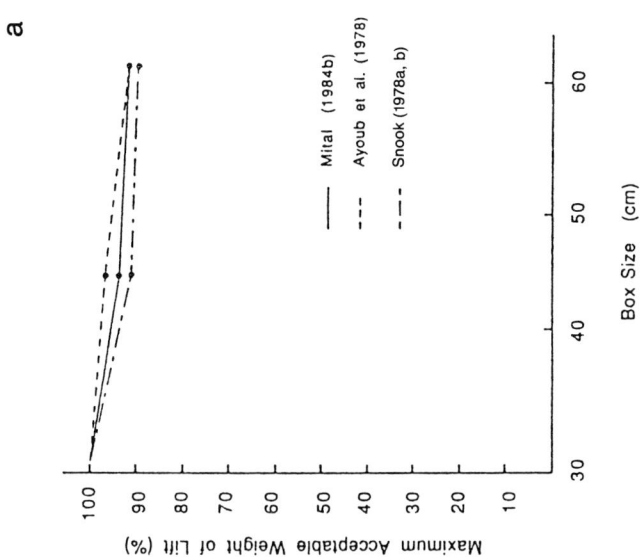

Figure 4.11 Effect of box-size on the maximum weight of lift acceptable to (a) industrial males, and (b) industrial females.

Table 4.30. Decrement (%) in maximum acceptable weight of lift with frequency, box size, and height of lift based on Snook (1978), Ayoub et al. (1978) and Mital (1984b) data

Task factor	Males			Females		
	Snook (1978a)	Ayoub et al. (1978)	Mital (1984b)	Snook (1978a)	Ayoub et al. (1978)	Mital (1984b)
Box size (cm)						
30-48	100*	100	100	100	100	100
45-72	91	97	94	89	95	98
60-96	89	92	92	87	91	95
Frequency (lifts/min)						
1	100	100	100	100	100	100
4	89	93	92	92	94	93
8	79	84	81	86	91	83
12	59	†	71	72	†	78
Height of lift						
Floor to knuckle	100	100	100	100	100	100
Knuckle to shoulder	98	94	96	90	86	92
Shoulder to reach	84	71	87	73	59	84

*Numbers rounded to the nearest integer value. † Experimental data not collected.

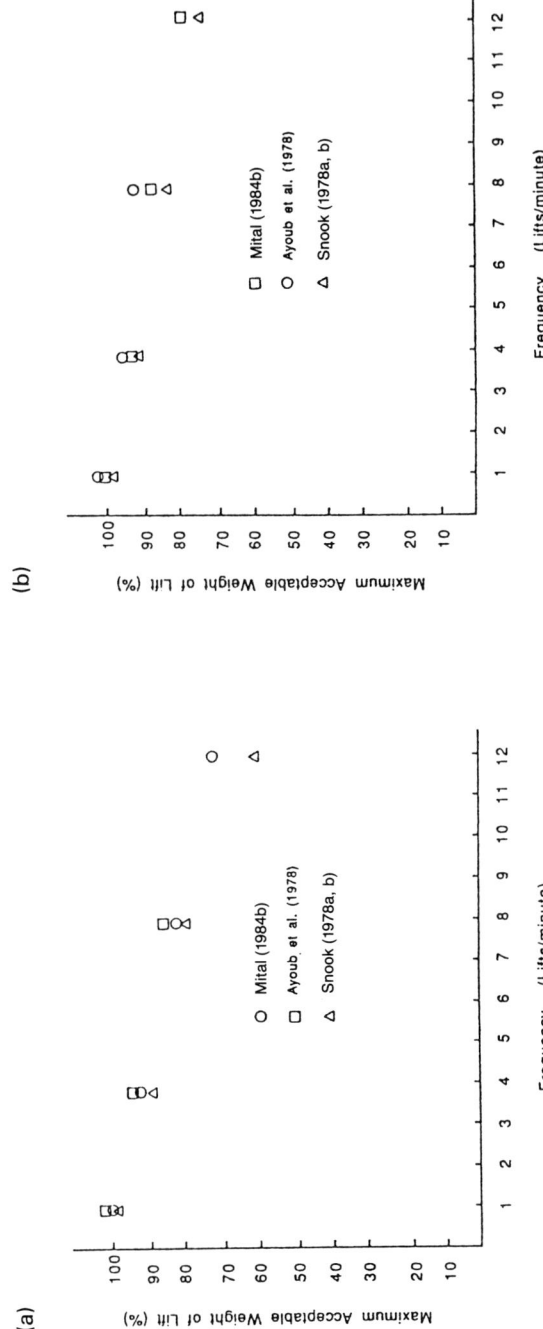

Figure 4.12 Effect of lifting frequency on the maximum weight of lift acceptable to (a) industrial males, and (b) industrial females.

4.6.2 Frequency Effect

Figure 4.12 shows reductions in the maximum acceptable weight of lift with frequency for males and females, respectively. In most cases the differences between data from the three studies were small. The exception being for males lifting at 12 lifts/min. In this case Snook's data showed a much sharper decrement than Mital's data.

In general, effects of frequency on the maximum acceptable weight of lift were more profound than box size effects. It is reasonable to expect such sharp decrease in weight with frequency since with every lift, individuals also lift a part of their body-weight. In some cases, this extra weight may be as much as 60% of the individual's body-weight (Ayoub et al., 1978).

4.6.3 Effect of Height Level

As shown in Figure 4.13, less weight was lifted as the starting height increased, and data from all three studies are in close agreement. For the shoulder to reach height level, however, the differences were large. Male subjects employed by Ayoub et al. lifted substantially lower weight for the shoulder to reach height level compared to male subjects of the other two studies. Females who participated in Mital's study lifted more weight for the shoulder to knuckle height level than the other two studies. Overall, as can be seen from these figures, in most cases the data from the three studies were in close agreement. The differences in some instances were large, but they were considered to be due to variation in population capabilities.

4.6.4 Comprehensive Maximum Acceptable Weight of Lift Data Base

Overall, the data from all the three studies (Ayoub et al., 1978; Snook, 1978a, b; Mital, (1984b) were in agreement. Minor differences existed due to population variations. The data from all the three studies, therefore, could be integrated to develop a comprehensive 'maximum acceptable weight of lift' database.

Psychophysical estimate data from all three studies were integrated by taking weighted means; largest of the standard deviation values was used (since standard deviation from each of the three studies was a measure of capacity dispersion, lower values would not include the majority of the population). The integrated data were then adjusted for the effect of the shift duration with the help of multipliers proposed in the literature (Mital, 1983c; section 4.3). Table 4.31 shows the means and standard deviations of the weight recommendations for males and females for 8-h work shifts.

4.7 Performance Ceiling for One- and Two-Handed Lifting Tasks

The performance criteria for evaluating repetitive lifting tasks, one-handed or two-handed, can be grouped under the physiological criteria and the psychophysical criteria. The physiological criteria mostly include the following:

1. Metabolic energy expenditure rate or oxygen uptake criterion;

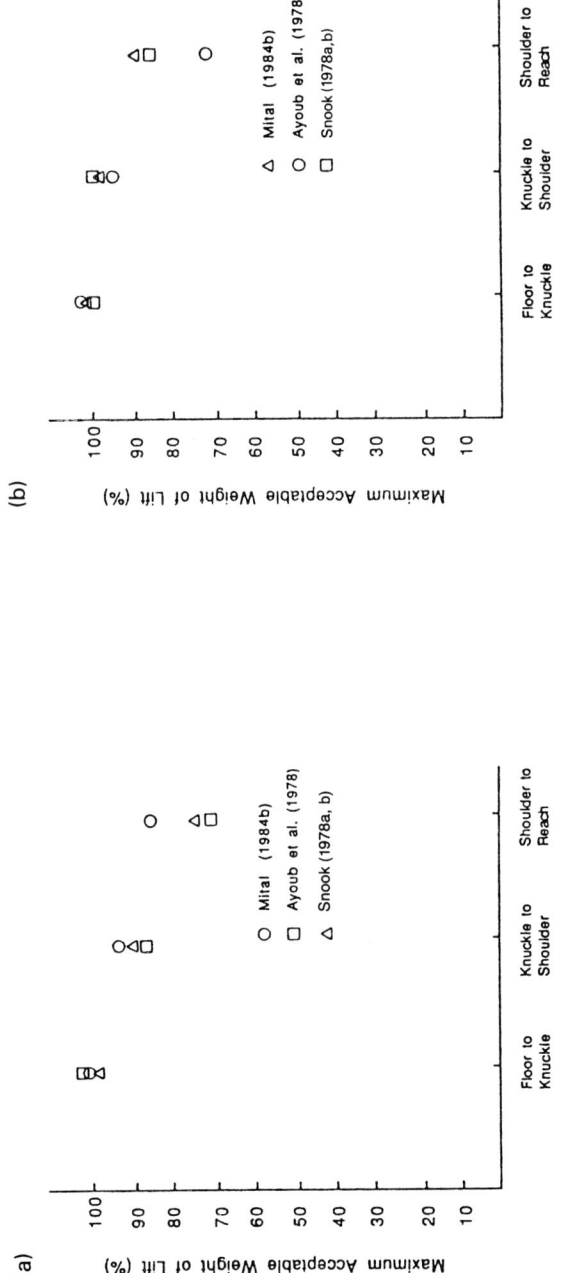

Figure 4.13 Effect of height level on the maximum weight of lift acceptable to (a) industrial males, and (b) industrial females.

Table 4.31. Means and standard deviations of the maximum acceptable weight of lift (kg) for two-handed lifting tasks, for males and females, as a function of task variables*

		Frequency of lift (lifts/min)											
		1			4			8			12		
Box size (cm)		30·48	45·72	60·96	30·48	45·72	60·96	30·48	45·72	60·96	30·48	45·72	60·96
Floor to knuckle	$\bar{x}$†	20·56	18·21	17·22	16·81	16·38	14·99	15·10	13·72	13·33	12·28	10·93	10·05
		(15·70)	(14·18)	(12·60)	(12·88)	(10·70)	(10·72)	(10·90)	(9·52)	(9·25)	(9·24)	(8·85)	(8·50)
	S‡	8·56	7·78	7·78	6·70	9·46	7·47	7·83	6·29	5·16	5·45	3·89	3·89
		(3·11)	(3·11)	(3·11)	(3·47)	(4·89)	(2·74)	(3·76)	(2·07)	(2·67)	(3·11)	(2·88)	(2·72)
Knuckle to shoulder	$\bar{x}$	16·82	18·65	15·13	15·08	16·53	13·75	13·30	13·84	11·62	11·23	10·77	10·21
		(11·88)	(11·26)	(10·78)	(10·92)	(10·78)	(10·07)	(9·83)	(9·37)	(8·83)	(8·54)	(7·91)	(7·78)
	S	7·00	6·23	6·23	7·09	8·74	7·47	5·39	5·34	7·29	3·89	3·89	3·89
		(2·33)	(2·73)	(2·38)	(2·09)	(2·38)	(3·21)	(3·62)	(3·03)	(2·35)	(1·55)	(2·04)	(2·31)
Shoulder to reach	$\bar{x}$	15·48	13·99	14·36	14·32	12·97	12·50	11·06	11·12	11·40	8·95	8·44	8·64
		(10·44)	(9·28)	(9·41)	(9·17)	(8·53)	(8·68)	(7·62)	(7·32)	(6·82)	(6·09)	(5·77)	(5·92)
	S	7·00	5·45	5·45	6·52	5·32	4·99	3·85	4·12	4·12	3·89	2·33	2·47
		(2·05)	(1·61)	(1·55)	(2·09)	(1·48)	(1·82)	(2·17)	(2·14)	(2·39)	(1·03)	(1·55)	(1·26)

*Numbers in parentheses are for females.
†$\bar{x}$ = mean; ‡S = standard deviation.

2. Heart rate or pulse rate criterion; and
3. Electromyographic (EMG) activity criterion.

The psychophysical performance criteria are mainly based on:

1. Maximum acceptable weight of lift;
2. Maximum acceptable frequency of lift;
3. Maximum acceptable workload (combinations of weight, frequency and lifting height), and
4. Ratings of perceived exertion.

Literature reviewed in sections 4.3 and 4.5 provides performance ceilings for two- and one-handed manual lifting tasks, respectively, which can maximize productivity. An individual's productivity is maximized when the task loading is optimum, not maximum or minimum (overloading usually results in exhaustion, fatigue and overexertion while underloading causes boredom and monotony (Mital, 1985b). The performance ceilings, based on the criteria discussed earlier, which maximize productivity, are summarized below.

4.7.1 One-Handed Lifting Tasks

1. For one-handed lifting tasks, the continuous work capacity should be based on a heart rate of approximately 90 beats/min (Mital and Asfour, 1983; Mital, 1985b).
2. For optimum performance, the tasks should elicit a rating of perceived exertion of about 12 on the Borg scale (Garg and Saxena, 1982; Mital and Asfour, 1983; Mital, 1985b).
3. For optimum performance, the frequency of lift should not exceed 50% of the maximum acceptable frequency for a 4-min period (Garg and Saxena, 1981, 1982; Mital and Asfour, 1983; Mital, 1985b).
4. For maximizing work output, the load should be between 4·5 and 5·5 kg (Mital and Asfour, 1983; Mital, 1985b).

4.7.2 Two-Handed Lifting Tasks

1. The continuous work capacity for two-handed lifting tasks should be based on a heart rate of approximately 100 beats/min and oxygen consumption of approximately 28% of the physical work capacity. For males, the oxygen consumption ceiling translates to approximately 4 kcal/min; for females, the corresponding number is 3 kcal/min (Mital, 1983c, 1984a, b).
2. The psychophysical ceilings for two-handed lifting tasks are given in Table 4.31.

Chapter 5
Pushing, Pulling, Carrying and Holding Data Bases

5.1 Strength Data for Two-Handed Pushing and Pulling Activities

Pushing and pulling constitute a different kind of materials handling activity. In this case, objects are generally pushed or pulled on floors. Often pulling or pushing forces are exerted on points above the floor. Pushing and pulling activities may not be as strenuous as lifting activities but are quite prevalent in industry.

Troup and Chapman (1969b) observed the maximum pulling force (for 1 s) exerted by 132 male and 98 female subjects. Forces were measured in the sitting and standing positions using a strain-gauge dynamometer. It was found that forces were greater in the sitting position than in the standing position. However, the forces transmitted by the erector spinae were of similar magnitude for sitting and standing. When standing, the lumbar spine is extended, and the erector spinae shortened; this provides a relatively greater mechanical advantage than does a sitting posture. In the sitting position, the smaller mechanical advantage is compensated for by a shorter distance between the pelvis and the upper limbs, resulting in a turning movement similar to that in the standing position. This leads to similar force transmission by the erector spinae as in the case of standing postures.

Warwick *et al.* (1980), using 29 adult males, measured push/pull forces that could be exerted at shoulder height (142 cm) and knee height (60 cm) for various foot positions (see section 4.2 for experimental details). Substantially greater push force was exerted at shoulder height when subjects were facing the handles (Table 5.1). The pulling force, however, was maximized in the anterior foot position at knee height. Table 5.1 shows the average push/pull forces in various postures at shoulder and knee heights. Push/pull to the right or left were in directions parallel to the frontal plane when the body was facing the work piece (anterior position). As shown in Table 5.1, push and pull forces are affected by the height of force application as well as the foot position.

Martin and Chaffin (1972) have proposed push/pull strength limits for the 50th percentile male as a function of horizontal and vertical distances of the hands from the ankles. The recommendations were made on the basis of isometric strength simulation which utilized maximum isometric strengths of 41 men (19–52 years age) and 48 women (18–48 years age). Table 5.2 shows the push/pull force recommendations after conversion from graphical representation.

Table 5.1. Average push/pull forces (kg) exerted by two hands at shoulder and knee heights (Warwick et al., 1980)

Foot position	Shoulder height				Knee height			
	Push forward	Pull backward	Push right	Push left	Push forward	Pull backward	Push right	Push left
Anterior	29·76	17·33	15·90	17·02	21·61	22·93	19·77	18·25
Right	26·20	19·67	11·93	14·37	28·34	18·25	13·86	15·70
Right posterior	24·06	17·94	11·42	11·93	—	—	—	—
Left	—	—	—	—	28·34	22·22	17·53	14·17

Table 5.2. Pushing-pulling capabilities* (kg) of 50% male for various hand locations (Martin and Chaffin, 1972)

Horizontal distance (m)	Vertical distance (m)	Force (kg)	
		Push	Pull
0·51	0·25	24	32
	0·51	23	54
	0·76	19	45
	1·02	14	34
1·02	0·25	0	0
	0·51	45	1
	0·76	48	5
	1·02	36	2
1·52	0·25	0	0
	0·51	0	0
	0·76	0	0
	1·02	1	0

The strength simulation predicted the maximum pushing force for horizontal distance between 0·89 and 1·27 m and vertical distance between 0·46 and 0·84 m. The pulling force was maximized between the horizontal distance of 0·13 and 0·51 m and the vertical distance between 0·38 and 0·61 m.

Kroemer (1969) measured the maximal isometric static horizontal push forces which could be exerted in many common working postures using 73 male subjects. Of these, 45 participated in the experiments with wall or foot rest, and 28 took part in the experiment with slippery floors. The experiment simulated all kinds of body supports and body postures which might occur in daily life when one has to exert a maximal horizontal push force. Subjects were instructed to maintain a maximum push force steadily over a 5-s period. The force exerted increasing during the first second, then stayed constant, and, finally, during the last second, it dropped. The mean force applied during the third second was used.

The results of this study indicated that the pushing or pulling force capabilities increased with the traction (friction). It was concluded that body posture, body support and the chain of body members through which the force is transmitted, have significant effects on the amount of force exerted. Body-weight, alone or in combination with body size, is not very useful in predicting force output. A physically strong subject may be able to exert only weak push or pull forces due to lack of body support.

In order to test the pushing force of an individual, Strindberg and Peterson (1972) measured the force required to start a trolley rolling. It was found that a person can exert, on an average, a pushing force of about 80% of his body-weight. Body-weight of the individual, thus, contributes significantly to the pushing force.

Caldwell (1964) studied the body position and the strength and endurance of manual pull. Ten male subjects were used (mean weight 77·46 kg, mean height 175 cm, age range 22–39 years). Strength (response) and endurance were measured for 20 different body positions (five elbow angles: 95, 110, 125, 140 and 155°; two thigh angles: 0 and 20°; and two knee angles: 10 and 150°. Each subject was required to pull maximally for a period of 8 s for the strength test. For the endurance test, the subject was asked to apply 80% of the maximal pull for as long as possible. A green light, which would turn on at 80% pull force, was used to assist subjects in maintaining this pull force.

An increase in the knee angle resulted in an increase in strength. The same was true for an elbow angle increase. Trunk stabilization improved with an increase in the thigh angle. These responses occurred during endurance tests also. It can be concluded that a change in body position, control placement or body stabilization, which increases pull strength, will reduce the effort required to maintain a given force on the control, and the endurance (time) of the holding response will be proportionately increased. In other words, static pull strength can predict static endurance (time) with reasonable confidence.

However, Smith and Edwards (1968) indicated that muscular endurance (dynamic endurance) may not be predicted from static strength. Forty male students participated in their study. A grip dynamometer was used to record static strength, static (isometric) endurance and dynamic (isotonic) endurance. Subjects were given two 45-min training sessions per week for 14 weeks. There was a poor correlation between strength and pre- and post-training isometric and isotonic endurance. These results directly contradict Caldwell's findings.

The Materials Handling Research Unit of the University of Surrey (Davis and Stubbs, 1980) has made recommendations for two-handed pushing and pulling forces that can be exerted by males in standing and kneeling positions. The recommendations (Table 5.3) account for worker's age and location of hands in front of the body (hori-

Table 5.3. Two-handed horizontal push/pull forces (kg) in standing and kneeling positions (Davis and Stubbs, 1980)

Position	Age (years)	Hand location					
		45° above horizontal		Horizontal		45° below horizontal	
		Push	Pull	Push	Pull	Push	Pull
Standing	Up to 40	12	18	20	35	30	50
	41–50	11	16	18	35	25	45
	51–60	10	16	17	34	24	40
Kneeling	Up to 40	11	15	20	38	22	50
	41–50	11	12	20	27	16	42
	51–60	10	11	19	24	15	30

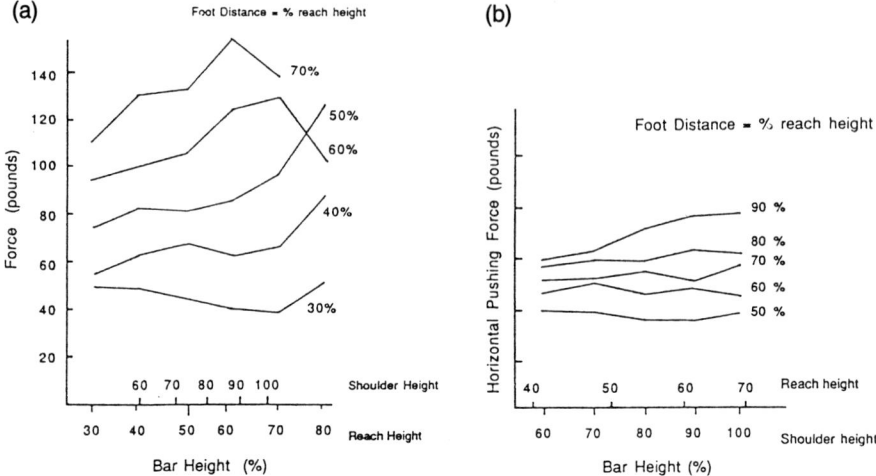

Figure 5.1. Effect of bar height and foot position interaction on horizontal pushing forces. (a) Males, n = 36. (b) Females, n = 11.

zontal in the plane of the shoulder and in planes 45° above and below the shoulder) and are applicable for occasional push and pull (once per minute or less frequent).

Ayoub and McDaniel (1974) studied the maximum isometric pushing and pulling forces exerted by 35 males and 11 females as a function of foot and hand placement. The effects of both the bar height and foot distance on the horizontal pushing force were significant for both males and females (Figure 5.1).

These figures indicate that the appropriate foot distance should be approximately 60–70% reach height at a bar height of 50–60% reach height. The efficiency of these positions was determined by plotting the time for a 10% drop in horizontal force with bar height. It was concluded that the optimum bar height should be as low as possible.

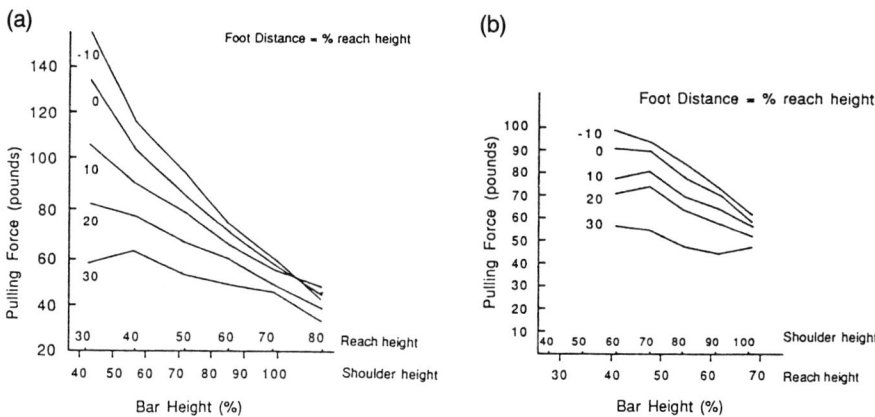

Figure 5.2. Effect of body configurations on pulling forces. (a) Males, n = 35. (b) Females, n = 11.

The maximum fatigue was observed to occur when the foot distance was between 50–55% reach height.

Figure 5.2 shows the pulling force for males and females as a function of foot distance and bar height.

The pulling force increased as the bar height decreased to 30% reach height for males and 40% reach height for females while the foot distance was at the −10% reach height. A negative value of the foot distance means the foot is located on the other side of the vertical plane of the bar.

Subject body-weight also had a significant effect on pushing and pulling forces. Its interaction with both foot distance and bar height resulted in considerably higher

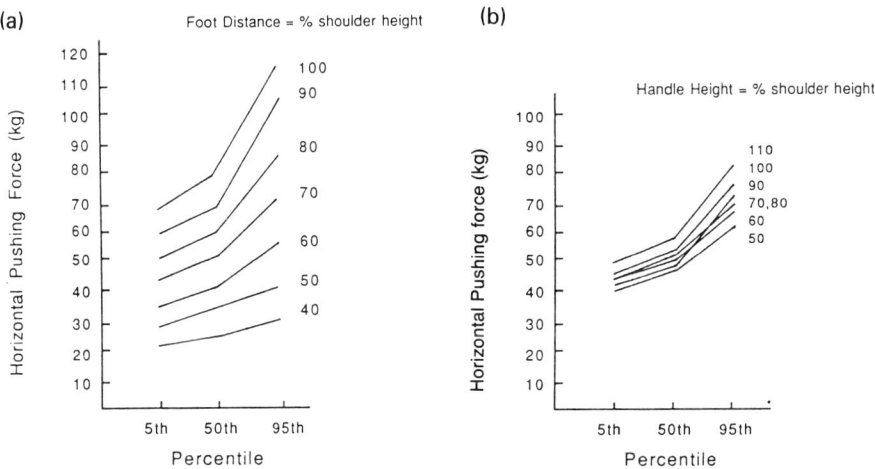

Figure 5.3. Effect of subject weight on pushing force (constant height) for different body configurations (Ayoub and McDaniel, 1974).

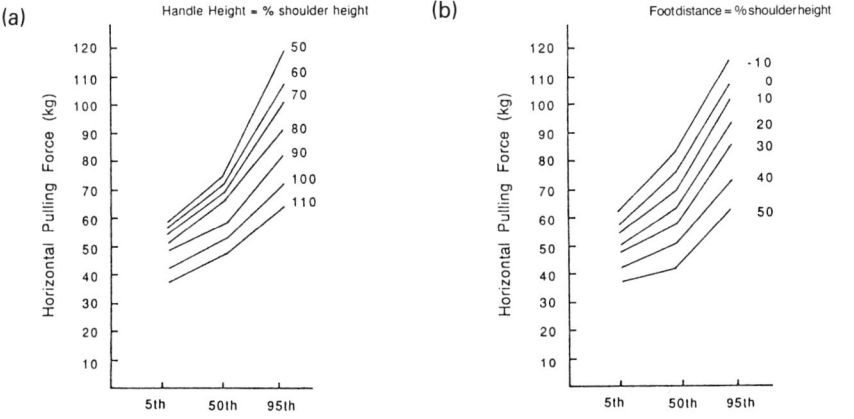

Figure 5.4. Effect of subject weight on pulling force (constant height) for different body configurations (Ayoub and McDaniel, 1974).

pushing forces at high bar height and larger foot distance (Figure 5.3) and higher pulling forces at the smallest foot distance and lowest bar height (Figure 5.4).

Chaffin *et al.* (1983) extended Ayoub and McDaniel's work and measured push/pull forces in unrestricted, but known postures chosen by three males and three females (21–23 years of age, 50–87 kg body-weight, and 160–190 cm height) in three different handle positions (68, 109 and 152 cm above the floor). The average push/pull forces exerted by males and females are given in Table 5.4.

Table 5.4. *Average push/pull forces (standard deviations) in kg for varied handle heights (Chaffin et al., 1983)*

Handle height (cm)	Pushing		Pulling	
	Male	Female	Male	Female
68	40·67 (9·68)	16·11 (6·22)	38·33 (7·44)	18·25 (7·44)
109	34·86 (9·99)	17·94 (6·93)	26·30 (2·65)	17·43 (3·36)
152	28·95 (8·46)	15·29 (4·89)	17·74 (1·43)	14·58 (3·46)

Subjects replicated the postures assumed during various pushing and pulling tasks and as indicated by postural data given in Table 5.5, the variation in angles and distances (Figure 5.5) were quite small.

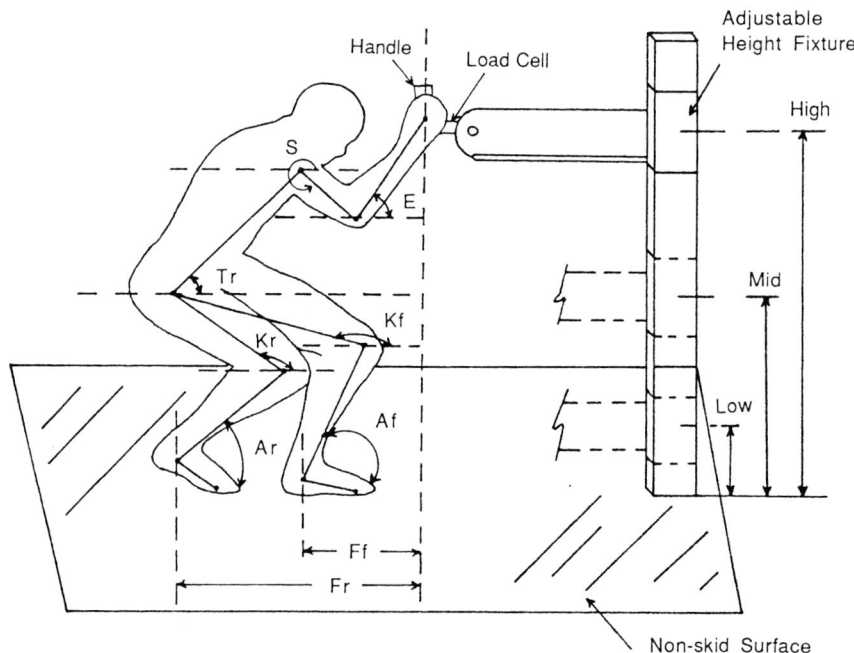

Figure 5.5. Test apparatus and postural variables for push/pull strength study (Chaffin et al., 1983). From *Human Factors*, Vol. 25. © 1983 by The Human Factors Society, Inc., and reproduced by permission

Table 5.5. Postural data* and its repeatability (Chaffin et al., 1983)

Foot Position†	Pushing with feet symmetric		Pulling with feet symmetric		Pushing feet apart		Pulling feet apart	
	Trial 1	Trial 2	Trial 1	Trial 2	Trial 1	Trial 2	Trial 1	Trial 2
AR (deg)	54	53	102	98	51	51	57	57
KR (deg)	84	88	139	136	65	69	120	117
AF (deg)	—	—	—	—	63	68	105	103
KF (deg)	—	—	—	—	124	128	139	138
TR (deg)	61	59	64	64	58	58	64	64
S (deg)	−62	−59	−31	−39	−50	−47	−32	−37
E (deg)	16	20	14	15	16	15	14	15
FR (cm)	81	80	38	33	104	105	76	74
FF (cm)	—	—	—	—	56	54	33	31

*Means for all subjects.
†See Figure 5.5.

Snook (1978a, b) summarized previous studies carried out at the Liberty Mutual Insurance Company and made recommendations for push and pull forces that could be exerted initially and sustained by male and female industrial workers for occasional performance. For pushing activities, the recommendations made are a function of handle height, frequency and distance. All pulling force recommendations are for a pulling distance of 2·1 m (experimental details of the studies are given in section 4.3.1). Tables 5.6 and 5.7 give pushing force capabilities of males and females for initial forces, which are forces required to get an object in motion, and sustained forces, to keep an object in motion. Table 5.8 gives the pulling force capabilities.

Both initial and sustained pushing capabilities decreased with distance. The pushing capabilities were greatest at 95 cm vertical height for males, and at 89 cm height for females. The differences due to height, however, were not large. The pulling capability decreased with vertical height. Peak forces were exerted by males and females at 64 cm and 57 cm vertical height, respectively.

Ciriello and Snook (1983) tested the assumptions made in generating data reported in Tables 5.6 and 5.7. Twelve females and 10 males participated in the study. Pushing was performed at a height midway between knuckle and elbow heights (approximately the middle height value in Tables 5.6 and 5.7). Two pushing distances (7·6 m and 30·5 m) were studied. The results are shown in Table 5.9. While the forces for males were higher, in general, forces for females were lower overall. Differences in 30 min and 480 min capabilities were also significant. The authors suggest using these trends to revise the capability data given in Tables 5.6 and 5.7.

Among the various investigations of pushing and pulling activities undertaken by various researchers, the investigations by Ayoub and McDaniel (1974) and Snook (1978a, b) are perhaps the most extensive. While Snook's investigation is a more realistic simulation of pushing and pulling actual objects placed on the floor, Ayoub and McDaniel's investigation appears to be more relevant to force exertions (push or pull) at a particular height. Depending upon the application, therefore, appropriate recommendations should be used.

Table 5.6. Initial push force (kg) capabilities of males and females (Snook, 1978a, b)

Sex	Handle height(cm)	Pushing distance (m)	Frequency Once/30 min		Once/480 min	
			$\bar{x}^*$	S†	$\bar{x}$	S
Male	144	2.1	52	12·76	56	13·37
		7.6	44	10·94	47	11·55
		15.2	40	9·12	42	9·73
		30.5	31	6·08	33	6·08
		45.7	29	6·68	31	6·68
		61.0	26	6·08	28	6·08
	95	2.1	56	13·37	60	13·98
		7.6	50	12·76	53	13·37
		15.2	45	10·33	48	10·94
		30.5	36	6·08	38	6·68
		45.7	33	7·29	35	7·90
		61.0	30	6·68	32	6·68
	64	2.1	51	12·16	54	13·37
		7.6	43	10·94	46	11·55
		15.2	39	8·51	41	9·73
		30.5	31	6·68	32	6·08
		45.7	28	6·68	30	6·68
		61.0	26	5·47	27	6·08
Female	135	2.1	35	6·68	38	6·68
		7.6	31	6·08	33	6·08
		15.2	27	4·86	28	4·86
		30.5	22	3·04	23	3·65
		45.7	21	2·43	22	3·05
		61.0	18	3·05	20	2·43
	89	2.1	38	6·68	40	7·29
		7.6	36	6·08	38	6·68
		15.2	30	5·47	32	6·08
		30.5	25	3·65	27	3·65
		45.7	23	3·65	25	3·04
		61.0	21	3·04	22	3·04
	57	2.1	34	6·68	37	6·68
		7.6	31	5·47	33	5·47
		15.2	26	4·86	28	4·86
		30.5	22	3·04	23	3·04
		45.7	20	2·43	21	3·04
		61.0	18	2·43	19	3·04

*$\bar{x}$ = mean; †S = standard deviation.

5.2 Capacity Data for Two-Handed Pushing and Pulling Activities

Pushing and pulling capacity is generally defined as the force that can be sustained over a certain distance. Snook et al. (1970), Snook and Ciriello (1974a, b), Snook (1978a, b) and Ciriello and Snook (1978, 1983), studied the pushing and pulling capacity, along with the lifting and lowering capacities described in the previous section. Pushing and pulling heights used were approximately the handle heights 'commonly' found in industry. Force was the dependent variable and was varied by changing the resistance against which the force was applied. In the study by Snook et al. (1970) and Snook and

Table 5.7. Sustained push force (kg) capabilities of males and females (Snook, 1978a, b)

Sex	Handle height (cm)	Pushing distance (m)	Once/30 min $\bar{x}^*$	Once/30 min $S\dagger$	Once/480 min $\bar{x}$	Once/480 min S
Male	144	2.1	36	12.76	39	13.37
		7.6	31	8.51	33	9.73
		15.2	27	7.29	29	7.90
		30.5	19	4.25	21	4.25
		45.7	18	3.65	20	3.65
		61.0	16	3.04	17	3.65
	95	2.1	38	12.76	41	13.37
		7.6	30	9.12	33	9.12
		15.2	27	7.29	29	7.29
		30.5	19	3.65	21	3.65
		45.7	18	3.65	19	4.25
		61.0	16	3.04	17	3.04
	64	2.1	38	12.16	40	13.98
		7.6	29	8.51	31	9.12
		15.2	26	6.68	27	7.90
		30.5	19	3.04	20	3.65
		45.7	17	3.65	19	3.65
		61.0	15	3.04	16	3.65
Female	135	2.1	26	7.29	28	7.29
		7.6	23	4.86	25	4.86
		15.2	19	3.65	21	3.65
		30.5	16	2.43	18	2.43
		45.7	14	1.82	15	1.82
		61.0	12	1.82	13	1.82
	89	2.1	27	7.29	29	7.90
		7.6	23	4.25	25	4.86
		15.2	19	3.04	21	3.04
		30.5	16	2.43	17	2.43
		45.7	14	1.82	15	1.82
		61.0	12	1.82	13	1.82
	57	2.1	27	7.29	29	7.29
		7.6	22	4.25	24	4.25
		15.2	19	2.43	20	3.04
		30.5	16	1.82	17	2.43
		45.7	13	1.82	14	2.43
		61.0	11	1.82	12	1.82

*$\bar{x}$ = mean; †S = standard deviation.

Ciriello (1974a), subjects were encouraged to make force adjustments by starting them with either a very light or very heavy force. The light initial force was: 4·53–15·83 kg for a low rate of work, 4·53–13·59 kg for an intermediate rate, and 4·53–11·32 kg for a high rate. The heavy initial force was: 36·24–40·77 kg for a low rate, 33·97–40·77 for an intermediate rate, and 31·71–40·77 kg for a high rate. The initial and sustained forces acceptable to male and female workers decreased with the distance moved.

Snook and Ciriello (1974b) investigated the effect of heat stress on pushing tasks. The workload for pushing decreased significantly (by about 16%) as the environmental temperature changed from 17·2° C WBGT to 27·0° C WBGT. There was

Table 5.8. Initial and sustained pull forces* (kg) acceptable to males and females (Snook, 1978a, b)

Sex	Vertical height (cm)	Condition	Frequency			
			Once/30 min		Once/480 min	
			$\bar{x}$†	S‡	$\bar{x}$	S
Male	144	Initial force	30	5·47	31	6·08
		Sustained force	22	6·68	23	5·47
	95	Initial force	42	7·29	44	7·90
		Sustained force	29	7·29	30	7·29
	64	Initial force	47	8·51	49	9·12
		Sustained force	31	7·29	32	7·90
Female	135	Initial force	23	4·25	24	4·25
		Sustained force	17	3·65	18	3·65
	89	Initial force	32	5·47	34	5·47
		Sustained force	23	4·25	23	5·47
	57	Initial force	36	6·68	38	6·68
		Sustained force	24	4·86	25	4·86

*Pulling distance = 2·1 m. †$\bar{x}$ = mean; ‡S = standard deviation.

Table 5.9. Maximum acceptable push force (kg) for males and females (Ciriello and Snook, 1983)

Distance (m)	Condition	Sex	Frequency			
			Once/30 min		Once/480 min	
			$\bar{x}$*	S†	$\bar{x}$	S
7·6	Initial force	Male	44	10·0	52	13·0
		Female	27	5·1	29	5·4
	Sustained force	Male	29	7·6	35	9·6
		Female	15	4·7	19	5·4
30·5	Initial force	Male	37	9·3	47	11·4
		Female	24	3·7	28	4·7
	Sustained force	Male	25	7·1	29	4·9
		Female	13	3·4	17	4·7

*$\bar{x}$ = mean; †S = standard deviation.

also an increase of 9–10 beats/min in the heart rate while the rectal temperature increased by 0·2–0·3°C.

Snook (1978a, b) and Ciriello and Snook (1978a) studied three pushing and pulling heights (64, 97 and 164 cm). A treadmill was used and powered (by the worker), as the force was exerted (push and pull) against a stationary box. Initial push/pull forces and sustained push/pull forces were recorded with the help of strain gauges. The results showed no significant difference among the sustained push forces at the three height levels. Initial push forces, though, were significantly different. Between pushing and pulling, sustained and initial forces were significantly different only at 147 cm height. Snook (1978a) integrated data from all previous studies to recommend pushing/pulling capability data for industrial males and females. Tables 5.10 and 5.11 show the initial and sustained pushing force capability of males and females, respectively. The pulling capabilities are given in Table 5.12.

Table 5.10. Maximum acceptable initial and sustained push forces (kg) for males (Snook, 1978a, b)

Force	Height (cm)	One push every (second)	2·1		7·6		15·2		30·5		45·7		61·0	
			$\bar{x}$*	S†	$\bar{x}$	S	$\bar{x}$	S	$\bar{x}$	S	$\bar{x}$	S	$\bar{x}$	S
Initial	144	6	31	7·9	24	4·2	22	4·9	—	—	—	—	—	—
		12	34	8·5	29	7·3	26	6·1	—	—	—	—	—	—
		60	38	9·1	32	7·9	29	6·7	23	4·2	22	4·9	—	—
		120	40	9·7	34	8·5	31	6·7	25	5·5	23	4·9	21	4·9
		300	48	12·1	41	10·3	37	8·5	29	5·5	27	5·5	24	5·5
	95	6	33	8·5	28	4·2	24	6·1	—	—	—	—	—	—
		12	37	8·5	33	8·5	29	6·7	—	—	—	—	—	—
		60	41	9·7	36	9·7	33	7·3	26	4·9	25	5·5	—	—
		120	43	10·3	38	9·7	35	7·9	29	5·5	27	4·9	24	5·5
		300	52	12·7	46	12·1	42	9·7	33	6·1	31	6·1	28	6·1
	64	6	30	7·9	22	3·6	20	4·2	—	—	—	—	—	—
		12	34	7·9	26	7·3	24	4·9	—	—	—	—	—	—
		60	37	9·1	31	7·9	28	6·7	21	3·6	20	4·2	—	—
		120	39	9·7	33	8·5	30	6·7	25	4·9	23	4·2	20	4·2
		300	47	11·5	40	9·7	36	9·7	28	5·5	26	6·1	24	4·9
Sustained	144	6	16	5·5	13	3·0	13	3·6	—	—	—	—	—	—
		12	21	7·3	16	4·9	16	4·2	—	—	—	—	—	—
		60	25	8·5	21	6·1	18	4·9	13	2·4	12	2·4	—	—
		120	27	8·5	22	6·7	20	4·9	16	3·0	15	3·0	12	2·4
		300	33	11·5	28	8·5	25	6·7	18	3·6	17	3·0	14	3·0
	95	6	17	5·5	14	3·0	14	3·6	—	—	—	—	—	—
		12	22	7·3	17	4·2	16	4·9	—	—	—	—	—	—
		60	26	9·1	21	6·1	18	5·5	13	3·0	13	2·4	—	—
		120	28	9·1	22	6·7	19	5·5	15	3·0	15	3·0	12	3·0
		300	35	11·5	28	7·9	24	5·5	18	3·0	17	3·0	14	3·0
	64	6	17	5·5	14	3·0	14	3·6	—	—	—	—	—	—
		12	22	4·9	17	4·9	16	4·9	—	—	—	—	—	—
		60	26	8·5	20	6·1	18	4·9	13	3·0	13	2·4	—	—
		120	27	9·1	21	6·7	19	4·9	15	3·0	14	3·0	13	2·4
		300	34	12·1	27	7·9	23	6·7	17	3·0	16	3·0	14	2·4

*$\bar{x}$ = mean; †S = standard deviation.

Ciriello and Snook (1983), in a series of two experiments studied pushing and pulling tasks. In the first series of experiments, 12 industrial females and 10 industrial males performed pushing at a height midway between knuckle and elbow heights (mid-height region). Pushing was carried out for 7·6 and 30·5 m distances for very low (once every 8 h) to very high (once every 15 s) frequency. The second series of experiments involved 12 female industrial workers. Five distances of push and pull were investigated: 2·1, 7·6, 30·5, 45·7 and 61·0 m. Frequencies ranged from once every 15 s to once every 180 s. The results indicated that forces for the 5-min frequency were overestimated by 10–15% (Tables 5.10 and 5.11). The initial force was significantly influenced by the pushing height, but not the pulling height (knuckle height–low, elbow height–mid and shoulder height–high). The sustained forces were not affected by height except for the 2·1 m push where the force at low height was significantly lower than the force at the other two heights. Heart rates and oxygen uptake values were also not affected significantly.

Table 5.11. *Maximum acceptable initial and sustained push forces (kg) for females (Snook, 1978a, b)*

Force	Height (cm)	One push every (second)	2·1 $\bar{x}^*$	2·1 S†	7·6 $\bar{x}$	7·6 S	15·2 $\bar{x}$	15·2 S	30·5 $\bar{x}$	30·5 S	45·7 $\bar{x}$	45·7 S	61·0 $\bar{x}$	61·0 S
Initial	135	12	23	3·0	—	—	—	—	—	—	—	—	—	—
		22	—	—	19	2·4	—	—	—	—	—	—	—	—
		24	25	3·6	—	—	—	—	—	—	—	—	—	—
		35	—	—	—	—	18	3·6	—	—	—	—	—	—
		44	—	—	20	3·6	—	—	—	—	—	—	—	—
		60	26	4·2	23	4·2	19	3·6	17	2·4	16	1·8	—	—
		120	27	4·9	24	4·2	20	3·6	18	3·6	16	2·4	15	1·8
		300	33	5·5	29	5·5	25	4·2	20	3·0	19	2·4	17	2·4
	89	12	24	3·6	—	—	—	—	—	—	—	—	—	—
		22	—	—	21	3·6	—	—	—	—	—	—	—	—
		24	27	4·2	—	—	—	—	—	—	—	—	—	—
		35	—	—	—	—	21	3·6	—	—	—	—	—	—
		44	—	—	23	4·2	—	—	—	—	—	—	—	—
		60	28	4·9	26	4·9	22	4·2	19	3·0	18	2·4	—	—
		120	29	5·5	27	5·5	23	4·2	21	3·6	18	2·4	17	2·4
		300	35	6·7	33	6·1	28	4·9	23	3·6	22	2·4	19	2·4
	57	12	22	3·6	—	—	—	—	—	—	—	—	—	—
		22	—	—	17	2·4	—	—	—	—	—	—	—	—
		24	24	4·2	—	—	—	—	—	—	—	—	—	—
		35	—	—	—	—	17	2·4	—	—	—	—	—	—
		44	—	—	19	3·0	—	—	—	—	—	—	—	—
		60	25	4·9	22	4·2	18	3·0	15	2·4	14	2·4	—	—
		120	26	4·9	24	4·2	20	3·6	18	3·0	16	2·4	14	1·8
		300	32	6·1	28	5·5	24	4·2	20	3·0	19	3·0	17	1·8
Sustained	135	12	15	2·4	—	—	—	—	—	—	—	—	—	—
		22	—	—	12	1·8	—	—	—	—	—	—	—	—
		24	17	3·0	—	—	—	—	—	—	—	—	—	—
		35	—	—	—	—	11	1·2	—	—	—	—	—	—
		44	—	—	14	1·8	—	—	—	—	—	—	—	—
		60	19	4·2	16	3·0	13	1·8	11	1·8	9	1·2	—	—
		120	20	4·9	17	3·0	14	2·4	12	1·8	11	1·8	9	1·8
		300	25	6·1	21	4·2	18	3·0	15	1·8	13	1·8	11	1·2
	89	12	16	2·4	—	—	—	—	—	—	—	—	—	—
		22	—	—	12	1·8	—	—	—	—	—	—	—	—
		24	17	3·0	—	—	—	—	—	—	—	—	—	—
		35	—	—	—	—	12	1·2	—	—	—	—	—	—
		44	—	—	14	2·4	—	—	—	—	—	—	—	—
		60	19	4·2	16	3·0	13	2·4	11	1·8	10	1·2	—	—
		120	20	4·9	17	3·0	14	2·4	12	1·8	11	1·8	9	1·8
		300	25	6·1	21	4·2	18	2·4	15	1·8	13	1·8	11	1·2
	57	12	16	2·4	—	—	—	—	—	—	—	—	—	—
		22	—	—	13	1·8	—	—	—	—	—	—	—	—
		24	17	3·0	—	—	—	—	—	—	—	—	—	—
		35	—	—	—	—	12	1·2	—	—	—	—	—	—
		44	—	—	14	2·4	—	—	—	—	—	—	—	—
		60	19	4·2	15	3·0	13	2·4	11	1·8	10	1·2	—	—
		120	20	4·9	16	3·0	14	1·8	12	1·2	11	1·2	10	1·2
		300	25	6·1	20	4·2	17	3·0	14	1·8	12	1·8	10	1·8

*$\bar{x}$ = mean; †S = standard deviation.

Table 5.12. Maximum acceptable initial and sustained pull forces* (kg) for males and females (Snook, 1978a, b)

Force	Sex†	Vertical height (cm)	One pull every (second)											
			6		12		24		1		2		5	
			$\bar{x}$‡	S§	$\bar{x}$	S	$\bar{x}$	S	$\bar{x}$	S	$\bar{x}$	S	$\bar{x}$	S
Initial	M	144	19	4·2	22	4·9	—	—	25	4·9	26	4·9	29	5·5
		95	27	5·5	31	6·1	—	—	35	6·1	36	6·7	40	7·3
		64	31	6·1	35	6·7	—	—	39	7·3	41	7·3	45	8·5
	F	135	—	—	19	3·6	19	4·2	19	3·6	20	3·0	22	3·6
		89	—	—	26	5·5	27	5·5	27	4·2	28	4·2	31	5·5
		57	—	—	29	6·1	30	6·1	30	5·5	31	5·5	35	6·1
Sustained	M	144	12	2·4	16	3·6	—	—	19	4·2	19	4·9	21	5·5
		95	15	3·0	20	4·9	—	—	25	6·1	26	5·5	28	6·7
		64	16	3·6	22	4·9	—	—	27	6·1	27	6·7	30	6·7
	F	135	—	—	14	3·0	14	3·0	15	3·0	15	3·0	17	3·0
		89	—	—	18	4·2	19	3·6	19	4·2	20	4·2	22	4·2
		57	—	—	18	4·2	20	4·2	20	4·2	20	4·2	23	4·9

*For 2·1 in pull. †M = male, F = female; ‡$\bar{x}$ = mean; §S = standard deviation.

The studies conducted by Snook and his colleagues at the Liberty Mutual Insurance Company are the only ones that have provided pushing/pulling capability data. In the absence of any other data, it is advisable to use the recommendations given in Tables 5.10–5.12.

5.3 Strength Data for One-Handed Pushing and Pulling Activities

The pushing and pulling capability recommendations have been made by Warrick et al. (1980) and the University of Surrey (Davis and Stubbs, 1980). Warrick et al. measured push forward, pull backward and push left and right forces in direction parallel to the anterior position of the subject (see Figure 4.5). The recommendations are given in Table 5.13.

The recommendations for one-handed pushing tasks for occasional application, made by the University of Surrey, are given in Table 5.14. These are applicable for hori-

Table 5.13. Mean magnitude of forces (kg) for males for one-handed push/pull tasks at shoulder and knee heights (Warwick et al., 1980)*

Hand	Height	Foot position	Push forward	Pull backward	Push right	Pull left
Left	Shoulder	Right	15·0	13·8	9·0	8·9
		Right posterior	14·6	10·7	7·5	10·3
	Knee	Left	19·2	17·8	14·0	11·5
		Right	17·2	16·3	11·6	11·5
Right	Shoulder	Right	19·4	15·0	11·4	11·5
		Right posterior	18·7	14·1	8·7	9·7
	Knee	Left	17·3	16·6	14·4	11·0
		Right	20·4	15·0	13·8	16·5

*See Figure 5.6.

zontal push in the plane of the shoulder, when the hand is in front of the body, and the individual is standing.

Clearly, very limited data are available for one-handed pushing and pulling strengths. Further investigations are needed to establish a more comprehensive data base.

Table 5.14. Push force values (kg) for one-handed tasks for males (Davis and Stubbs, 1980)*

Acromial-grip distance (cm)	Age group (years)		
	Up to 40	41–50	51–60
70	15	14	11
60	18	17	13
50	20	19	14
45	22	21	16
20	26	24	19
5	30	28	22

*Values converted from graphical representation.

5.4 Capacity Data for One-Handed Pushing and Pulling Activities

The only recommendations for pushing tasks come from the University of Surrey's Material Handling Research Unit (Davis and Stubbs, 1980). For one-handed pushing tasks that are performed more frequently than once per minute, the recommendations in Table 5.14 should be reduced by 30%. The reduced forces can also be applied for one-handed tasks.

5.5 Strength Data for Two-Handed Carrying Activities

Investigations of carrying strength of workers are relatively few (Snook, 1978a, b; Ciriello and Snook, 1983a; Mital and Ilango, 1983; Mital, 1986c). The summary of studies reported by Snook (1978a, b) is the most comprehensive and includes maximum acceptable weight of carry data for male and female industrial workers as a function of carrying distance and carrying weight. The data are reported for elbow and knuckle heights for distances of 2·1, 4·3 and 8·5 m. Table 5.15 present the data for an occasional carry.

Substantially more weight was carried at the knuckle height than weight carried at the elbow height. The carrying capability for an occasional lift decreased with distance and with frequency. The differences due to frequency, however, were not very large. For females only, differences due to distance were also small.

Ciriello and Snook (1983), in a subsequent study, determined that differences due to frequency are much larger than had previously been reported. Using 12 females and 10 males, they determined carrying capabilities at knuckle height for 4·3 m distance. Males accepted 56 kg for carrying once every 30 min, and 66 kg for carrying once every 8 h (standard deviations, 17·1 and 15·2 kg, respectively). Females accepted 20 and

Table 5.15. Maximum acceptable weight of carry (kg) for occasional performance (Snook, 1978a, b)

Sex	Carrying height (cm)	Carrying distance (m)	Frequency			
			Once/30 min		Once/480 min	
			$\bar{x}$*	S†	$\bar{x}$	S
Male	Elbow (111)	2·1	38	9·73	41	9·73
		4·3	35	8·51	37	9·12
		8·5	30	7·29	32	7·90
	Knuckle (79)	2·1	48	11·55	50	12·76
		4·3	42	10·33	44	10·94
		8·5	40	9·73	42	10·94
Female	Elbow (105)	2·1	23	3·04	24	3·65
		4·3	22	3·65	24	3·65
		8·5	20	3·65	22	3·65
	Knuckle (72)	2·1	27	4·25	29	4·86
		4·3	25	3·04	26	3·65
		8·5	25	3·65	26	4·25

*$\bar{x}$ = mean; †S = standard deviation.

27 kg for one carry every 30 min and every 8 h, respectively (standard deviations, 4·3 and 4·7 kg, respectively). These numbers are substantially different from those in Table 5.15; the capability of males being much higher. The 8-h capability of females, however, was not much different. These trends are more realistic since they are based on actual data rather than on the assumption that males and females have similar trends. The data in Table 5.15, therefore, should be adjusted by these trends.

As noted in cases of lifting, lowering, pushing and pulling, subjective perception of acceptable weight is strongly influenced by frequency even though the time interval between successive attempts is large enough to permit complete recovery from fatigue.

Mital and Ilango (1983) investigated the influence of material density, load asymmetry, hand preference and carrying distance on maximum acceptable weight of carry at knuckle height for occasional performance. Using 10 male students (mean age 23·2 years, mean weight 79·1 kg and mean height 181·4 cm), the effect of three different materials (water, dry sand and lead shot), three asymmetric loads (c.g. offset of 0, 12·7 and 25·4 cm from the sagittal plane in the frontal plane), two carrying distances (3·05 and 9·14 m), and two load orientations (heavier end held by the preferred hand and non-preferred hand) on carrying capabilities was investigated. Table 5.16 gives the average carrying capability for various conditions. In general, carrying capability was not influenced by either the carrying distance or the load orientation. Material density and asymmetric loading effects, however, were relatively large.

The carrying capability of students was also found to be substantially lower than that of industrial males (Table 5.15). Since students are, by and large, inexperienced in carrying materials, it can be concluded that carrying capability of experienced workers is substantially greater than inexperienced workers.

Mital (1986c) determined the carrying capability of male and female students for one performance every 8 h. Twenty-seven males and 11 females (21–32 years age, 157·7–187·2 cm height and 48·5–89 kg weight) carried a load through open and

Table 5.16. Maximum acceptable weight of lift* (kg) for occasional performance (Mital and Ilango, 1983b).

Main effect	Mean	Standard deviation
Material density (g/cc)		
1·0 (water)	19·52	2·80
2·02 (dry sand)	21·00	4·62
10·68 (lead shot)	22·20	4·89
Load c.g. location (cm)		
0	21·66	4·50
12·7	20·82	4·30
25·4	20·21	4·12
Carrying distance (m)		
3·05	21·10	4·14
9·14	20·70	4·52
Load orientation (hand preference)		
Stronger (preferred hand)	21·04	4·49
Weaker (non-preferred hand)	20·77	4·18

*Male students carried the load at the knuckle height.

confined spaces (0·56 m wide and 1·83 m high). The carrying distance was 7·62 m. The load was carried at knuckle and elbow heights, with and without handles, in five different boxes. The box dimensions (length × width × height) were: 30·48 × 30·48 × 25·4, 45·72 × 30·48 × 25·4, 69·96 × 30·48 × 25·4 and 30·48 × 30·48 × 30·48 cm; the c.g. of the 30·48 cm³ box was offset by 0 and 7·61 cm in the direction of the preferred hand (to obtain the fifth box). Table 5.17 shows the maximum acceptable weight of carry for occasional performance.

Males carried 54·6% more weight, on average, than females. The carrying capability declined by 13%, for both males and females, when load was carried through confined spaces. Males carried 14·1% more weight and females 11·7% more weight at

Table 5.17. Maximum acceptable weight of carry (kg) for occasional performance (Mital, 1986c).

	Males		Females	
Main effect	$\bar{x}$*	S†	$\bar{x}$	S
Space				
Open	28·7	5·2	18·6	4·0
Confined	24·9	4·8	16·1	3·8
Couplings				
Handles	29·6	5·1	19·1	4·1
No-handles	24·3	4·3	15·7	3·4
Height				
Knuckle	28·6	5·2	18·3	4·2
Elbow	25·1	4·9	16·4	3·8
Box size (cm)				
30·48 × 30·48 × 25·4	27·8	5·0	18·9	4·1
45·72 × 30·48 × 25·4	27·9	5·3	17·5	3·7
60·96 × 30·48 × 25·4	26·8	5·3	16·4	3·9
30·48 × 30·48 × 30·48 − 0 cm offset	26·2	5·4	17·4	4·2
30·48 × 30·48 × 40·48 − 7·61 cm offset	25·5	5·4	16·5	4·1

*$\bar{x}$ = mean; †S = standard deviation.

the knuckle height compared to the weight they carried at the elbow height. Approximately 21% more weight was carried by subjects when boxes were held by handles. Box size effects, even though statistically significant, were not very large. For each condition, loads carried resulted in an average heart rate of 96 beats/min. For more stressful conditions, subjects accepted lower weight to maintain the same levels of circulatory strain.

5.6 Capacity Data for Two-Handed Carrying Activities

Parts of the studies by Snook et al. (1970) and Snook and Ciriello (1974a) were to determine the maximum acceptable weight of carry. Subjects determined their own workloads at different walking speeds. Snook (1978a, b) integrated the results (Table 5.18). The results indicated that the maximum acceptable weight of carry decreases with carrying distance and carrying frequency. More weight is carried at the knuckle height than at the elbow height.

Table 5.18. Maximum acceptable weight of carry (kg) for males and females (Snook, 1978a)

			One carry every (second)									
	Carrying height (cm)	Carrying distance (m)	6		12		60		120		300	
Sex			$\bar{x}^*$	$S\dagger$	$\bar{x}$	S	$\bar{x}$	S	$\bar{x}$	S	$\bar{x}$	S
Males	111	2·1	18	4·2	24	6·1	29	7·3	31	7·3	36	9·1
		4·3	19	4·2	22	5·5	26	6·7	28	6·7	33	7·9
		8·5	17	3·6	19	4·9	23	5·5	24	6·1	28	7·3
	79	2·1	23	5·5	29	7·3	36	9·1	38	9·1	45	10·9
		4·3	23	5·5	28	6·7	32	7·9	33	8·5	39	9·7
		8·5	22	5·5	26	6·1	30	7·9	32	7·9	38	9·1
Females	105	2·1	14	2·4	15	3·0	17	3·0	18	3·0	21	3·6
		4·3	14	3·0	15	3·0	17	3·0	18	2·4	21	3·0
		8·5	13	1·8	14	2·4	15	3·0	16	3·0	19	3·0
	72	2·1	17	2·4	19	3·0	21	3·0	22	3·0	26	3·6
		4·3	17	2·4	19	2·4	19	2·4	19	3·0	23	3·0
		8·5	16	2·4	17	2·4	19	2·4	20	2·4	23	3·6

*$\bar{x}$ = mean; †S = standard deviation.

Snook and Ciriello (1974b) studied the effects of heat stress on carrying weights. Two different environmental conditions were used: 17·2°C WBGT and 27·0°C WBGT. There was a significant decrease (about 11%) in the weight carrying capability of workers due to increased temperature. In addition, the heart rate increased by about 9–10 beats/min, and the rectal temperature went up by 0·2–0·3°C.

Mital and Okolie (1982) investigated the effect of container shape (rectangular, cylindrical, barrel), carrying height (knuckle height, elbow height), frequency of carry (once per minute, once every four minutes), distance of carry (3·05 and 9·14 m), and presence and absence of partitions in a container. Thirteen college students (average age 22·6 years, average body-weight 74·4 kg, and average height 175·5 cm) performed 48 variations of carrying tasks (3 containers × 2 carrying heights × 2 frequencies × 2 carrying distance × 2 container designs). The results (Table 5.19) indicated that liquid carrying capabilities can be enhanced by as much as 10% if appropriate container designs (containers with partitions) are used. The maximum acceptable amount of liquid carried may be up to 30% lower than the maximum acceptable weight of solids that can be carried. Carrying capacity also decreased with frequency and distance.

Table 5.19. Average maximum acceptable amount of liquid (kg) carried by males (Mital and Okolie, 1982)

Frequency	Distance		Elbow height		Knuckle height	
			Partitions	No-partitions	Partitions	No-partitions
Once/4 min	3·05	$\bar{x}$*	24·4	23·9	29·3	26·6
		S†	4·1	3·7	5·1	4·5
	9·14	$\bar{x}$	24·3	22·7	27·4	25·6
		S	3·4	4·3	5·8	4·6
Once/min	3·05	$\bar{x}$	23·2	22·1	26·7	25·1
		S	3·6	2·4	5·1	4·9
	9·14	$\bar{x}$	21·8	21·3	25·3	23·5
		S	2·4	3·1	4·0	4·2

*$\bar{x}$ = mean; †S = standard deviation.

It is clear from these studies that more weight can be carried at the knuckle height than at elbow height. Also, if the material carried is a solid, more weight can be carried. Liquid carrying capacity can be enhanced if the movement of liquid (sloshing) in the container can be restricted.

5.7 Strength Data for One-Handed Carrying Activities

Carry capability of individuals is generally defined as the maximum weight individuals are willing to carry over a certain distance in a given container. Mital and Manivasagan (1983b) used 10 males and 5 females (21–23 years of age, 48–94 kg body-weight and 1·53–1·86 m height) to determine carrying capabilities for occasional performance. Four different shapes of containers (plastic bucket, galvanized iron bucket, tool box, and radiator can, Figure 2.12) and two different volumes (8·5 and 12·3 l) were used to carry loads for distances of 30·48, 60·96 and 91·44 m. Individuals first subjectively estimated how much weight they could carry in a given type of container, for the required distance, and then verified it by actual carrying. The preferred (stronger) hand was used to hold the containers. In addition to the weight acceptable for comfortable carrying, pulse rate and ratings of perceived exertion (RPE) of the arm and whole body were also used as response measures.

Results indicated that subjects could accurately estimate the amount of weight they could carry comfortably in one hand. RPE for one-handed carrying tasks was not one tenth of the pulse rate as is the case for whole body tasks. Instead, it was found to be one eighth of the heart rate. The average pulse was 100 beats/min for the weights selected. Shape of the containers significantly influenced the amount of weight subjects were willing to carry in one hand. The acceptable amount of weight decreased with distance, but was found to increase with volume. Table 5.20 gives the means and standard deviations of response measures for males and females.

Table 5.20. Mean ($\bar{x}$) and standard deviation (S) of response measures for males (Mital and Ilango, 1983b).

Factor	Description		Weight carried (kg)		Pulse (beats/min)		RPE (arm)		RPE (whole body)	
			Male	Female	Male	Female	Male	Female	Male	Female
Container shape	Plastic bucket	$\bar{x}$	9·8	7·3	99·1	96·5	12·7	12·6	11·3	10·8
		S	2·8	1·4	8·6	17·2	1·7	1·7	1·4	1·5
	Galvanized iron bucket	$\bar{x}$	9·3	6·7	101·1	100·7	12·3	12·5	11·1	10·9
		S	2·5	1·1	8·8	17·1	1·9	1·6	1·3	1·3
	Tool box	$\bar{x}$	10·6	8·4	100·8	98·9	12·4	12·6	11·2	11·2
		S	3·1	1·4	10·4	19·0	1·9	1·7	1·4	1·3
	Radiator can	$\bar{x}$	10·1	7·8	98·7	99·5	12·5	12·4	11·1	10·8
		S	3·0	1·6	11·5	17·6	1·6	1·6	1·2	1·3
Container volume (l)	8·5	$\bar{x}$	9·7	7·4	100·5	98·2	12·3	12·3	11·1	10·7
		S	2·8	1·5	10·0	18·0	1·8	1·7	1·3	1·4
	12·3	$\bar{x}$	10·2	7·7	99·4	99·6	12·7	12·7	11·2	11·2
		S	3·1	1·5	9·9	17·3	1·7	1·6	1·3	1·2
Distance (m)	30·48	$\bar{x}$	11·0	8·0	100·5	98·9	12·3	12·1	11·0	10·5
		S	3·4	1·9	9·1	17·1	1·8	1·8	1·2	1·3
	60·96	$\bar{x}$	9·7	7·5	98·8	98·8	12·4	12·6	11·0	11·0
		S	2·6	1·2	10·7	16·7	1·6	1·8	1·4	1·4
	91·44	$\bar{x}$	9·2	7·1	100·5	99·0	12·8	12·9	11·4	11·2
		S	2·4	1·3	9·9	19·2	1·9	1·3	1·4	1·2

5.8 Capacity Data for One-Handed Carrying Activities

While there are no weight/force recommendations available for frequent one-handed carrying tasks, it is possible to propose physiological limits (heart rate). As observed by Mital and Manivasagan (1983b), for infrequent tasks, subjects selected a weight that resulted in average heart rate of 100 beats/min. The heart rate did not change significantly with distance, container volume or container shape. It appears that subjects select a weight that results in the same circulatory strain, irrespective of task factors. For repetitive one-handed carrying tasks, weights that yield a heart rate of 100 beats/min are therefore recommended.

5.9 Holding Data Base

There exists a small amount of literature pertaining to holding tasks. Ayoub et al. (1987) investigated holding in an attempt to determine performance of holding and unusual postures, determine performance on scores of strength tests, determine holding time and object weight relationships, and to develop and validate prediction equations based on performance on one or more of the strength tests and/or one or more anthropometric characteristics.

Subjects, who were Texas Tech University students and whose ages ranged from 19–35 years, simulated 55 holding tasks based on how the container was held, the posture assumed during the lift and the location of the vertical target and ceiling. The container was (1) held against a wall with no barrier, (2) held against a wall with a vertical restraining barrier, (3) held against a wall with an overhead barrier, and (4) held against the ceiling. Lifting postures were (1) standing, (2) sitting (12-in high seat), (3) squatting, (4) kneeling on left knee, and (5) lying on side or supine. The location of the ceiling and the target were determined by the subject's standing vertical reach.

Table 5.21 details the postures and target heights. Figure 5.6 shows two examples of holding activities.

Subjects were instructed to hold a $10 \times 10 \times 10$ in container against a target for as long as they could without exceeding 60 s. Subjects lifted the container to the target using both hands and held it in place with the left hand only. (It was assumed that the right hand would be used to perform precision movements such as bolting the container into place.) Dependent measures were (1) time on target, (2) time that the container was misaligned (time off target 1), (3) time that the container was not in contact with the target (time off target 2), (4) total time off target (time off target 1 plus time off target 2), and (5) the number of times that the container was off target.

Typical results are given by Figure 5.7(a) and (b) which correspond to Figures 5.6(a) and (b), respectively.

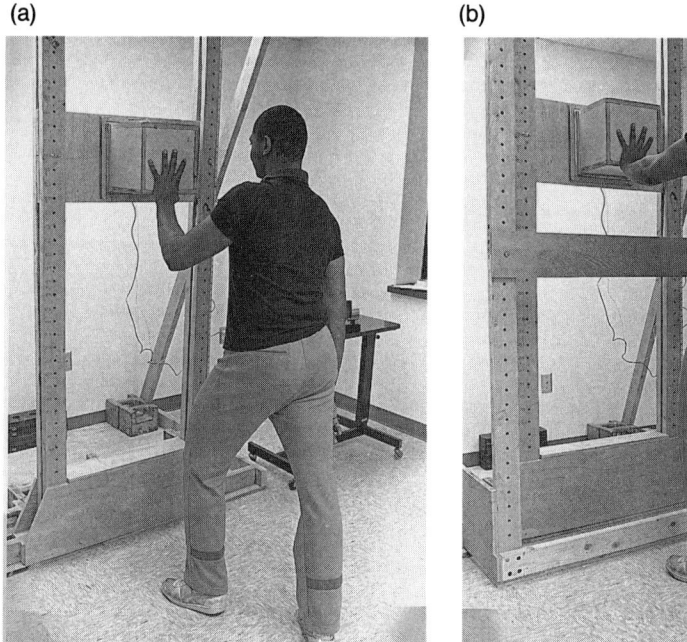

Figure 5.6 (a) Hold task WNUST6 (container held against wall, no ceiling, no constraint, standing and target at 60% of SVR) (b) Holding task WNCST6 (container held against wall, no ceiling, constraint, standing and target at 60% of SVR) (Ayoub et al., 1987)

Table 5.21. Holding tasks against wall with no barrier (WNU tasks) (Ayoub et al., 1987)

Holding task	Posture	Ceiling height*	Target height*
Against wall – no barrier (WNU tasks)	Standing	—	40, 60, 80
	Sitting	—	20, 40, 60
	Squatting	—	20, 40, 60
	Kneeling	—	20, 40, 60
	Lying on side	—	20, 30
Against wall with vertical restraining barrier (WNC tasks)	Standing	—	40, 60, 80
	Sitting	—	20, 40, 60
	Squatting	—	20, 40, 60
	Kneeling	—	20, 40, 60
Against wall with overhead barrier (WCU tasks)	Standing	60	40
		60	50
	Sitting	30	20
		40	30
		40	20
	Squatting	30	20
		40	30
		40	20
	Kneeling	30	20
		40	30
		40	20
Against ceiling (OVU tasks)	Standing	—	60, 70, 80
	Sitting	—	30, 40, 50, 60
	Squatting	—	30, 40, 50, 60
	Kneeling	—	30, 40, 50, 60, 70
	Lying on side	—	20, 30

*Percentage of standing vertical reach.

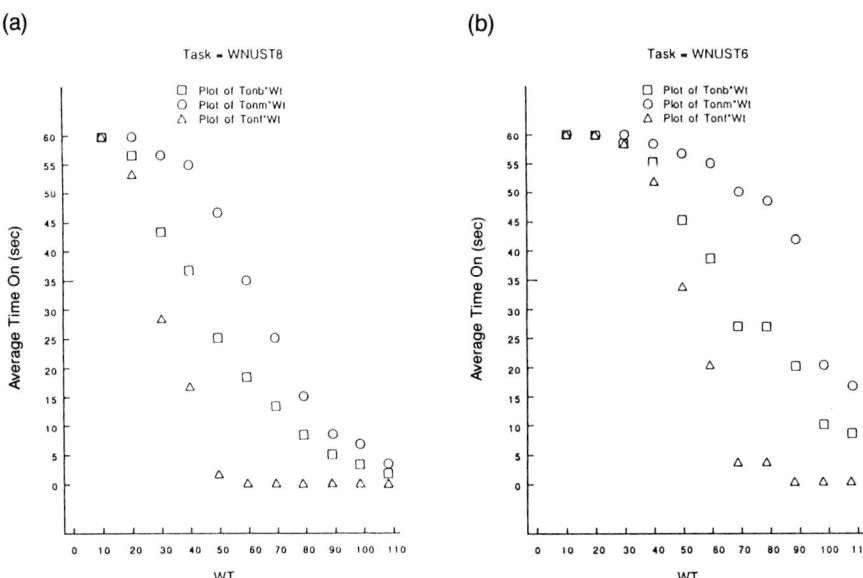

Figure 5.7. Plot of Average Holding Time vs Weight for (a) Holding Task WNUST6, and (b) Holding Task WNCST6. (Ayoub et al., 1987)

As expected, the data revealed significant differences between male and female holding task performance. Comparing all four holding activities (container held against a wall with no barrier, container held against a wall with a vertical restraining barrier, container held against a wall with an overhead barrier, and container held against the ceiling) data revealed that overhead holding tasks resulted in the lowest average time on target values. Containers held against the wall with no barriers had the highest time on target values. It is thought that this is due to the posture assumed for this task.

Chapter 6
Job Design/Redesign and Screening Procedure

6.1 Principles of Job Design/Redesign

There are certain risks that the individual takes when he performs an MMH activity. Individual, task and environmental variables can affect MMH. Figure 6.1 delineates these variables and their place in the job design/employee placement paradigm. Risks can be greatly reduced if the jobs are designed taking the human component into consideration. Eliminating or reducing MMH, decreasing job demands and minimizing body movements are three goals of job design. Each are discussed below.

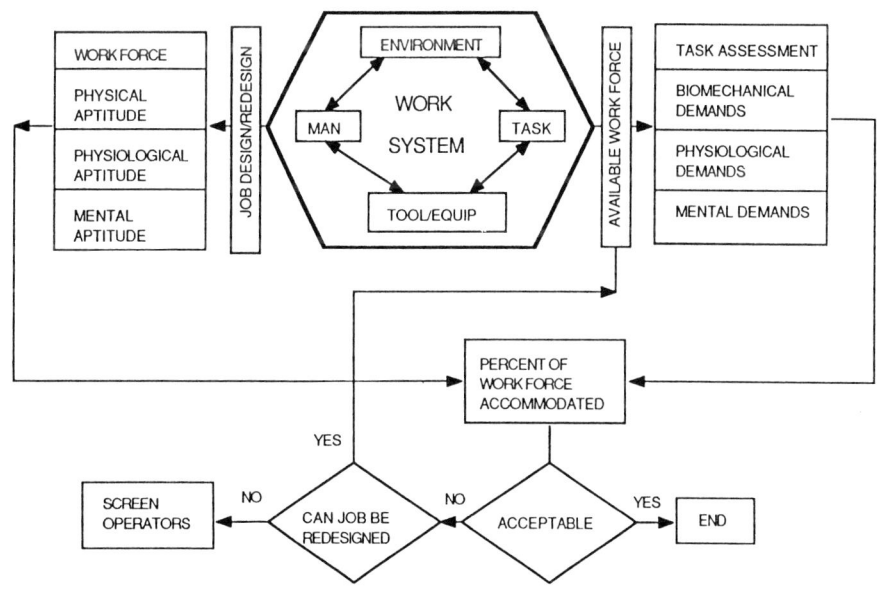

Figure 6.1. A general lifting model (modified from Ayoub et al., 1983a)

6.1.1 Eliminating MMH

An optional solution to MMH-related problems is to eliminate heavy MMH. This can be accomplished by (1) using mechanical aids (hoists, lift trucks, lift tables, cranes, elevating conveyors, gravity dumps, chutes) to eliminate some MMH stress, and (2) altering work area layout such that all material is at work level. Altering the area layout can involve either change in the height of either the work level or the worker level. See Figure 6.2 for a summary of eliminating heavy MMH (Ayoub et al., 1987).

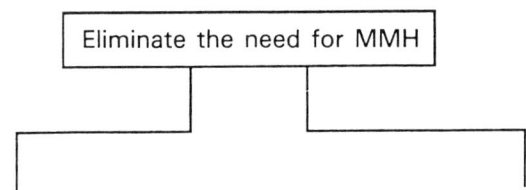

1. Use mechanical aids: lift table, lift trucks, hoists, cranes, conveyors, drum & barrel dumpers, gravity dumps, chutes.

2. Change work area layout, change height of work level or worker level, provide all material at work level.

Figure 6.2. Methods to eliminate the need for MMH (Ayoub 1982, and Ayoub et al., 1983a)

6.1.2 Decreasing Job Demands

If MMH cannot be eliminated realistically, attempts should be made to decrease MMH demands of a given job. The following are means by which to satisfy this principle:

1. Decrease the weight of the object handled by assigning the handling to two or more people, by distributing the load into two or more containers, by reducing the capacity of the container, or by using a lightweight container.
2. Decrease job demands by changing the type of MMH activity. Lifting, lowering, pushing, pulling, carrying and holding are MMH activities, each having a relative job demand. It is suggested that it is better to lower rather than lift, pull rather than carry and push rather than pull.
3. Decrease job demands by changing the work area layout. This can be accomplished by reducing the horizontal distance between the starting and ending points of a lift, limiting stacking height to the shoulder height of operators, and keeping heavy objects at the knuckle height of operators. Distance travelled should be minimized for pushing, pulling and carrying activities.
4. Decrease job demands by maximizing the time available to perform the job. The means by which to accomplish this are to reduce the frequency of lift, and/or

incorporate workrest schedules or job rotation programmes into the work design (Ayoub *et al.*, 1987)

See Figure 6.3 for a summary of methods to decrease job demands.

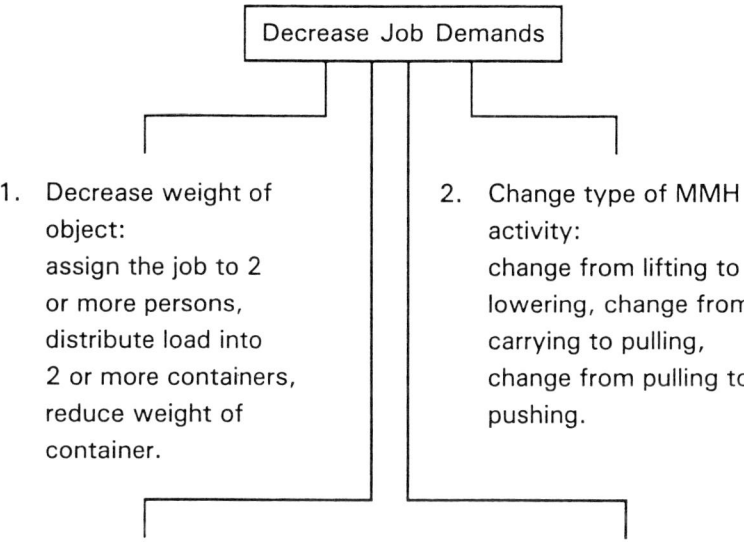

Figure 6.3. *Methods to decrease job demands (Ayoub* et al., *1987)*

6.1.3 Minimizing Body Movements

Minimizing stressful body movements should also be strived for. This can be accomplished by the following:

1. Placing the object to be handled within the operator's arm reach envelope whereby bending will be reduced. Bending can also be reduced by providing all material at

work level of the operator, and by avoiding deep shelves that require the operator to bend and reach for objects placed toward the back of the shelf.
2. Twisting motions can be reduced by locating objects within the arm reach envelope. It is important to provide the worker with sufficient space for the entire body to turn, and to provide seated operators with adjustable swivel chairs.
3. Design considerations should allow the operator to lift objects safely. Safe lifting practices involve the allocation for the object to be handled close to the body, better control of the handled object when handles are used, balancing of the contents of the containers, using rigid containers which increase operator control of the object, and avoiding lifting wide objects from the floor level (Ayoub et al., 1987). See Figure 6.4 for a summary of methods to minimize stressful body movements.

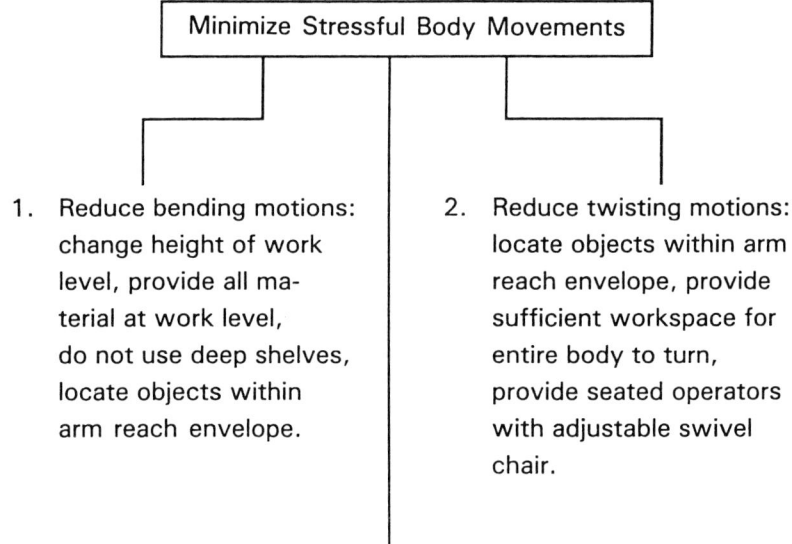

Figure 6.4. *Methods to minimize stressful body movements (Ayoub et al., 1987; and Chaffin and Baker, 1970)*

6.2 Principles of Screening Procedures

Oftentimes, it is not feasible or it is impractical to design or redesign a job. Therefore, in order to control manual lifting hazards, the strategy of pre-employment screening must be used. The pre-employment screening provides information of the potential employee's capacity to work. Once this potential employee's capacity has been estimated, an attempt is made to assign him/her to a job whereby his/her capacity is not exceeded by task demands.

In the past, back X-ray films, strength testing, medical examinations, psychological tests, job simulators and rating methods have been used for employee placement. Justification for using the various methods resides in the fact that age, gender, stature and body-weight are personal risk factors (Ayoub, 1982b). However, Chaffin et al. (1977a) found that neither age, gender, stature nor body-weight correlate with increased incidence or severity rates of later medical problems of any kind. Thus, suggesting that the aforementioned employee placement techniques are inadequate sole measures of a worker's capacity to work. Although taken as sole measures of work capacity, the aforementioned screening procedures are not adequate, they can be useful in some aspects of pre-employment screening. In fact, several of these procedures have become common practice for pre-employment screening.

The kernel of today's pre-employment screening procedures come from the rationale and structure of pre-employment screening programmes that had been devised in the past. The following are brief descriptions of screening alternatives and their respective functions of the past and present.

6.2.1 Back X-Ray Films

It was thought that back abnormalities may predispose an individual to low-back injury. Consequently, prospective workers had their backs X-rayed to disclose the status of the spine. Whether back X-rays are useful predictors of back injury is much debated. Steward (1947), Becker (1955), Kosiak et al. (1968), and McGill (1968) support the predictive validity of back X-rays. However, more recent research of LaRocca and MacNab (1970) and Redfield (1971) suggests the contrary.

At one extreme, Montgomery (1976) states that evidence does not substantiate the relationship between developmental abnormalities and an increased incidence of low-back injury. Part of the problem is that no criteria for radiological screening have been defined (NIOSH, 1981). Furthermore, NIOSH (1981) states that interpretation problems make it impossible to quote a value for the increased probability of back trouble for each abnormality.

Two partial exceptions to this statement involve sacralization and spondylolisthesis which have been extensively researched. However, still no reliable screening data resulted from these investigations. Another aspect in the interpretation of lumbar spinal radiographs is that of size and shape of the spinal canal and intervertebral foramina. There is some evidence suggesting that these variables may be related to back injuries, but how they are related is not known.

At the other extreme, Rowe (1969) reports that a complete physical examination accompanied with a lumbosacral radiographic evaluation predicts approximately 10%

of later low-back sufferers. Additionally, according to Leggo and Mathiasen (1973), gross skeletal abnormalities that are detected by X-rays can lead to low-back injury in the manual materials handlers.

Generally speaking, today's use of back X-ray examinations in pre-employment testing is to establish the present health status of the applicant, not to predict back injury risks of a physically stressful job (American College of Radiology, 1973; Ayoub, 1982b). The controversy over back X-ray examinations predictive validity and the additional concern over excessive radiation exposure led researchers to search for new methods of predictive injury risk tools.

6.2.2 Strength Testing

Strength testing evolved with the hypothesis that there is a direct relationship between probability of injury and percentage of strength capacity utilized by the worker when performing his job. While strength testing can be a very useful tool, it neglects other factors contributing to the individual's ability to perform work, such as cardiac and pulmonary functions. Therefore, strength testing alone is not a sufficient measure of worker capacity (Ayoub, 1982b).

The usefulness of strength testing for predicting lifting capacity should not be overlooked however. Many jobs require occasional or frequent lifting where large amounts of force are needed. This force requirement increases the frequency and severity of musculoskeletal problems and low-back pains. In support of strength testing, it has been established that abdominal strength is correlated with reducing the compressive force acting on the lumbar spine while lifting (Bartelink, 1957; Davis, 1969; Rowe, 1971). Furthermore, strength of back extensors have been found to protect the back during MMH activities (Troup and Chapman, 1969b; Poulsen and Jorgensen, 1971). Thus, the need for careful selection and placement of employees on jobs requiring high forces based on their strength (NIOSH, 1977). To choose the appropriate strength test, a job analysis should be conducted. From this, minimum strength scores can be obtained with the knowledge of the required job capacities for job placement (Selan *et al.*, 1986).

Isometric strength tests have been the preferred strength tests because they are safe and reliable (Chaffin, 1974). In isometric strength tests, the worker is required to slowly increase the force exerted until he has reached an acceptable level. Thus, isometric strength testing presents no challenges during testing which, in turn, reduced the chance of injury during testing. Assessments of strength are discussed further in chapter 7.

6.2.3 Medical Examinations

Pre-employment medical examinations used as screening devices have been and still are commonplace. In this context, the purpose of pre-employment medical examinations is to match the individual's capacity to task demands, both physically and psychologically. Basic anthropometric data are taken such as body-weight and height. Body-weight and height can lead to assumptions about metabolic energy expenditure and predispositions to low-back pain. However, there is no evidence to link these variables directly to low-back pain (NIOSH, 1981). Thus, the question of how a physician is to match the individual's capacity to task demands remains. In order to accomplish this end, the physician must know all job demands and have very accurate measurements of the individual's capacity. However, in practice, the physician probably has only vague knowledge about job demands, and therefore has no basis to screen workers.

According to Ayoub (1982b), at best the physician can screen those workers with gross abnormalities and reveal the present day's health status. Some researchers are more optimistic about the predictive validity of medical examinations. As stated previously, a complete physical examination accompanied by lumbosacral radiographic evaluation predicts approximately 10% of low-back injury sufferers (Rowe, 1969). NIOSH (1977) also advocates stringent medical examinations under certain conditions.

NIOSH (1977) suggests that stringent medical examinations be required for any job that exceeds suggested maximum load lifted as delineated in Figure 6.5. That is, if the maximum load lifted on the job is only lifted occasionally, a stringent medical examination is less likely needed than if the maximum load lifted on the job is frequently lifted. Size and centre of gravity are also considered in the graph.

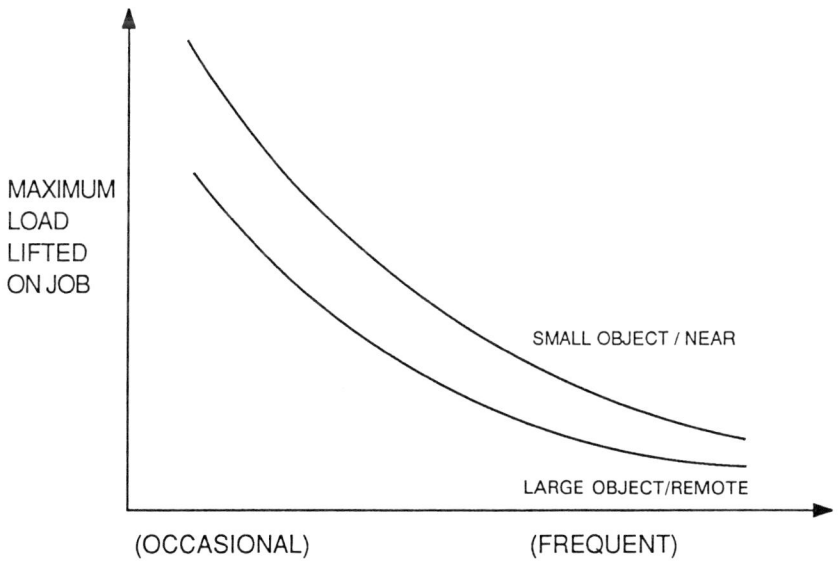

Figure 6.5. Proposed medical action level

The following findings warranted the need for medical action levels:

1. A more severe and higher frequency of musculoskeletal and contact type injuries is correlated with jobs requiring high strengths.
2. Age, gender, body-weight and stature are not good predictors of the strengths of the working population.
3. Jobs populated by weaker persons had higher frequencies and more severe incidences of musculoskeletal problems.

6.2.4 Psychological Tests

Psychological tests have been used as screening devices in the USA since the beginning of the 20th century and continue to be used today. The rationale behind psychological testing is that they can predict success/failure of the individual's job performance. Popular attributes to examine are motivation, maturity and aptitude on relevant tasks. It is noted that personality characteristics may increase the worker's susceptibility to hazards on the MMH job. However, several confounding variables prevent clear interpretations of this data (NIOSH, 1981). From test outcomes, employers can select a more favourable work-force. While this type of information may improve work-force quality, it does not incorporate physical work capability, and is therefore an inadequate sole selection device (Ayoub, 1982b).

6.2.5 Job Simulators

Job simulators do as their name implies, simulate actual working conditions. Thus, the applicant's ability to perform the job can be approximated easily. However, this method has several serious drawbacks. First, to build and maintain simulators can be very expensive for the company. Secondly, the screening process could become very lengthy since most applicants could fill several different positions performing several different tasks. The third and most serious limitation is that screening is based on the lack of physical ability for a certain task. The situation may arise where an applicant strains him/herself in the hopes of acquiring a job. Thus, this method of screening is potentially injurious.

Additionally, job simulators reveal little information about job capacity since time constraints prevent replications of working conditions. Therefore, it will not be known whether an applicant can perform tasks continually as the job will demand (Ayoub, 1982b).

6.2.6 Rating Methods

Rating methods attempt to incorporate physical and psychological assessments. The GULHEMP method rates applicants on (1) general physique, (2) upper extremities, (3) lower extremities, (4) hearing, (5) eyesight, (6) mentality, and (7) personality type. This rating scale is based on a complete job analysis to define job demand and corresponding scaling categories, and subjective categorization by a physician of the applicant's capability. Another rating method, the Hanman method also rates both job demands and the applicant's capacity to do work. The primary difference between the two methods is that the Hanman method details the job demand attributes of weight, distance and time. Both of these rating methods have been quite successful in practice. The major criticism is their lack of objectivity (Ayoub, 1982b).

In brief summary, it is concluded that the past selection devices, many of which have become common practice, are inadequate in today's work world. Screening programmes should separate fit and unfit workers for the task such that the worker will not be working beyond safe limits. According to Ayoub (1982b), common scales and determinants should be used to describe the job as well as the individual. The next section discusses some criteria on which to base pre-employment selection procedures.

6.3 Criteria for Screening of Personnel

Up to this point, inadequate pre-employment screening procedures have been discussed, but adequacy has yet to be defined. There are certain medical, social, economic and legal criteria that must be met for each assessment. However, when it comes to choosing between methods, NIOSH's (1981) *Work Practices Guide for Manual Lifting* offers some criteria to follow:

1. Pre-employment screening procedures should be safe to administer. Some pre-employment screening procedures, such as X-rays, may be potentially more dangerous to the applicant than the valuable screening information they may give. In the case of physical strength and endurance type assessments, there is a risk of over-exertion injury since the promise of a job may motivate the applicant to attain some lifting goal beyond his lifting capacity. It is recommended that the inclusion of this type of test proceed after the applicant has passed a physical examination and has no history of musculoskeletal or coronary problems.
2. Pre-employment screening procedures should give reliable, quantitative values. Any reliable test is a test that produces consistent results. The coefficient of variation, the standard deviation of the repeated values divided by their mean value, is a common measure of test reliability. According to Chaffin *et al.* (1977a), it should be possible to achieve a coefficient of variation of less that 15% for physical capacity tests.
3. Pre-employment screening procedures should be related to specific job requirements. Physical attributes do not correlate well with one another. That is to say, that if an applicant scores highly on grip strength it does not mean that he will score highly on some other strength measure. Job physical requirements must be assessed for each required physical task. Lifting capability may be important for one task, while pulling capability is important for another. Legally, job requirements must not discriminate against women, minority groups, aged workers or the physically disabled.
4. Pre-employment screening procedures should be practical. What is necessary for one industry may not be necessary for another industry, therefore selection procedures should be tailored to each industry. Typically, small industries are able to construct a set of carefully controlled tests that can be completed by the worker

within the first few days of employment. Larger industries usually use standardized tests so that a large number of applicants can be processed at once. According to NIOSH (1981), practical conditions for pre-employment screening procedures should:
   (a) Require minimum hardware expense;
   (b) Have hardware capable of simulating different job conditions;
   (c) Require minimum time to administer;
   (d) Require minimum instruction and learning time.
5. Pre-employment screening procedures should predict risk of future injury or illness. In order to do this, the chosen pre-employment selection procedure must be supported by injury and illness data. The effects of injury or illness data should be assessed to ensure the goal of reducing health and saftey problems. Additionally, this evaluation must include the degree to which the worker's physical capabilities match the job's physical demands.

6.4 Tools for Job Design/Redesign and Screening with Examples

6.4.1 Job Severity Index

Job severity index (JSI) is a tool used to design jobs/tasks requiring lifting or lowering (Ayoub et al., 1978). JSI is the ratio of job demand to the capacity of the person or population working under the job conditions. The following is a conceptual representation of JSI:

$$JSI = \frac{job\ demand}{operator\ capacity}; \text{ for the given job conditions.}$$

Job demand is determined through job analysis while capacity of worker(s) is determined through psychophysical method. Job conditions include frequency of lift, range of lift, container size, and any of these coupled with a few anthropometric measurements.

6.4.1.1 Using JSI as a Job Design/Redesign Device

To describe a job, preliminary definitions must be made, and tasks must be organized. A job is organized into two subcategories, subtask groups and number of tasks for each group. Figure 6.6 describes a job which has two subtask groups. Subtask group 1 carries out three tasks, while subtask group 2 carries out two tasks. Decisions of job description are based on task type (lifting or non-lifting task), range of motion, and time spent on the task.

A lifting task is defined as lifting or lowering of an item from one point in space to another. Ranges of lifting space are listed in Table 6.1. A single lifting task can include several slightly different weights and sizes. Two or more lifting tasks should be defined if (1) the range of item weights exceeds 25 lbs or (2) item weights are not uniformly distributed. A non-lifting task is defined as the negation of a lifting task; an item is not lifted or lowered from one point in space to another.

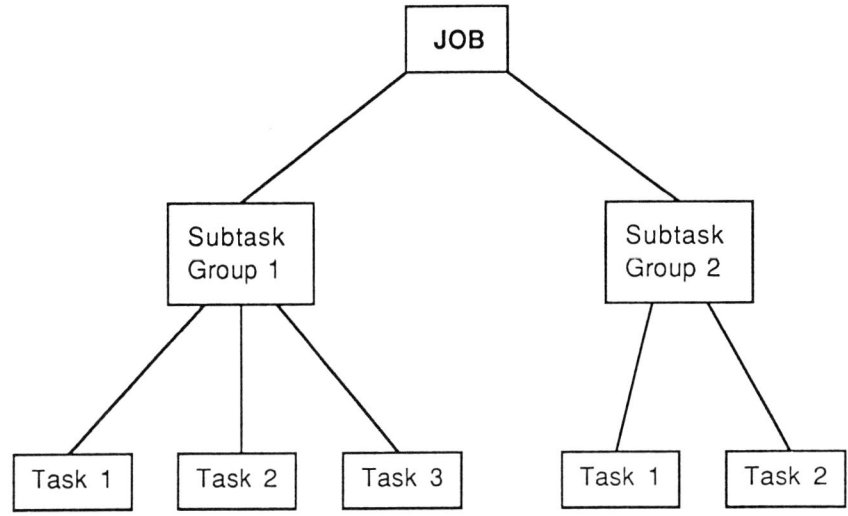

Figure 6.6. Job organization for JSI analysis

Table 6.1. Lifting range assignment (Ayoub et al., 1978)

Point of Lift Initiation	Point of Lift Termination	Range† Assignment
0" to KL/2	0" to KL* + 10	#1
	KL + 10" to KL + 30"	#2
	KL + 30" and above	#3
KL/2 to KL	KL/2 to KL	#1
	KL to KL + 30"	#4
	KL + 30" and above	#5
KL to KL + 10"	KL to KL + 30"	#4
	KL + 30" and above	#5
KL + 10" to KL + 20"	KL + 10" to KL + 20"	#4
	KL + 20" and above	#6
KL + 20" and above	KL + 20" and above	#6

*KL = knuckle level.
†Range #1 = floor to knuckle; #2 = floor to shoulder; #3 = floor to reach;
#4 = knuckle to shoulder; #5 = knuckle to reach; #6 = shoulder to reach.

Lifting and non-lifting tasks, which account for more than 1 h of the workday, are organized into task groups. A task group is defined as one or more tasks that are performed within the same time frame.

Once preliminary job descriptions are made, job demand, job demand parameters and operator capacity are determined. Job demands are acquired through the following job analysis:

Step 1: Determine the following:
  (A) Average length of the work week (days);
  (B) Average length of the workday or shift (hours);
  (C) Number of shifts per day (one, two or three);
  (D) Average time exposure (based on A–C);
Step 2: Describe job demand parameters in terms of:
  (A) Actual weight of lift (lbs);
  (B) Average frequency of lift (lifts/min);
  (C) Container size (inches in sagittal plane);
  (D) Range of lift (lift initiation and termination points).
Note: Each job may be divided into component tasks because of the inconsistent job parameters.

Operator capacity is determined using the psychophysical method. That is to say, the worker adjusts the weight of the load such that the maximum acceptable weight of lift is reached. Thus, the criterion is the worker's perceived stress. The psychophysical model is used in this context to determine operator capacity because it is a measure of the integration of both biomechanical and physiological stresses (Karwowski, 1982; Ayoub, 1982a).

More explicitly defined, JSI is the exposure-time and lifting-frequency weighted average of the ratio of maximum required weight of lift to worker capacity, as described by the following equation:

$$JSI = \sum_{i=1}^{n} \frac{hours_i \times days_i}{hours_t \times days_t} \sum_{j=1}^{m_i} \frac{F_j}{F_i} \times \frac{WT_j}{CAP_j}$$

where n = number of subtask groups; m = number of tasks in group i; $days_i$ = exposure days per week for group i; $days_t$ = total days per week for job; $hours_i$ = exposure hours per day for group i; $hours_t$ = total number of hours per day that a job is performed; F_j = lifting frequency for task j; F_i = total lifting frequency for group i; WT_j = maximum weight of lift required by task j; CAP_j = the smallest applicable maximum acceptable weight of lift adjusted for frequency of lift and box size.

Lifting capacity (*CAP*) of a group will have to be calculated before the JSI calculation can be completed. The information that is needed to calculate lifting capacity is the range of lift (Table 6.1), the gender of the workers, frequency of lift (lift/min), box size, and the percentile of the target population. First, adjustment for frequency of lift (Tables 6.2 and 6.3) is made, then the adjustment for box size (Tables 6.4 and 6.5), and then the adjustment for the population percentile (Tables 6.6, 6.7, and 6.8).

The following is an example of the simplest case using JSI. Given:

(A) range — knuckle to reach (K–R);
(B) population — male;
(C) frequency of lift (*FY*) — 5 lifts/min;
(D) box size (*BX*) — 20 in;
(E) population percentage to design for — 85%;
(F) number of subtask groups (*n*) — 1;

Table 6.2. Prediction equations for lifting capacity based on frequency of lift – males

Range of Lift, Male	Capacity
	$0.1 < FY < 1.0$
F–K†	$57.2‡(FY)^{**}(-0.184697)$
F–S	$51.2\ (FY)^{**}(-0.184697)$
F–R	$49.1\ (FY)^{**}(-0.184697)$
K–S	$52.8\ (FY)^{**}(-0.138650)$
K–R	$50.0\ (FY)^{**}(-0.138650)$
S–R	$48.4\ (FY)^{**}(-0.138650)$
	$1.0 \leq FY \leq 12.0$
F–K	$57.2 - 2.0(FY - 1)$
F–S	$51.2 - 2.0(FY - 1)$
F–R	$49.1 - 2.0(FY - 1)$
K–S	$52.8 - 2.0(FY - 1)$
K–R	$50.0 - 2.0(FY - 1)$
S–R	$48.4 - 2.0(FY - 1)$

FY = frequency of lift (lifts/min).
†K, knuckle; F, floor; S, shoulder; R, reach.
‡57.2 = mean capacity for lift based on data from Ayoub et al. (1978) and Snook (1978a) for the various ranges of lift for the 50th percentage and 1.0 lift/min.
**, Exponentiation (e.g., FY to the power of -0.184697).

Table 6.3. Prediction equations for lifting capacity based on frequency of lift – females

Range of Lift†, Female	Capacity
	$0.1 < FY < 1.0$
F–K†	$37.4‡(FY)^{**}(-0.187818)$
F–S	$31.1\ (FY)^{**}(-0.187818)$
F–R	$28.1\ (FY)^{**}(-0.187818)$
K–S	$30.8\ (FY)^{**}(-0.156150)$
K–R	$27.3\ (FY)^{**}(-0.156150)$
S–R	$26.4\ (FY)^{**}(-0.156150)$
	$1.0 \leq FY \leq 12.0$
F–K	$37.4 - 1.1(FY - 1)$
F–S	$31.1 - 1.1(FY - 1)$
F–R	$28.1 - 1.1(FY - 1)$
K–S	$30.8 - 1.1(FY - 1)$
K–R	$27.3 - 1.1(FY - 1)$
S–R	$26.4 - 1.1(FY - 1)$

FY = frequency of lift (lifts/min).
†, See Table 6.2 for legend.
‡37.4 = mean capacity for lift based on data from Ayoub et al. (1978) and Snook (1978a) for the various ranges of lift for the 50th percentage and 1.0 lift/min.
**, Exponentiation (e.g., FY to the power of -0.187818).

(G) number of tasks in group i (m_i) — 1;
(H) exposure days per week for group i (days $_i$) — 5;
(I) total days per week for job (days $_t$) — 5;
(J) exposure hours per day for group i (hours $_i$) — 7;
(K) number of hours per day that a job is performed (hours $_t$) — 8;

Table 6.4. Adjustment of lifting capacity based on box size – males

Range of Lift†, Male	Capacity
	$12'' \leq BX \leq 18''$
F-K	$CAP + 1\cdot 65(18-BX)$
F-S	$CAP + 1\cdot 65(18-BX)$
F-R	$CAP + 1\cdot 65(18-BX)$
K-S	$CAP + 1\cdot 10(18-BX)$
K-R	$CAP + 1\cdot 10(18-BX)$
S-R	$CAP + 1\cdot 10(18-BX)$
	$BX > 18''$
F-K	$CAP + 0\cdot 8(18-BX)$
F-S	$CAP + 0\cdot 8(18-BX)$
F-R	$CAP + 0\cdot 8(18-BX)$
K-S	$CAP + 0\cdot 8(18-BX)$
K-R	$CAP + 0\cdot 8(18-BX)$
S-R	$CAP + 0\cdot 8(18-BX)$

CAP = Capacity of lift as determined in Table 6.2.
BX = Box size (in).
†, See Table 6.2 for legend.

Table 6.5. Adjustment of lifting capacity based on box size – females

Range of Lift†, Female	Capacity
	$12'' \leq BX \leq 18''$
F-K	$CAP + 1\cdot 1\ (18-BX)$
F-S	$CAP + 1\cdot 1\ (18-BX)$
F-R	$CAP + 1\cdot 1\ (18-BX)$
K-S	$CAP + 0\cdot 55(18-BX)$
K-R	$CAP + 0\cdot 55(18-BX)$
S-R	$CAP + 0\cdot 55(18-BX)$
	$BX > 18''$
F-K	$CAP + 0\cdot 4(18-BX)$
F-S	$CAP + 0\cdot 4(18-BX)$
F-R	$CAP + 0\cdot 4(18-BX)$
K-S	$CAP + 0\cdot 2(18-BX)$
K-R	$CAP + 0\cdot 2(18-BX)$
S-R	$CAP + 0\cdot 2(18-BX)$

CAP = Capacity of lift as determined in Table 6.3.
BX = Box size (in).
†, See Table 6.2 for legend.

(L) lifting frequency for group i (F_j) — 5 lifts/min;
(M) total lifting frequency for group i (F_i) — 5 lifts/min;
(N) maximum weight of lift required by task j (WT_j) — 30 lbs.

Step 1. Calculate lifting capacity of the group.
    A. Frequency adjustment
        (1) Since this example specifies male workers, Table 6.2 will be used, not Table 6.3 which is designated for female workers. The frequency of lift given is 5 lifts/min.

Table 6.6. Adjustment of lifting capacity based on percentile and frequency – males

Range of Lift†, Male	Capacity
	$0 \cdot 1 \leq FY < 1 \cdot 0$
F–K	$CAP + Z \times 16 \cdot 86(FY)^{**}(-0 \cdot 174197)$
F–S	$CAP + Z \times 15 \cdot 09(FY)^{**}(-0 \cdot 174197)$
F–R	$CAP + Z \times 14 \cdot 47(FY)^{**}(-0 \cdot 174197)$
K–S	$CAP + Z \times 14 \cdot 67(FY)^{**}(-0 \cdot 156762)$
K–R	$CAP + Z \times 13 \cdot 89(FY)^{**}(-0 \cdot 156762)$
S–R	$CAP + Z \times 13 \cdot 45(FY)^{**}(-0 \cdot 156762)$
	$1 \cdot 0 \leq FY \leq 12 \cdot 0$
F–K	$CAP + Z(16 \cdot 86 - 0 \cdot 5964(FY-1))$
F–S	$CAP + Z(15 \cdot 09 - 0 \cdot 5338(FY-1))$
F–R	$CAP + Z(14 \cdot 47 - 0 \cdot 5119(FY-1))$
K–S	$CAP + Z(14 \cdot 67 - 0 \cdot 5534(FY-1))$
K–R	$CAP + Z(13 \cdot 89 - 0 \cdot 5240(FY-1))$
S–R	$CAP + Z(13 \cdot 45 - 0 \cdot 5074(FY-1))$

CAP = capacity of lift as determined in Table 6.4.
Z = Z-score of population percentage (from normal tables).
FY = Frequency of lift (lifts/min).
†, see Table 6.2 for legend.
**, Exponentiation.

Locate the group of equations that pertain to 5 lifts/min. Within that group under the heading 'Range of lift', locate K–R which is the given range of lift. The equation corresponding to that range which is found under the heading 'Frequency of lift' is the equation to use for this adjustment.

$$CAP = 50 \cdot 0 - 2 \cdot 0 \times (FY - 1)$$
$$CAP = 50 \cdot 0 - 2 \cdot 0 \times (5 - 1)$$
$$CAP = 42 \cdot 0 \text{ pounds}$$

B. Box adjustment

(1) Next, based on the gender of workers, Table 6.4 will be used for the adjustment of lifting capacity based on box size. The given box size is 20 in. Locate the group of equations that pertain to 20 in boxes. Within that group under the heading 'Range of lift', locate K–R. The equation corresponding to that range which is found under the heading 'Box Size' is the equation to use for this adjustment. CAP refers to the lifting capacity calculated in the previous step.

$$CAP = CAP + 0 \cdot 8 \times (18 - BX)$$
$$CAP = 42 \cdot 0 + 0 \cdot 8 \times (18 - 20)$$
$$CAP = 40 \cdot 4 \text{ lifts/min.}$$

C. Population Percentage Adjustment

(1) Lastly, based on the gender of workers, Table 6.6 will be used for the adjustment of lifting capacity based on population percentile used. As done before, locate the appropriate group of equations based on frequency of lift and range of lift. The equation corresponding to K–R is the equation to use for this adjustment. CAP refers to the lifting capacity calculated in the previous step.

Table 6.7. *Adjustment of lifting capacity based on percentile and frequency – females*

Range of Lift†, Female	Capacity
	$0 \cdot 1 \leq FY < 1 \cdot 0$
F-K	$CAP + Z \times 6 \cdot 87(FY)^{**}(-0 \cdot 251605)$
F-S	$CAP + Z \times 5 \cdot 71(FY)^{**}(-0 \cdot 251605)$
F-R	$CAP + Z \times 5 \cdot 16(FY)^{**}(-0 \cdot 251605)$
K-S	$CAP + Z \times 5 \cdot 66(FY)^{**}(-0 \cdot 258700)$
K-R	$CAP + Z \times 5 \cdot 01(FY)^{**}(-0 \cdot 258700)$
S-R	$CAP + Z \times 4 \cdot 85(FY)^{**}(-0 \cdot 258700)$
	$1 \cdot 0 \leq FY \leq 12 \cdot 0$
F-K	$CAP + Z(6 \cdot 87 - 0 \cdot 1564(FY-1))$
F-S	$CAP + Z(5 \cdot 71 - 0 \cdot 1300(FY-1))$
F-R	$CAP + Z(5 \cdot 16 - 0 \cdot 1175(FY-1))$
K-S	$CAP + Z(5 \cdot 66 - 0 \cdot 1289(FY-1))$
K-R	$CAP + Z(5 \cdot 01 - 0 \cdot 1141(FY-1))$
S-R	$CAP + Z(4 \cdot 85 - 0 \cdot 1104(FY-1))$

CAP = capacity of lift as determined in Table 6.5.
Z = Z-score of population percentage (from normal tables).
FY = Frequency of lift (lifts/min).
†, see Table 6.2 for legend.
**, Exponentiation.

(a) Table 6.8 lists Z-scores for various population percentages. The Z-score corresponding to population of 85 is $-1 \cdot 0364$ (Z).

$$CAP = CAP + Z(13 \cdot 89 - 0 \cdot 5240 (FY - 1))$$
$$CAP = 40 \cdot 4 + -1 \cdot 0364 (13 \cdot 89 - 0 \cdot 5240 (5 - 1))$$
$$CAP = 40 \cdot 4 + -12 \cdot 22$$
$$CAP = 28 \cdot 18 \text{ lifts/min.}$$

Step 2. Calculate JSI

$$JSI = \frac{hours_i \times days_i}{hours_t \times days_t} \times \frac{F_j \times WT_j}{F_i \times CAP_j}$$

$$JSI = 0 \cdot 9315$$

Since JSI value is 0·9315, this job is within acceptable limits. The risk of injury is minimal.

Ayoub et al. (1983a) reported a critical JSI value of 1·5; the occurrence of injuries increases substantially when the JSI is above 1·5. Figure 6.7(a) describes the relationship between JSI and cumulative injury rate. A substantial increase in cumulative injury rate is evident beyond the JSI value of 1·5. Situations where the JSI is between 0·0 and 0·75 results in the fewest number of back injuries, approximately three back injuries per 100 full-time equivalent employees (FTE), while the JSI range of 0·75 to 1·5 results in approximately 5 back injuries per FTE (Ayoub et al., 1983c). Figures

Job Design and Screening

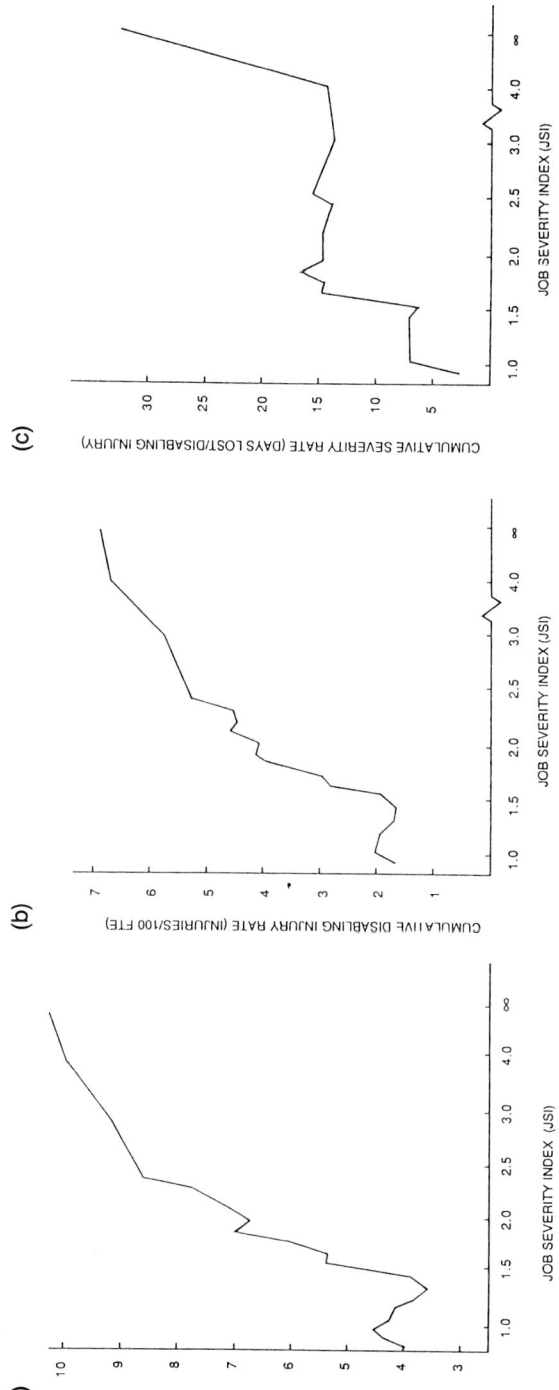

Figure 6.7(a). *Relationship between JSI and cumulative injury rate (b).* *Relationship between JSI and cumulative disabling injury rate (c).* *Relationship between JSI and cumulative severity rate*
From *Human Factors, Vol. 26.* © 1983 by The Human Factors Society, Inc., *and reproduced by permission*

Table 6.8. Z-scores for various population percentages

Population Percentage	Z-score
95	−1·6449
90	−1·2816
85	−1·0364
75	−0·6745
50	0·0
25	0·6745
15	1·0364
10	1·2816
5	1·6449

6.7(b) and (c) describe the relationship between JSI and cumulative disabling injury rate, and the relationship between JSI and cumulative severity rate, respectively. Once again, a large increase in cumulative injury rate and cumulative severity rate is evident beyond the JSI value of 1·5.

JSI is not only a tool used for job design/redesign, but also a tool to judge whether the individual's capacity (or group of worker's capacities) will be exceeded by job demands. JSI is unique in its flexibility to be used as an individual employee screening procedure. JSI can be used when the job requires relatively constant, measurable requirements, or the job requires at least 25 lifts/day of objects weighing 10 lbs or more and lifting activities occupy at least 25% of time on the job.

6.4.1.2 Using JSI as a Job Screening/Placement Device

As discussed in the previous section, JSI is an exposure time and lifting frequency weighted average of the job demands to the capacity of the worker or work-force. Thus, job demand, operator capacity and given job conditions define JSI. To use JSI as a screening device, job demands and operator capacity are determined in the same way as for job design/redesign. However, the JSI employee placement procedure requires some additional steps so that the individual company can tailor the use of the JSI to its demands. The following steps outline the general procedure:

1. The maximum acceptable injury rate and its corresponding JSI are determined for the company based on management policy.
2. Job analysis is performed.
   (a) Determine and organize tasks.
   (b) Determine job exposure information.
   (c) Determine lifting task information.
3. Predict the individual's lifting capacities based on the job analysis using a chosen capacity equation. The chosen capacity equation is based on how well that equation describes the job.
4. Determine the individual's JSI if that individual is to be placed in the given job.
5. Use the JSI from step 4 to determine an expected injury frequency rate for that individual.

6. Make employee placement decision based on acceptability in terms of JSI from step 5.

The following is an example of employment placement using JSI procedure:

Given:
1. Maximum acceptable JSI = 0·8032.
2. Job description is as follows:
    Subtask group 1:
    (a) Exposure hours per day — 2;
    (b) Total number of hours worked per day — 7;
    (c) Exposure days per week — 5;
    (d) Total days per week for job — 5;
    (e) Frequency of lift for task — 6;
    (f) Total lifting frequency for subtask group — 14.
    Subtask group 2:
    (a) Exposure hours per day — 3;
    (b) Total number of hours worked per day — 7;
    (c) Exposure days per week — 5;
    (d) Total days per week for job — 5;
    (e) Frequency of lift for task — 8;
    (f) Total lifting frequency for subtask group — 14.
3. Jiang and Ayoub's (1986) prediction model which requires the following information:
    (a) Shoulder strength (SHO) = 45 kg;
    (b) Arm strength (ARM) = 40 kg;
    (c) Back strength (BACK) = 66 kg
    (d) Leg strength (LEG) = 150 kg;
    (e) Composite strength (COM) = 130 kg;
    (f) Body-weight (BW) = 81 kg;
    (g) Abdominal depth (ABD) = 20 cm.

For lifting task 1, the employee's capacity (*CAP*) would be computed as follows:

$$CAP. = C + 0\cdot238 \times (0\cdot211 \times SHO + 0\cdot212 \times ARM + 0\cdot216 \times BACK$$
$$+ 0\cdot255 \times LEG + 0\cdot226 \times COM) + 1\cdot595 \times (0\cdot483 \times BW$$
$$+ 0\cdot561 \times AD) - 0\cdot303 \times F$$
$$= 2\cdot962 + 0\cdot238 \times (0\cdot211 \times 45 + 0\cdot212 \times 40 + 0\cdot216 \times 66$$
$$+ 0\cdot225 \times 150 + 0\cdot226 \times 130) + 1\cdot595 \times (0\cdot483 \times 81$$
$$+ 0\cdot561 \times 20) - 0\cdot303 \times 6$$

$CAP = 96\cdot05$ (lifting capacity plus body-weight)
$CAP - BW = 15$ kg (33 lb).

Table 6.9 gives lifting capacity equations for all lift ranges which has been adjusted to 5 lifts/min and an 18-in box size.

In a similar fashion, the individual's capacity for lifting task 2 is computed as 13 kg (28·6 lb). If the weight for task 1 is 50 lb and the weight for task 2 is 70 lb, the JSI for

Table 6.9. *Regresion coefficients for maximum acceptable weight of lift plus body weight (pounds) for both males and females* (Mital and Ayoub, 1980)

Regression terms	Lifting ranges†					
	F-K	F-S	F-R	K-S	K-R	S-R
Constant term	-19·944	-108·770	-850·993	-148·125	-194·777	-904·215
Sex code#	-15·630	-7·109	—	-6·688	-6·679	—
Arm strength	1·051	1·372	1·449	1·521	1·394	0·964
(Arm strength)²	-0·004	-0·007	-0·007	-0·008	-0·006	-0·004
Age	-0·661	-1·894	-2·253	-2·161	-2·011	-0·759
(Age)²	—	0·016	0·018	0·020	0·027	—
Shoulder height	—	—	10·687	—	—	-0·031
(Shoulder height)²	0·004	0·005	-0·034	—	—	-0·031
Back strength	0·448	—	0·078	—	—	0·110
(Back strength)²	-0·001	0·0003	—	0·0004	-0·0004	—
Abdominal depth	—	7·685	8·026	7·658	7·129	15·026
(Abdominal depth)²	0·146	—	—	—	—	-0·174
Dynamic endurance	10·250	14·809	18·193	14·881	18·460	18·078
(Dynamic endurance)²	-1·213	-1·682	-1·945	-1·598	-1·988	-1·915

*Adjusted to 5 lifts/min, box size of 18 in in sagittal plane. #sex code = 0 for males, 1 for females.
†See Table 6.2 for legend.

this employee if placed on this particular job is computed as follows:

$$\text{JSI} = \left(\frac{2 \times 5}{7 \times 5} + \frac{3 \times 5}{7 \times 5}\right) \left(\frac{6 \times 50}{14 \times 33} + \frac{8 \times 70}{14 \times 29}\right) = 0\cdot714 \times (0\cdot65 + 1\cdot38) = 1\cdot45$$

Given that the computed JSI of this individual does not exceed the target JSI of 1.5, the decision should be made to place this individual in the job.

6.4.1.3 JSI Validation

Two field validation studies were conducted to establish the relationship between MMH injury and JSI. Jointly, 101 jobs, 385 male and 68 female industrial workers from 28 private companies and government entities were evaluated. Lifting requirements of the job and injury data were analysed. It was concluded that JSI is a useful tool for job design (Ayoub et al., 1983c). Additional research supports that the application of the JSI method does result in a decrease in the MMH problem (Liles et al, 1983).

6.4.2 Job Design/Redesign in Conjunction with JSI

Job design/redesign procedures used in conjunction with JSI are described in detail elsewhere (Liles, 1986; Ayoub et al., 1983c). Briefly, the procedure involves the following:

1. Determine the lifting capacities of the design population and specify the percentage

of the population for which the job is to be designed. Lifting capacities can be described biomechanically, physiologically or psychophysically.
2. Describe the existing or proposed job design first identifying distinct tasks that comprise the job. Two or more tasks should be defined if the range of item weights exceeds approximately 10 kg. Second, organize the task into task groups. Task groups are all tasks performed during the same time frame. Lastly, describe job and task parameters necessary to compute JSI. The following parameters need to be included for each task:
    (a) hours spent per day (in percentage);
    (b) number of days spent per week;
    (c) required lifting range(s);
    (d) frequency of lift;
    (e) weight of load.

(JSI is designed such that the most stressful task component is isolated. Therefore, the job description need only describe the worst case).
3. Calculate the JSI indices for each lifting task and for the job.
4. Determine the acceptability of the job by comparing the calculated JSI to critical JSI of 1·5. If the calculated JSI falls between 0 and 1·5 risk is low and relatively constant. JSI values greater than 1·5 indicate high risk, problem areas. Redesign considerations should be made for these tasks.
5. If the job is determined to be unacceptable, job requirements can be adjusted until an acceptable JSI level is achieved.

6.4.3 NIOSH Guidelines for Job Design/Redesign

Another tool used for job design is the NIOSH lifting guideline concept. NIOSH established two load limits, the action limit (AL) and the maximim permissible limit (MPL). These limits were designed to identify hazardous lifting jobs provide recommendations to alleviate the hazardous elements associated with lifting jobs (see Figure 6.8).

The AL is based on the following assumptions:

1. Increased musculoskeletal injury incidence and severity rates are associated with populations that had been exposed to AL lifting conditions.
2. Tolerable compressive forces of 350 kg on the L5-S1 disc of most young, healthy workers is created by AL lifting conditions.
3. Metabolic rates would not exceed 3·5 kcal/min under AL lifting conditions.
4. Lift loads described by AL account for more than 75% of women and over 99% of men.

The MPL attempts to meet the following criteria:

1. Musculoskeletal injury and severity rates increase in populations exposed to lifting conditions above MPL.
2. Intolerable compressive forces of 650 kg are created on the L5-S1 when lifting conditions are above the MPL.

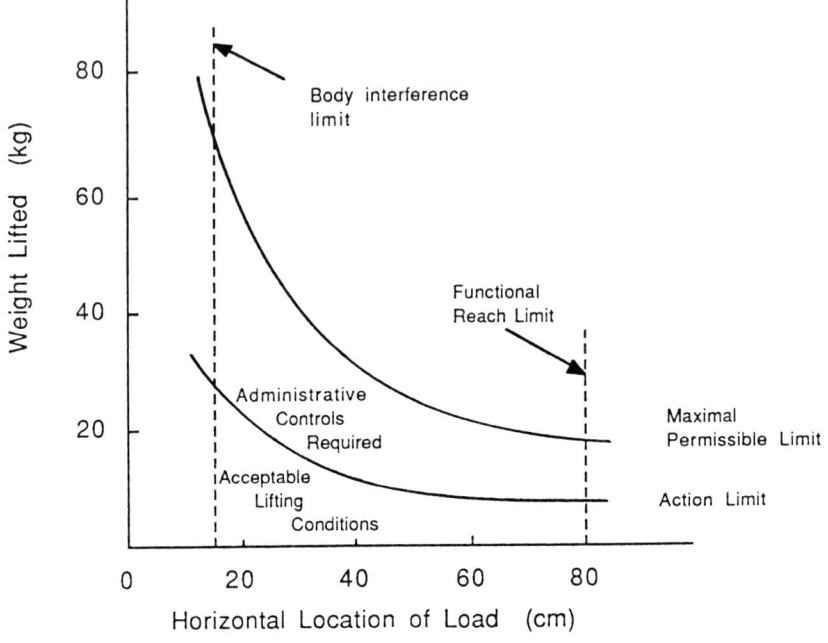

Figure 6.8. Lifting regions established by NIOSH Guidelines

3. Metabolic rates exceed 5·0 kcal/min under lifting conditions above the MPL.
4. Work above the MPL can be performed by only a minority of workers; 25% of men and less than 1% of women.

Three lifting regions are established by plotting AL and MPL (see Chapter 4 for details). The lifting region falling below AL identifies acceptable lifting conditions and, therefore, is a low injury risk region. Lifting tasks falling above MPL identifies unacceptable lifting conditions and demands engineering controls. Lastly, lifting tasks falling between AL and MPL are acceptable with administrative or engineering controls.

Using NIOSH lifting guidelines, an ideal weight is given and then adjusted by factors used to improve job design. Modifiable factors include the horizontal location of the object, vertical location of the object, vertical distance lifted and lifting frequency (see Chapter 4 for details).

6.4.4 Comparing JSI Method and NIOSH Guidelines for Job Design/Redesign

Liles *et al.* (1983) conducted an experiment comparing NIOSH lifting guidelines to the JSI procedure. The goal of this experiment was to compare each procedure's assessment of risk for agreement, and then to compare these assessments to actual injury data.

Of 101 jobs, 19 jobs were assessed as being a higher risk using the NIOSH guidelines rather than using JSI. Additionally, 10 jobs were assessed as being a lower risk using the NIOSH guidelines rather than JSI. Differences in assessment have been attributed to the different assumptions defining the two assessment procedures (Liles et al., 1983). More specifically, the JSI method ignores lowering, while NIOSH guidelines assumes no difference between lifting and lowering. The JSI method defines the required weight of lift more conservatively and differs in its method to define task geometry than NIOSH guidelines.

In comparing risk assessments to injury statistics, Liles et al. (1983) found that for number of back injuries, number of lost time per back injuries, and number of days lost per back injury, the two methods were equally effective at identifying low risk jobs. However, it was found that the JSI method was more effective at differentiating moderate and high risk jobs (Table 6.10). When medical expenses were predicted according to the risk assessment models, both models predicted low risk jobs well, however, neither differentiated well between moderate and high risk jobs (Table 6.11).

Table 6.10. *Maximum sustainable frequency of lift*

Period (h)	Average vertical location	
	$V > 75$ cm (30 in) Standing	$V \leq 75$ cm (30 in) Stooped
1	18	15
8	15	12

As job analysis and assessment tools, both the JSI method and the NIOSH guidelines have equal potential according to Liles et al. (1983). An added bonus to the JSI method is that it offers procedures for employee placement based on individual capacities. NIOSH guidelines have no such provision.

Table 6.11. *Number of jobs observed in each risk category* using JSI or ALR†*

ALR	JSI			
	L	M	H	Total
L	38	7	0	45
M	13	27	3	43
H	1	5	7	13
Total	52	39	10	101

*L, low risk category; M, moderate risk; H, high risk.
†ALR, action limit ratio; ratio of the averaged weight lift to the calculated action limit.

6.4.5 Lift Strength Ratio

Yet another guide to engineering design is the lift strength ratio (LSR). The underlying assumptions of the lift strength ratio are:

1. The LSR is based on the predicted lifting strength capability of the large/strong male (Chaffin, 1974).
2. The predicted strength graph describes symmetrical lifting of relatively compact objects in the sagittal plane (Figure 6.11).

Although these assumptions may seem quite restrictive, they are effective for the incorporation of load location when assessing strength requirements of a job.

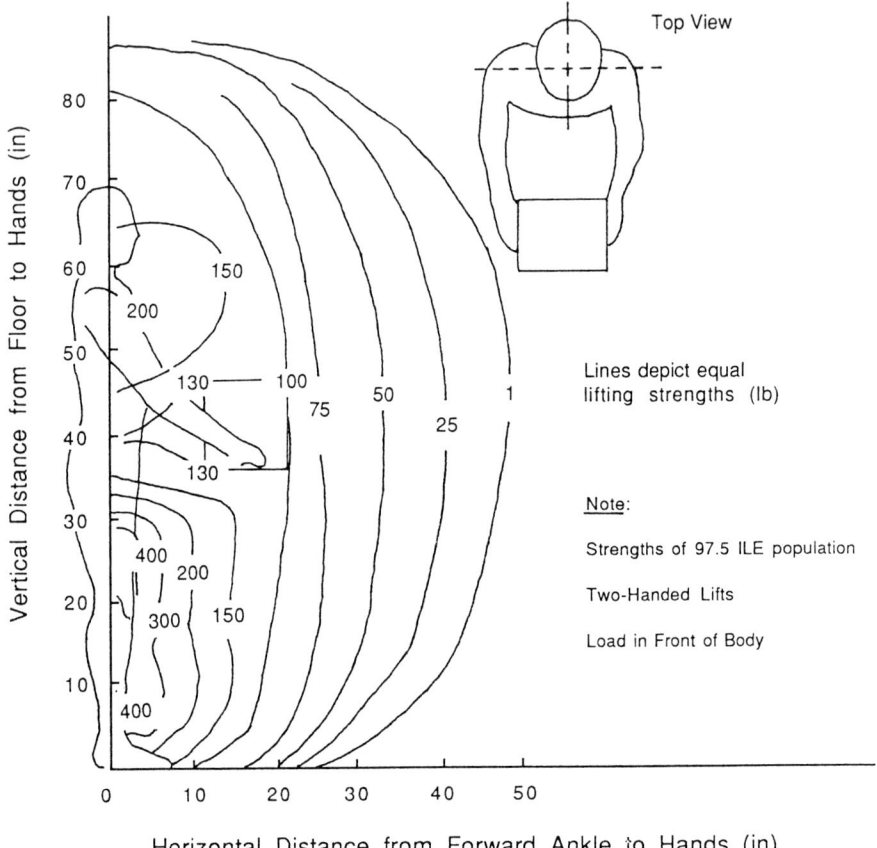

Figure 6.9. *Predicted lifting strength of large/strong male (Chaffin, 1974)*

6.4.5.1 Using the Lift Strength Ratio as a Job Design/Redesign Tool

In order to compute LSR, the following information must be collected:

(a) The origin and destination of the object to be lifted.
(b) The weight of the object (lbs).
(c) The horizontal distance (in) from the front foot to the object.

(d) The vertical distance (in) from the floor to the object.
(e) The length, width and height of the object (in).
(f) The frequency of lift.
(g) Identify extraordinary workplace conditions that may affect a worker's performance.

Note that items (c) and (d) are used in conjunction with Figure 6.11 to define the denominator of the LSR equation.

The LSR is represented as the following:

$$\text{LSR primary} = \frac{\text{load (lbs) lifted on job}}{\text{predicted strength (lbs) in same position}}$$

LSR on job is computed by observing the job and determining the weights handled and their extreme locations from the ankle of the forward foot and the floor. The smallest distance between the individual and object to be lifted is used. Predicted strength depends on worker body position and hand position which corresponds to the LSR graph (Figure 6.11). The graph represents symmetric lifting of relatively compact objects in the sagittal plane. Although seemingly restrictive, the graph effectively incorporates load location when assessing a job's strength requirements (NIOSH, 1977). The maximum value of these comparisons becomes the primary LSR.

Quite obviously there is more than one stressful task in any given job. Thus, secondary LSRs can be computed. The information required for such a computation correspond to (a), (b), (c), (d) and (f) as described above. Secondary LSRs are simply lesser values than the primary LSR. Additionally, one-handed or asymmetrical loads may also be computed. The information required for such a computation also corresponds to (a), (b), (c), (d) and (f) as described above.

An example of using LSR for job design is as follows:

Situation:
A worker must lift an object onto a shelf.
Variables of the required task:
  Weight of the object = 50 lbs;
  Hands and centre of gravity;
  Inches in front of the ankle of the leading foot = 20 in;
  Inches above the floor = 45 in.
From Figure 6.11 and the information about the hands and centre of gravity, the denominator becomes 100 lbs.

$$\text{LSR} = \frac{50 \text{ lbs}}{100 \text{ lbs}} = 0 \cdot 5$$

The LSR value of 0·5 indicates that this particular job design would require 50% of a large/strong man's strength to successfully perform the job. If this LSR value had not been acceptable, the job could then be redesigned. Using this method, an engineer may consider the load location and its strength requirements when designing work tasks.

Measuring isometric strength (see Chapter 7 for details) of the individual has become an expected, popular way to describe populations of people. Isometric strength data was used for developing guides for engineering design. However, these data have been recognized as a useful screening device for employee placement. More specifically, isometric strength data can contribute to the assessment of personal risk injury to the individual assigned to MMH activities. Chaffin (1974) found a three-fold increase in the incidence of low-back pain for those workers having inadequate isometric strength required on their job compared to their stronger cohorts.

Jobs having high lifting strength ratings were jobs having an increased incidence of low-back pain. NIOSH (1977) concluded:

1. There is a greater severity of back problems and other types of musculoskeletal problems the heavier the load lifted regardless of the frequency of lift.
2. The severity rates of musculoskeletal problems in areas other than the back and severity of contact injuries increase with an increase of lifting of maximum loads.
3. The frequency and severity of musculoskeletal problems and contact injuries increase the further the load centre of gravity is from the body. Note that load centre of gravity can be further displaced from the body due to the bulk of the object being handled and/or the workplace layout.

Based on these findings, it is recommended that a medical action level be established. That is, qualified medical personnel should determine whether the worker's capacity can withstand the pressures of the job.

Medical action levels are established for the individual with the variables of the findings warranting the need for these levels as discussed in section 6.4.5.1. As a result, MMH will be safer for the individual. Calculations of the lift strength ratio are identical to those shown in section 6.4.5.1.

6.4.6 Mital's Model for Evaluating Manual Handling Jobs

Mital's model for evaluating manual handling jobs was designed to evaluate or aid in designing a dynamic MMH job based on task parameters and task duration. Task parameters and task duration are determined using the psychophysical approach. This model does account for subtasks, lifting, pushing, pulling or carrying activities, which comprise the dynamic MMH job (Mital, 1983a).

This model is based on the principle that operators will work within a subjectively safe limit. This model is based on the assumptions that the worker works under normal workday conditions, regular breaks are taken, water drinking and rest-room breaks are allowed for, and that the working population is normally distributed (Jiang and Mital, 1986). The model structure form developed by Mital is shown in Table 6.12.

Table 6.13 illustrates the various functions. Each capacity function was developed using stepwise regression on data generated and reported from several sources and based on the four different approaches: physiological, biomechanical, psychophysiological and epidemiological. Additionally, frequency of handling, vertical range of

Table 6.12. Model structure form developed by Mital (1983a)

(a) Population capacity
    to perform MMH jobs = function (lifting capacity, lowering capacity, pushing capacity, pulling capacity, and carrying capacity)

(b) Lifting capacity
    Lowering capacity
    Pushing capacity
    Pulling capacity
    Carrying capacity   = function (operator variables, task variables, environmental variables)
                          = function (sex, age, body size, strength,... frequency, object size, height range,... temperature, humidity,....)

(c) Population capacity   = A* [Lifting capacity function] +
                          = B [Lowering capacity function] +
                          = C [Pushing capacity function] +
                          = D [Pulling capacity function] +
                          = E [Carrying capacity function] +

*A, B, C, D and E are multipliers having the value of one or zero. One indicates the performance of a task and zero indicates the lack of the performance of that task.

handling, box sizes and free-style handling were included in these data bases.

The analysis is broken down into the following steps:

1. The job is broken up into subtasks using each item listed in option (c) listed above (lifting, lowering, pulling ...).
2. Determine acceptable weight/force and work-rate for each subtask that is appropriate for the gender and task variables used (refer to Tables 6.13 and 6.14).
2(a). Determine the acceptable capacity in accordance with the table of handling frequency (refer to Table 6.13) (Jiang and Mital, 1986).
2(b). Adjust value calculated in step 2(a) for the differences between job task variables and task conditions. Tables 6.15 and 6.16 give vertical distance and box size adjustments. To account for body twist, reduce the capacity by 5%, and to account for lack of handles, reduce the acceptable capacity by 7·2%. Table 6.17 is used to adjust distance for pushing, pulling and carrying (Jiang and Mital, 1986).
2(c). Adjust the acceptable capacity according to the desired population percentile (see Tables 6.18 and 6.19).
2(d). Find the acceptable work-rate using the weight/force that was determined in step 2(c) (Jiang and Mital, 1986).
3. Modify the work-rate in step 2 for the actual work durations using equations given in Table 6.20..
4. Determine the actual work-rate for each activity.
5. Calculate the risk potential for each subtask using the following equation (Jiang and Mital, 1986):

$$R = \frac{\text{required work-rate}}{\text{predicted work-rate}}.$$

6. Use the following criteria to evaluate the risk potential (Jiang and Mital, 1986):
   If $R > 1$ for any subtask then redesign the task.
   If $R \leq 1$ for all elements, then the job is acceptable.

See Figure 6.10 for schematic representation of these steps.

6.4.6.1 A Numeric Example

This example was taken from Jiang and Mital (1986) verbatim.

Example 1

A major producer of athletic goods received unwashed, knitted material that must be dyed and processed, then folded into bundles by a machine, called a calender. These bundles are about 38·1 cm wide and weigh about 20–45 kg. A worker removes each bundle from the calender which is 135 cm from the floor to a table 63·5 cm high and ties it. The worker then carries the bundle 6·1 m (at his knuckle height of 78·9 cm) to a scale, weighs it, then tags the bundle with a computer-generated card. Next, this bundle is loaded on a buggy for temporary storage (2·6 m away from the scale), from

Table 6.13. Acceptable capacity models and designated conditions for 50th percentile

Activity†	SEX	Capacity*	Conditions†	
LFK	Male	$27 \cdot 94 \exp(-0 \cdot 052)(F)$	BOX = 49 cm	V = 76 cm
	Female	$15 \cdot 96 \exp(-0 \cdot 040)(F)$	BOX = 49 cm	V = 76 cm
LKS	Male	$27 \cdot 51 \exp(-0 \cdot 043)(F)$	BOX = 49 cm	V = 51 cm
	Female	$14 \cdot 36 \exp(-0 \cdot 027)(F)$	BOX = 49 cm	V = 51 cm
LSR	Male	$22 \cdot 17 \exp(-0 \cdot 047)(F)$	BOX = 49 cm	V = 51 cm
	Female	$13 \cdot 06 \exp(-0 \cdot 054)(F)$	BOX = 49 cm	V = 51 cm
LOWKF	Male	$32 \cdot 42 \exp(-0 \cdot 060)(F)$	BOX = 49 cm	VL = 51 cm
	Female	$19 \cdot 64 \exp(-0.054)(F)$	BOX = 49 cm	VL = 51 cm
LOWSK	Male	$24 \cdot 25 \exp(-0 \cdot 027)(F)$	BOX = 49 cm	VL = 51 cm
	Female	$14 \cdot 71 \exp(-0 \cdot 005)(F)$	BOX = 49 cm	VL = 51 cm
LOWRS	Male	$20 \cdot 28 \exp(-0 \cdot 039)(F)$	BOX = 49 cm	VL = 51 cm
	Female	$13 \cdot 05 \exp(-0 \cdot 014)(F)$	BOX = 49 cm	VL = 51 cm
Carry	Male	$39 \cdot 86 \exp(-0 \cdot 056)(F)$	HC = 2 m	
	Female	$22 \cdot 86 \exp(-0 \cdot 036)(F)$	HC = 2 m	
Push	Male	$29 \cdot 32 \exp(-0 \cdot 055)(F)$	HP = 2 · 1 m	
	Female	$21 \cdot 08 \exp(-0 \cdot 067)(F)$	HP = 2 · 1 m	
Pull	Male	$28 \cdot 65 \exp(-0 \cdot 056)(F)$	VH = 0 · 6 m	
	Female	$21 \cdot 19 \exp(-0 \cdot 032)(F)$	VH = 0 · 6 m	

*F = Frequency of handling (No. of handlings/min) for push/pull: $0 \cdot 1 \leq F \leq 10$ for males, $0 \cdot 1 \leq F \leq 5$ for females; for lift/lower: $1 \leq F \leq 12$; for carry: $0 \cdot 1 \leq F \leq 10$.
†, See Table 6.14 for abbreviations.

Table 6.14. *Input variables and their units*

Activity	Input variables	Units
For all activities	Sex (S)	Male; Female
	Box length (L)	Centimetres
	Frequency (F)	No. of handlings/min
	Body twist (BT)	1, if BT exists 0, if no BT
	Handles (HAN)	1, if handles present, 0, if no handles
Lifting	Height level (H)	Floor to knuckle = 1 (LFK)
		Knuckle to shoulder = 2 (LKS)
		Shoulder to reach = 3 (LSR)
	Vertical distance (V)	Centimetres
Lowering	Height level (HL)	Knuckle to floor = 1 (LOWKF)
		Shoulder to knuckle = 2 (LOWSK)
		Reach to shoulder = 3 (LOWRS)
	Vertical distance (VL)	Centimetres
Carrying	Horizontal distance (HC)	Metres
	Frequency (FC)	No. of carries/min
Pushing	Frequency (FP)	No. of pushes/min
	Horizontal distance (HP)	Metres
Pulling	Frequency (FP)	No. of pulls/min
	Vertical height (VH)	Metres

Table 6.15. Vertical distance adjustment for lifting and lowering activities (% of adjustment)

Activity†	25–51 cm		51–76 cm	
	Male	Female	Male	Female
LFK	+0·38	+0·74	+0·14	0
LKS	+0·73	+0·88	−0·49	−0·30
LSR	+0·58	+0·72	−0·34	−0·33
LOWKF	+0·65	+0·53	−0·08	−0·19
LOWSK	+0·75	+0·75	−0·28	−0·34
LOWRS	+0·67	+0·74	−0·29	−0·39

†, 25 cm ≤ vertical distance ≤ 76 cm.
See Table 6.14 for abbreviations.

Table 6.16. Box size adjustment for lifting and lowering activities (% of adjustment)

Activity†	36–49 cm		49–76 cm	
	Male	Female	Male	Female
LFK	+0·70	0·01	−0·58	−0·71
LKS	+0·58	+0·54	−0·01	−0·01
LSR	+0·22	+0·60	−0·02	−0·01
LOWKF	+0·79	+0·93	−0·52	−0·52
LOWSK	+0·60	+0·41	−0·01	−0·01
LOWRS	+0·53	+0·44	−0·01	−0·01

*36 cm ≤ box size ≤ 76 cm.
†, See Table 6.14 for abbreviations.

Table 6.17. Distance adjustment for carry, push, and pull activities (percentage)

Activity	Sex	Percentage*
Carry	Male	$100·27 \exp(-0·018)(D)$
	Female	$99·70 \exp(-0·012)(D)$
Push	Male	$88·66 \exp(-0·011)(D)$
	Female	$90·91 \exp(-0·010)(D)$
Pull	Male	$131·49 \exp(-0·004)(VH)$
	Female	same as male)

*D Handling distance in m: VH, Vertical height in cm for carry: $2 \text{ m} \leq D \leq 8·5 \text{ m}$; for push: $2·1 \text{ m} \leq D \leq 60 \text{ m}$; for pull: $60 \text{ cm} \leq VH \leq 150 \text{ cm}$.

whence it is trucked to another plant for sewing. The worker is on the job 7·5 hours each day. The plant processes 470 bundles for each calender per day (1·044 bundles/min). The workplace layout (see Figure 6.11) causes a body twist to occur in each element of the job. The plant manager wishes this job designed to accommodate 75% of the male population.

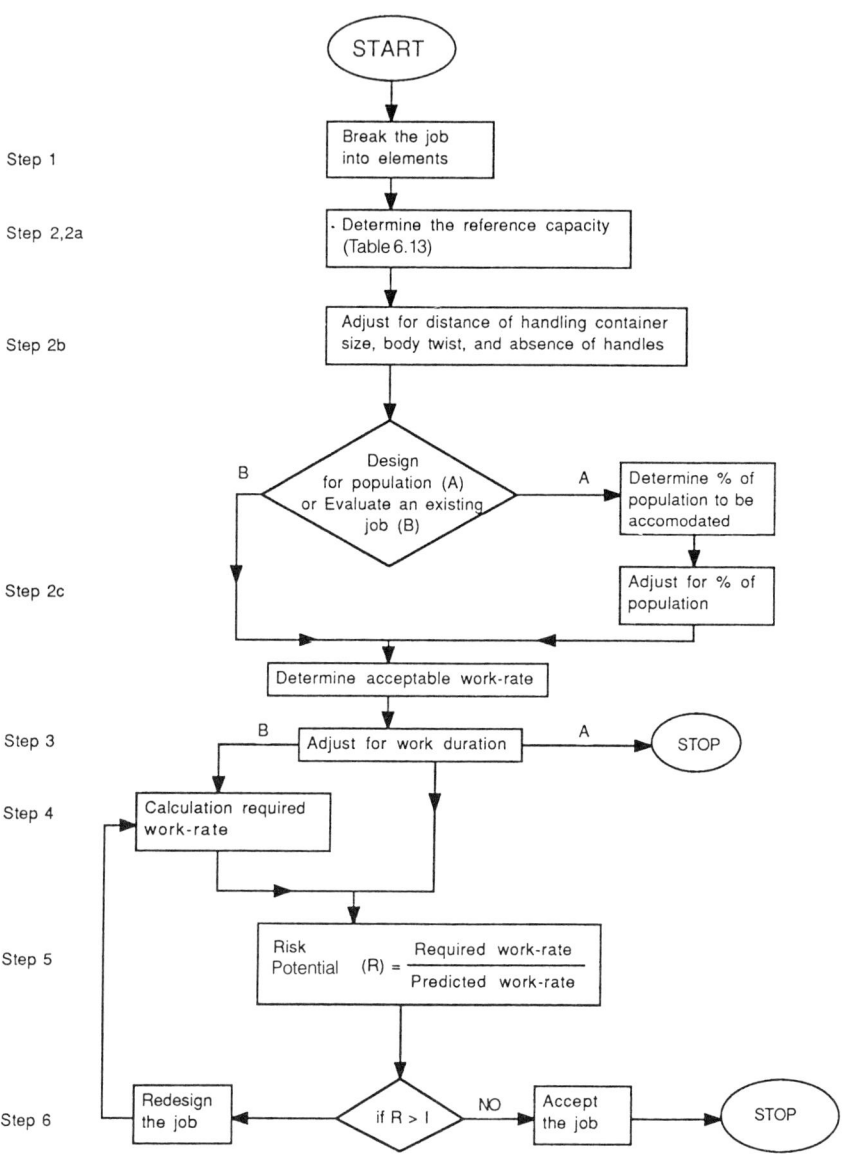

Figure 6.10. Flow diagram for MMH job design/redesign

Table 6.18. Standard deviation models for MMH capacity (kg)

Activity*	Male†	Female
LFK	6·52−0·30 F	3·48−0·12 F
LKS	6·75−0·26 F	2·52−0·07 F
LSR	5·56−0·18 F	1·58−0·02 F
LOWKF	9·60−0·43 F	3·22−0·08 F
LOWSK	7·22−0·16 F	3·12
LOWRS	6·04−0·19 F	1·88
Carry	12·71−0·56 F	4·00−0·04F
Push	12·57−0·56 F	6·63−0·84 F
Pull	8·27−0·36 F	5·72−0·06 F

*See Table 6.14 for abbreviations.
†F = frequency of handling (no. of handlings/min).

Table 6.19. Z-score for various population percentages

Population percentage	Z-score
95	−1·6449
90	−1·2816
85	−1·0364
75	−0·6745
50	0·0
25	0·6745
15	1·0364
10	1·2816
5	1·6449

Table 6.20. Equations to adjust the work-rate for actual work duration

Gender	Work-rate (%)
Male	= 101·42 − 3·4 DR
Female	= 100·81 − 1·94 DR

Note: DR = actual work duration in hours (0.42 < 0·42 DR ≤ 12).

Table 6.21 lists the input to the model structure, acceptable work-rates for the task duration, and actual work-rates for all elements. To determine the duration of each element, the time required for each element for each cycle was multiplied by the number of bundles (470) produced each day. The work duration is used to adjust the acceptable work-rate in Table 6.21.

The analysis in Table 6.21 indicates that the lowering element is far above the acceptable limit. Either the two operators should lower bundles from the calender or the table height should be raised. The first carry element was also above the acceptable limit. In the redesign, the scale table was moved closer to the calender.

Table 6.21. *Job analysis for Example 1*

Element	Input data†	Acceptable weight/force (kg)	Acceptable work-rate (kg m/min)	Actual work-rate (kg m/min)	Risk potential
Lower a bundle from calender to table	HL = 3 VL = 71·5 BT = 1 DR = 98·47 min	13·16	9·41	15.27	1·62
Tie the bundle	DR = 87·3 min				
Lift the bundle	H = 1 V = 15·4 cm BT = 1 DR = 27·51 min	25·31	4·06	3·29	0·81
Carry the bundle to scale	HC = 6.1 m BT = 1 DR = 145·75 min	21·6	128.15	130.23	1·02
Carry the bundle to buggy	HC = 2.6 m BT = 1 DR = 90·97 min	23·54	61·51	55·51	0·90

† Common variables for all elements: sex, male; bundle weight, 20·45 kg; bundle width, 38·1 cm; handling frequency, 1·044 bundles/min; designed population percentile, 75%; no handles. See Table 6.14 for abbreviations. DR is calculated duration of element.

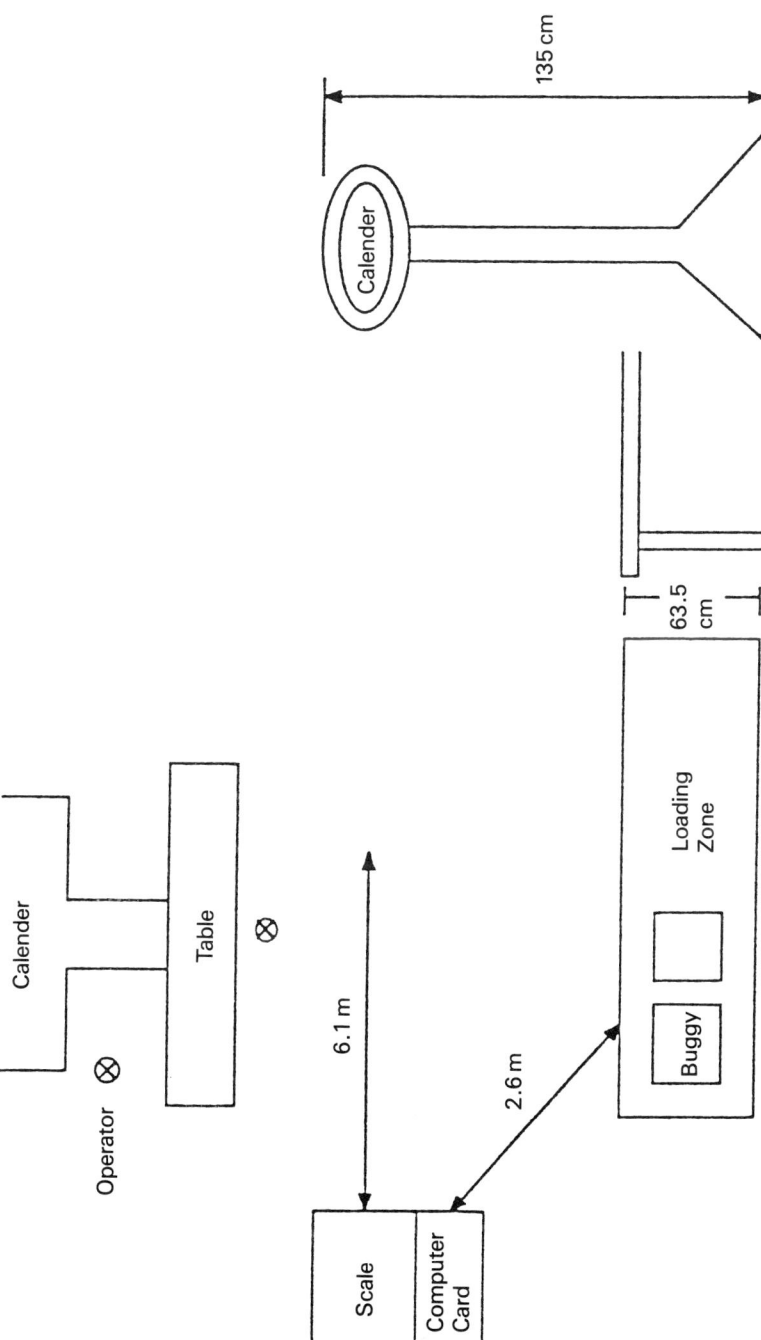

Figure 6.11. Schematic layout: Example 1

6.4.7 Lifting Optimization Model for Job Design/Redesign

The lifting optimization model attempts to optimize the individual's performance while remaining consistent with task constraints. The lifting optimization model is based on the assumption that the human body can be represented as five rigid links. These five rigid links, described in terms of angular positions, can represent any body movement or configuration. The five-link model is delineated by Figure 6.12. Notice that the links are located at the ankle, knee, hip, shoulder and elbow. As presented by the five-link model, principles of mechanics can be applied to the motions of the human body (Muth et al., 1978).

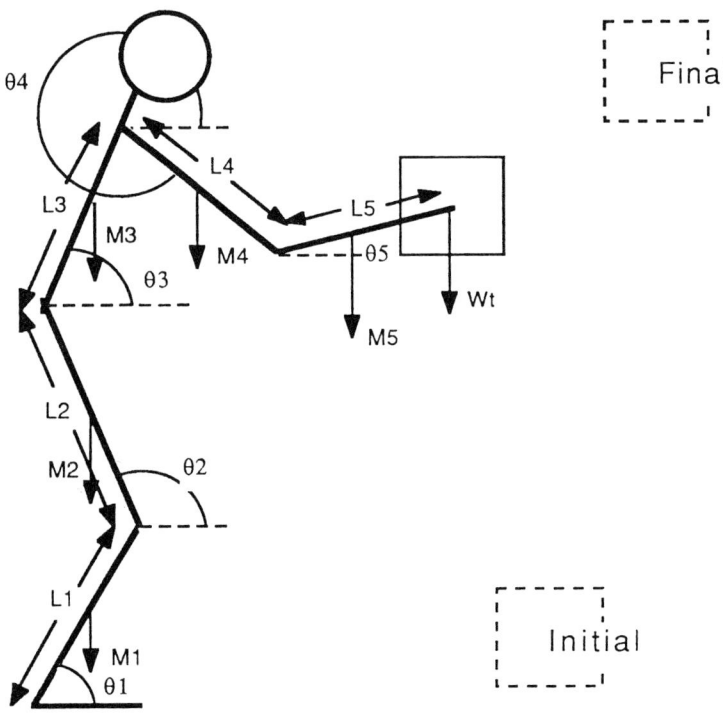

Figure 6.12 Five-link model of worker lifting

The following are underlying assumptions of the class of motions using this model:

(a) The worker does not walk with the object.
(b) The worker remains stationary, lifting the load between two set points in space.
(c) The object remains in arm reach.
(d) Movement is symmetrical.
(e) The ankle is in a fixed position.
(f) All motions occur in the sagittal plane.

(g) Four different types of lift exist; foot to shoulder, waist to shoulder, foot to waist, and knuckle to shoulder (Muth *et al.*, 1978).
(h) There are two methods of lift; straight knees/bent back, and bent knees/straight back.

As previously stated, this model attempts to optimize the movement of the worker while maintaining them within the constraints of the task by using the principles of mechanics. It does so through mathematical models which have been validated. This model is effective although it is limited by the physical characteristics of the workplace and the worker. Muth *et al.* (1978) suggest an expansion on their model to incorporate the effects of heat, illuminations and confined spaces.

Chapter 7
Pre-Employment Strength Testing

7.1 Why Measure Human Strengths?

Strength of an individual is his or her capability to exert force, generated by muscular contraction, on an external object. It is the maximum force that a muscle, or group of muscles, can develop when contracted voluntarily (maximum voluntary exertion or contraction — MVE or MVC). The exertion, which is the result of voluntary muscular contraction, may or may not result in the movement of body-segment(s)/object. Even though MVC is regarded as the maximum force that can be generated by an individual voluntarily, it is well below the failure limit of the muscle–tendon–bone–joint system (failure being defined as the failure of one or more of the system elements, for example, if the muscle ligament is torn or the joint is dislocated, the system fails); the failure limit strength maybe one-third greater than the MVC (Hettinger, 1961). Since the MVC represents the upper limit of force that can be exerted by an individual safely in a given situation, it is an indication of that person's capability to perform physical work.

Knowledge of what a person can or cannot do, under specified circumstances, is necessary for designing work, equipment, workplaces and tools. In the context of MMH activities, human strengths are used to ensure that job demands (muscular exertions required for satisfactory completion of the job) do not exceed the MVC of the person under similar conditions. If the strength requirements exceed MVC, risk for personal injury increases substantially (Chaffin *et al.*, 1977a, 1978; Keyserling *et al.*, 1980a, b). Strength measurement is, therefore, necessary to prevent incidences and severity of musculoskeletal injuries, especially those of the lower back and upper and lower extremities. Many experts agree that strength testing should be an integral part of pre-employment screening if such injuries are to be avoided (Chaffin *et al.*, 1977a, 1978; Keyserling *et al.*, 1980a, b; Kamon *et al.*, 1982; Kroemer, 1983, 1985).

7.2 Classification and Definition of Strengths

Human strengths can be broadly classified into two categories: (1) static and (2) dynamic. The static strength is also known as isometric muscle strength. Under the dynamic strength category, several different kinds of strengths are grouped: (i) isotonic strength, (ii) isokinetic strength, and (iii) isoinertial strength.

Static, or isometric, strength is the capacity of muscles to produce force or torque by a single maximal voluntary isometric exertion (Chaffin, 1975b; Roebuck *et al.*, 1975).

The internal muscular effort, resulting from muscular contraction, is amplified by the mechanical advantages of the body members involved. The net effect measured is the external static effect either in the form of force or torque × time. The length of the muscles involved remains constant and neither the body segment attached nor the object held moves.

In case of dynamic muscular exertions, both the body segments, to which contracting muscles are attached, and the objects held move. The muscle length and mechanical advantages also change continuously. With the exception of isotonic strength, the muscle tension also changes.

The various types of dynamic strengths are defined below:

Isotonic strength — is the measure of a person's maximum voluntary muscle contraction in which the tension developed in the muscle remains constant throughout the range of motion. In reality, however, the tension changes due to constantly changing muscle length and mechanical advantage. The isotonic effort and the isometric effort are frequently found in combination. This occurs, for example, when an object is held motionless.

Isokinetic strength — is the measure of a person's maximum voluntary muscle contraction when the body segments involved move at a constant speed. The speed control is achieved by either a mechanical or a hydraulic device. This device does not allow the body segment to move faster than the pre-selected speed.

Isoinertial strength — measures the ability of a person to overcome the initial static resistance by measuring the maximum amount of weight he or she can handle and

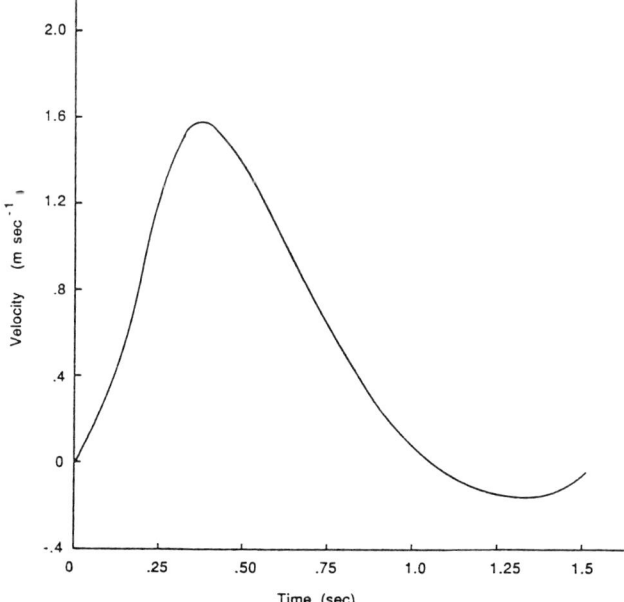

Figure 7.1. Typical velocity profile of load movement during lifting

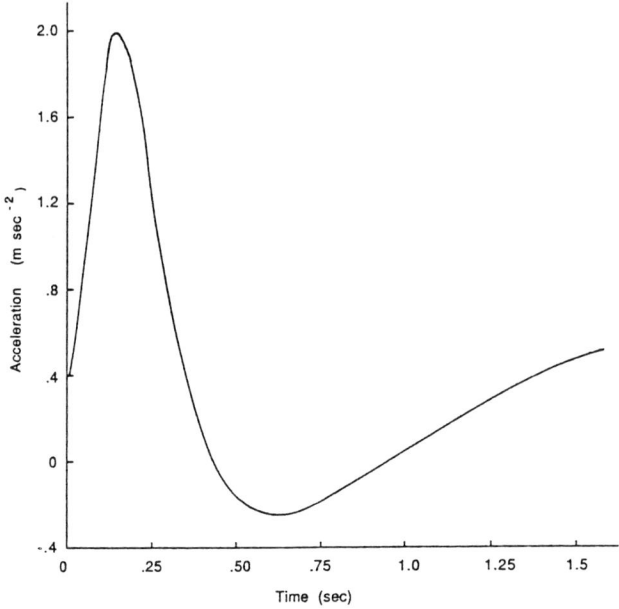

Figure 7.2. Typical acceleration profile of load movement during manual lifting

move to an assigned point at a freely chosen speed. The actual speed of movement does vary within the specified range of movement. Figures 7.1 and 7.2 show the typical velocity and acceleration profiles of the weight movement when lifting from the floor to approximately knuckle height.

The determination of the maximum weight is accomplished with the help of a lifting machine which permits incremental weights to be lifted and lowered along guides. Weight increments may be as small as 1·135 kg (Jiang, 1984) or as high as 11·4 kg (Kroemer, 1983, 1985). The method is a variation of the psychophysical methodology. The essential difference between the two (isoinertial strength measurement and the psychophysical capacity determination) is the manner in which the weights are added or subtracted. The other differences, such as the use of guides, are minor.

7.3 Measurement of Strengths

The measurement devices that are in use at the present time are either mechanical or hydraulic and generally convert the muscular exertion into an electrical signal proportional to the strain produced by the exertion. The electrical signal is conditioned, amplified and displayed (in designated units) on either a digital or analogue display. Figure 7.3 shows one such system which utilizes an electonic load cell to measure the exerted forces (push or pull type). The force measured is displayed on a digital display system.

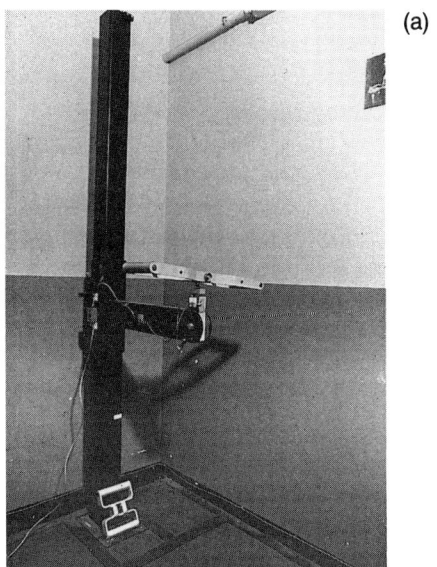

Figure 7.3(a) Isometric strength measuring device with load cell; (b) digital display unit for the isometric strength measuring device

7.3.1 Assessment of Isometric (Static) Strengths

The procedure to assess isometric strengths has been summarized by experts in considerable detail (Caldwell *et al.*, 1974; Chaffin, 1975b). According to these experts the following factors are important:

1. Exertion duration,
2. Strength measuring device,
3. Rest periods between repeated exertions,
4. Body position or posture, and
5. Reporting of test conditions, subject biographic data and strength data.

According to Kroemer (1970), the total duration of muscular exertion, for strength assessment purposes, should be less than 10 s if fatigue is to be avoided. In general, a 4–6 s time period is adequate for a person to reach peak exertion (isometric strength). Individuals can also maintain the peak exertion for 4–6 s period. This also permits recording of a 3-s mean value.

The characteristics of the strength measuring device have been summarized by Chaffin and Andersson (1984) from earlier recommendations (Chaffin, 1975b). The device should be capable of recording peak and 3-s time averaged exertions, should not generate discomforting localized pressure, should accommodate the population, and should permit easy adjustments for recording different types of exertions.

A break of at least 30 s should be provided between successive exertions if only a few measurements are to be made. It is necessary to increase the rest duration to 2 min if

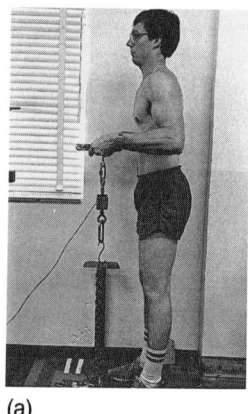 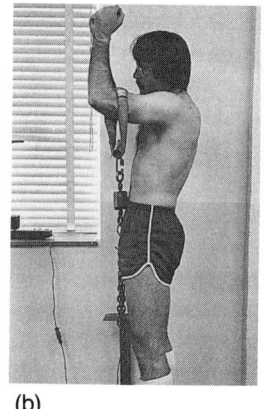 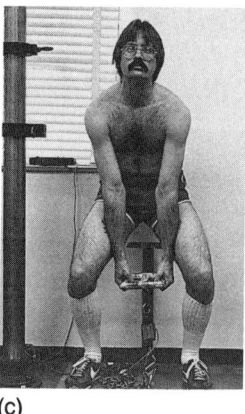

(a)   (b)   (c)

Figure 7.4. Body posture for isometric measurement of (a) arm strength, (b) shoulder strength, (c) composite strength

about 15 measurements are to be made in one test session. This additional rest is necessary to permit recovery from fatigue generated due to the isometric exertion. Longer rest breaks should be allowed if individuals being tested so desire.

The posture assumed by the individuals during isometric exertion influences strength (Chapter 2). It is necessary, therefore, to specify and control the body position if the corresponding strength values of the individuals are to be compared. For pre-employment screening for manual lifting activities, four different body positions are widely used. These positions are shown in Figure 7.4 and 7.5(a).

The isometric arm strength measurement requires that the long handle (Figure 7.3) be adjusted such that the forearms are flexed 90°, i.e., perpendicular to the individual's torso, and the upper arms are vertical, i.e., parallel and adjacent to the torso. The subject stands erect, with legs and back straight with the feet flat (Figure 7.4a). The handle is connected to an electronic load cell which sends the electrical signal to the digital display. The exerted force is upward, vertical and generated by only the arm muscles. The individual is instructed to avoid any shoulder movement.

The isometric shoulder strength measurement requires that the long handle be adjusted such that the forearms are vertical, i.e., parallel and adjacent to the torso and the upper arms are horizontal, i.e., perpendicular to the torso. Sometimes, two straps, joined by a bar, may be used by positioning the straps over the distal end of the humerus by inserting the arms outside-in (Figure 7.4b). The position of the feet, back, etc., remains the same as in the case of isometric arm strength measurements.

The isometric composite strength measurement, or leg lifting strength measurement, requires the short handle to be adjusted to approximately 38 cm height (the height is measured from the platform to the lower horizontal plane of the handle). The individual takes a semi-squat position, such that the handle is between the legs. The individual's elbows must be extended. Contact between the upper and lower extremities is not permitted. The heads of the first metatarsals should be placed opposite one another and intersect the vertical plane of the load cell. The feet remain flat on

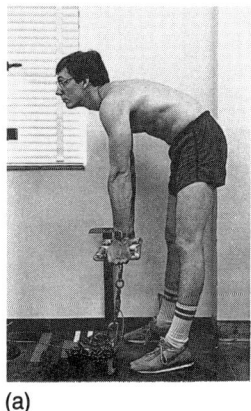
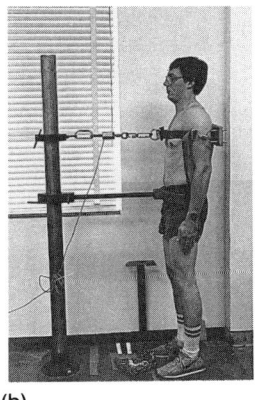

Figure 7.5(a) Body posture for isometric back strength measurement; (b) body posture for isometric back-extension strength measurement

the platform and an upward, vertical force is exerted by simultaneously extending the knees and the torso.

The isometric back strength measurement, or torso lifting strength measurement, requires that the long handle be located at 75% of the knee height (tibial height) and approximately 38 cm in front of the medial malleolus (the height measurement is the same as in the case of isometric composite strength). The individual's feet are separated at shoulder width. Both feet are kept at an equal distance from the chain anchor point (Figure 7.5a). The individual flexes the torso in order to grip the handle. The upward, vertical force is exerted by torso extension. The elbows and knees are fully extended and the eyes look straight ahead.

The isometric back extension strength measurement, or European method of back strength measurement, requires that the individual stand erect and rest the anterior surfaces of the pelvis and the abdominal muscles against a padded brace. The brace is adjusted to a height such that the subject can comfortably apply a horizontal pelvic/abdominal pressure. A padded board is placed on the individual's back at the level of the posterior surface of the crest of the spine on the scapulae. The load cell lies in the horizontal plane between a vertical column and the strap connected to the padded board (Figure 7.5b). The load cell is kept perpendicular to the torso. The measurement requires the individual to exert a horizontal, rearward force against the padded board, by extending the torso. The knees are fully extended, the upper extremities are kept parallel to the lateral surfaces of the torso and the feet are kept flat.

The procedure, in each case, requires the individual to build up to the maximum exertion, over a 3–4-s period, without jerking, and holding the exertion at the maximum for about 3 s.

Prior to strength measurements, individuals should be instructed as to the intent of the measurements, procedure and the risks involved. Coercion of the individual should be avoided and the individual should be told how the test data will be used.

Competition among individuals and environmental distractions (e.g., noise, presence of other individuals besides the experimenter, unusual temperature) should be avoided. The individual should be permitted to discontinue if and when he or she so desires. The anthropometric and biological characteristics, such as sex, age, height and weight, of the sample population must be recorded in order to describe the population. The mean, standard deviation and range of the strength data should be provided.

The procedure described here has been utilized by several researchers to measure various isometric strengths for pre-employment screening purposes (Chaffin et al., 1977a, 1978; Ayoub et al., 1978; Kamon and Goldfuss, 1978; Mital, 1984a). The data are tabulated in Tables 7.1 to 7.3.

Table 7.1. Isometric arm, composite (leg), and back (torso) lifting strengths (N) of male and female workers (Chaffin et al., 1977a)

Strength	Males ($n = 443$)*		Females ($n = 108$)	
	$\bar{x}$†	S‡	$\bar{x}$	S
Arm lifting	382	127	200	78
Composite lifting	942	340	416	198
Back lifting	545	243	266	138

*n = sample size; †$\bar{x}$ = mean; ‡S = standard deviation.

Table 7.2. Isometric arm, shoulder, composite, and back lifting strengths (N) of male and female workers (Ayoub et al., 1978)

Strength	Males ($n = 73$)*		Females ($n = 73$)	
	$\bar{x}$†	S‡	$\bar{x}$	S
Arm	362	95	230	59
Shoulder	490	125	270	66
Composite	1108	266	633	178
Back	827	234	527	138
Back extension	706	155	484	135

*n = sample size; †$\bar{x}$ = mean; ‡S = standard deviation.

Table 7.3. Isometric arm, shoulder, composite, and back lifting strengths (N) of male and female workers (Mital, 1984a)

Strength	Male ($n = 37$)*			Female ($n = 37$)		
	$\bar{x}$†	S‡	Range	$\bar{x}$	S	Range
Arm	341	103	137–598	187	56	89–343
Shoulder	436	142	226–991	216	67	108–373
Composite	975	274	539–1658	534	161	274–1001
Back	542	193	206–1138	353	102	196–716

*n = sample size; †$\bar{x}$ = mean; ‡S = standard deviation.

Kamon and Goldfuss (1978), in addition to the back extension and arm flexion strengths, also measured grip strength of 461 men and 138 women. The positioning of arms was slightly different from that described earlier for the measurement of arm lifting strength (forearms were kept vertical and upper arms were kept horizontal), but the force was generated by flexing the elbow. Table 7.4 shows the strength of the workers by age group. The study concluded that the younger workers were much stronger than older workers. Also, the strengths of workers assigned to very heavy, heavy, moderate, or light jobs, were not significantly different.

See Chapter 4 and 5 for other studies that have measured isometric strengths for other MMH activities such as pushing and pulling.

Table 7.4. Weight, height, and isometric strengths (N) of workers according to age (Kamon and Goldfuss, 1978)

	Weight (kg)			Height (cm)			Back extension			Elbow flexion			Grip		
	n^*	$\bar{x}$†	S‡	n	$\bar{x}$	S	n	$\bar{x}$	S	n	$\bar{x}$	S	n	$\bar{x}$	S
Men															
Above 31 years	300	84·3	13·9	296	175·8	6·8	245	562	174	302	261	82	307	431	103
Up to 31 years	149	77·3	13·1	149	177·6	7·0	130	658	169	152	311	90	152	487	97
Women															
Above 31 years	46	65·2	11·8	45	162·8	5·5	42	333	113	44	152	51	46	265	69
Up to 31 years	87	61·9	10·9	86	163·2	6·5	82	391	107	86	163	52	86	271	63

$^*n$ = sample size; †$\bar{x}$ = mean; ‡S = standard deviation.

7.3.2 Assessment of Isotonic Strengths

According to the definition (section 7.2), isotonic strength is the maximum voluntary muscle contraction in which the tension developed in the muscle remains constant. This means isotonic strength is the maximum resistance that can be sustained by voluntary contraction of muscles without change in the muscle tension when moving the maximum possible weight at a constant velocity. This, however, does not happen in reality because the tension changes due to changes in muscle length and mechanical advantage. Only if the body segment is kept motionless can constant tension be achieved. Such effort, however, includes isometric force exertion in combination with isotonic strength.

The measurement of static and dynamic endurance may be regarded as variations of isotonic strength measurement. The measurement of static endurance requires the individual to sit on a flat-top stool with scapulae and buttocks against a vertical wall. The forearms are flexed 90°, i.e., perpendicular to the torso. The individual holds, with both arms, a total weight equal to 25% of his/her average isometric arm strength.

The individual is required to hold the weight in that position as long as possible. Throughout the test, the upper arms are kept vertical and adjacent to, but not contacting the torso. The forearms are kept horizontal, i.e., perpendicular to the torso (Figure 7.6a). The individual assumes the posture prior to the measurement. The weight is then handed over and the duration for which the weight is held is recorded as the static endurance time.

The measurement of dynamic endurance requires that the individual stand erect but otherwise assume a posture similar to that required for the measurement of the static

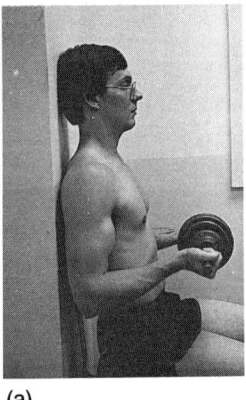  
(a) (b) (c)

Figure 7.6. Body posture for measuring (a) static endurance, (b) dynamic endurance (starting position of the movement), (c) dynamic endurance (end position of the movement)

endurance. Prior to the measurement, the individual is given the weight and is required to move it from a horizontal plane (Figure 7.6b) to the chest (Figure 7.6c) and vice versa, by 90° flexion of the elbow, 50 times/min for as long as possible. The pace is established by a metronome. The task duration is recorded as the dynamic endurance.

The static and dynamic endurance of industrial workers was determined by Ayoub et al. (1978). Table 7.5 gives the average and standard deviation values for males and females.

Table 7.5. Static and dynamic endurance* (min) of industrial workers (Ayoub et al., 1978)

Endurance	Males ($n = 73$)†		Females ($n = 73$)	
	Average	S‡	Average	S
Static	3·82	1·85	3·89	2·27
Dynamic	2·55	1·42	2·63	1·62

*When handling a weight equivalent to 25% of the average isometric arm strength.
†n = sample size; ‡S = standard deviation.

7.3.3 Assessment of Isokinetic Strengths

The prime requirement of isokinetic strength measurement is that the body segment velocity should remain constant during the maximum voluntary contraction. Using this requirement as the primary basis, several researchers have measured isokinetic strengths of industrial and non-industrial workers (Pytel and Kamon, 1981; Kamon et al., 1982; Aghazadeh and Ayoub, 1985; Mital, 1985e; Mital and Karwowski, 1985; Mital et al., 1985a, 1986c). The initial efforts were made by Pytel and Kamon (1981) and Kamon et al. (1982). They measured isokinetic strength by using a modified Mini Gym (Figure 7.7). The modified unit consisted of a load cell and a speed sensor to

Figure 7.7. Isokinetic strength dynamometer (Mini-Gym Model Super 2)

measure the force output and the speed of the motion, respectively. These data were recorded on a pen recorder. The peak force (the highest recorded value) was representative of the isokinetic dynamic strength. The speed of motion of 0·73 m/s was found more useful in predicting the maximum weights people were willing to lift infrequently (Pytel and Kamon, 1981) and, therefore, was used in subsequent analysis and presentation (Pytel and Kamon, 1981; Kamon *et al.*, 1982).

The isokinetic strength was exerted by pulling on the rope wrapped around a shaft. The resistance was present to maintain constant shaft rotation during pulling. The mechanical clutch ensured that the prescribed speed was not exceeded.

In the initial study (Pytel and Kamon, 1981), 10 males and 10 females applied maximum voluntary exertions on the rope. Three different isokinetic strengths were recorded: dynamic lift strength (DLS — pulling the handle from the floor to chest height), dynamic back extension strength (DBES — pulling the handle from the floor to approximately knuckle height), and dynamic elbow flexion strength (DEFS — pulling from the knuckle height to chest height). In all cases, the rope movement was in the sagittal plane. In the subsequent study (Kamon *et al.*, 1982), isokinetic lift strength (DLS) and isokinetic elbow flexion strength (DEFS) were measured on 48 male steelmill workers. The procedure and equipment employed were identical. Table 7.6 gives the isokinetic strengths measured in the two studies.

Table 7.6. Isokinetic strengths (N) of males and females (Pytel and Kamon, 1981; Kamon et al., 1982)

Study	Strength DLS		DBES		DEFS		Age (years)		Height (cm)		Weight (kg)	
	$\bar{x}$*	S†	$\bar{x}$	S	$\bar{x}$	S	$\bar{x}$	S	$\bar{x}$	S	$\bar{x}$	S
Pytel and Kamon (1981)												
Men (n = 10)‡	601	129	540	101	323	55	26	3	177	5	75·2	7·6
Women (n = 10)	379	95	315	87	167	33	30	10	165	6	59·7	11·5
Kamon et al. (1982)												
Men (n = 48)	601	100	—	—	324	46	39	12	173	7	81	14

*$\bar{x}$ = mean; †S = standard deviation; ‡n = sample size.

The equipment used by Pytel and Kamon (1981) and Kamon et al. (1982) was modified by Mital and Vinayagamoorthy (1984) to permit measurement of isokinetic strengths at almost any point within the working space. The device is shown in Figure 7.8a.

The basic equipment cell consists of a Super-2 Mini Gym, which allows easy speed control. As such, the basic cell lacks the degrees of freedom to make it versatile. The degrees of freedom of this basic equipment cell were increased to make it useful for measuring isokinetic muscle strengths in space. The major components of the device are also shown in Figure 7.8a. Figures 7.8b, c and d show the device in detail. The whole set-up consists of: (1) Super-2 Mini Gym (D); (2) vertical column (A); (3) horizontal rails on the base plate (F); (4) horizontal support column (C); (5) pivot and slider assembly; (6) support structure connected to column (A) and the platform; and (7) platform (wooden top with steel base).

The design shown in Figure 7.8 allows the Mini Gym to be oriented to any position without any major adjustments. Such orientation enables isokinetic strength measurements at any height, angle or position. The horizontal support column (C) can be rotated about the pivot and slider assembly (B). Five different angular positions, from the horizontal to the vertical, are possible (Figures 7.8b and c).

In each position, the support column can be tightened. The vertical movement of the Mini Gym is accomplished with the help of a balance weight inside of column (A). A locking secures the position of pivot and slider assembly (B) with respect to the vertical column (A). The vertical column (A) can also be moved horizontally along the guide rails on the base plate (F). This permits the entire unit to be fixed in any horizontal position. These adjustments make possible (1) horizontal movement of the entire unit, (2) vertical movement of the Mini Gym along the vertical column (Figures 7.8a and b), and (3) angular positioning about the horizontal plane (Figure 7.8d).

Thus, the device permits measurement of isokinetic strength in vertical (Figures 7.8a and b), horizontal (Figure 7.8c) and transverse (Figure 7.8d) planes. The pulley and rope support permit a still wider choice of positions. Suitable handles attached to the end of the rope (Figures 7.8a and b) can be used to simulate lifting or other material handling tasks such as short distance carrying or pulling in the horizontal, vertical or transverse planes.

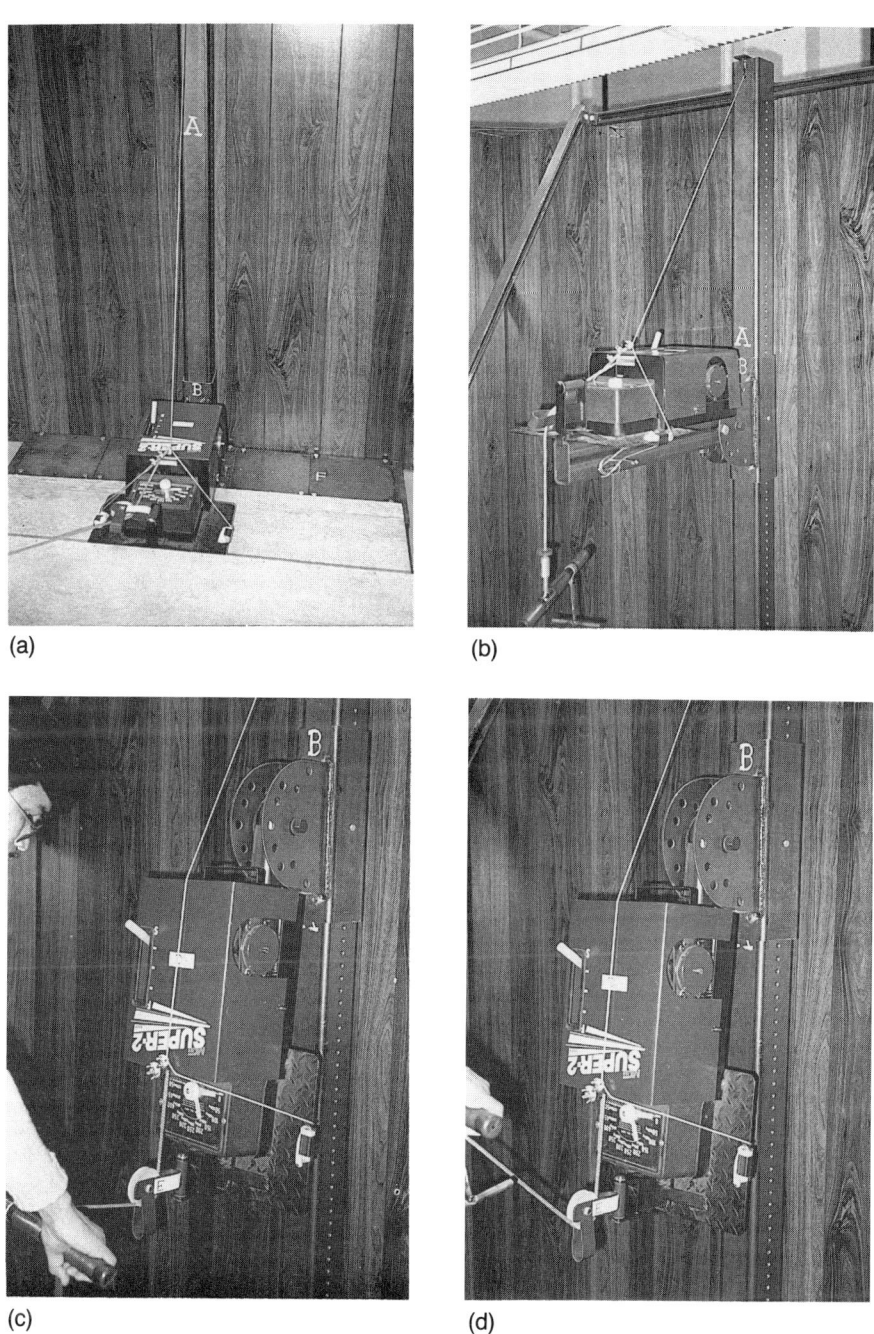

Figure 7.8. Three-dimensional isokinetic strength simulator (Mital and Vinayagamoorthy, 1984)
(a) overall view, (b) mounting, (c) measuring horizontal pull strength, (d) measuring pull strength in an oblique plane

Table 7.7. Isokinetic strengths (N) of males and females in the vertical and horizontal planes (Mital, 1985c; Mital and Karwowski, 1985; Mital et al., 1985a, 1986c)

Study	Age (years)		Height (cm)		Weight (kg)		Strength							
							DLS		DEFS		DS81		DS452	
	x̄†	S‡	x̄	S	x̄	S	x̄	S	x̄	S	x̄	S	x̄	S
Mital et al. (1985a)														
Men (n = 30)§	22·8	2·7	178·8	5·8	78·6	8·3	1121	247	557	196	—	—	—	—
Women (n = 30)	21·2	1·5	161·5	7·3	59·0	10·4	493	139	196	104	—	—	—	—
Mital et al. (1986c)														
Men (n = 19)	22·9	1·6	175·1	7·0	76·7	18·5	1083	297	741	327	344	93	594	172
Women (n = 6)	23·0	2·5	164·9	3·7	59·4	16·3	632	251	269	132	223	122	312	156

*Measured at the speed at which weight is lifted and/or carried.
†x̄ = mean; ‡S = standard deviation; §n = sample size.

The equipment described above was used to measure isokinetic lifting and elbow flexion strengths of male and female civilians (Mital et al., 1985a). Subsequently, isokinetic lifting and elbow flexion strengths and isokinetic strengths at 81 cm and 152 cm height were also measured on 19 male and 6 female college students (Mital, 1985e; Mital and Karwowski, 1985; Mital et al., 1986c). Individuals were instructed to exert as hard as possible, but without jerking. Table 7.7 shows the isokinetic strengths measured in the vertical and the horizontal planes.

Aghazadeh and Ayoub (1985) modified a CYBEX II Isokinetic Dynamometer, shown in Figure 7.9, to measure isokinetic lifting and elbow flexion strengths.

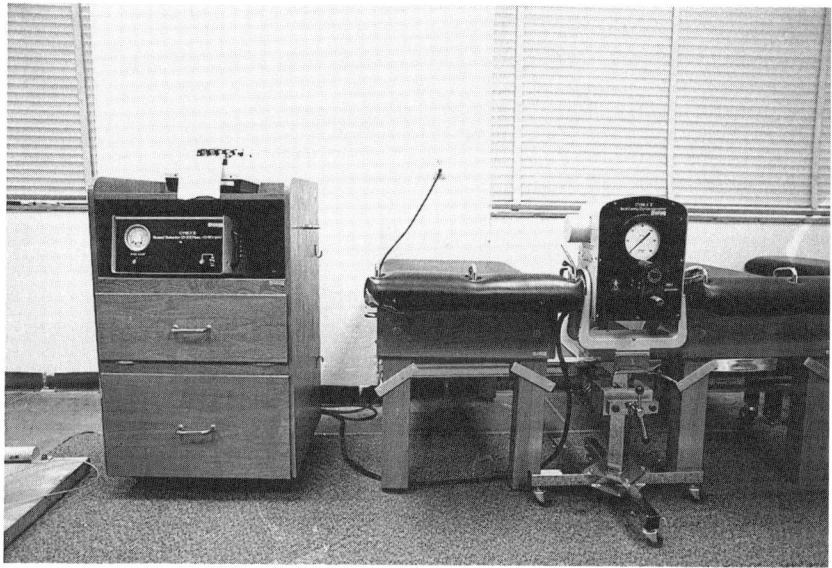

Figure 7.9. Cybex isokinetic dynomometer

CYBEX consists of a shaft that different handles can be attached to measure torque. Torque can be measured if the activity follows a circular path. Since lifting requires upward movement, it was necessary to modify CYBEX. A wheel (48·26 cm radius) was designed and built so that the vertical upward movement could be transformed into circular motion. The wheel permitted vertical movement in excess of 1·5 m. The wheel was attached to the CYBEX. A cable connected the wheel and the handle which was lifted upward. Figure 7.10 shows the final set up.

Using the modified CYBEX, isokinetic lifting and elbow extension strengths of nine male students were determined.

Subjects were instructed to exert as much force as possible, without jerking. The speed of movement was maintained at 75 cm/s. Figures 7.11a and b show the subject posture for lifting and elbow flexion strengths, respectively.

Figure 7.10.   Modification of Cybex isokinetic dynamometer for measuring lifting strengths (Aghazadeh and Ayoub, 1985)

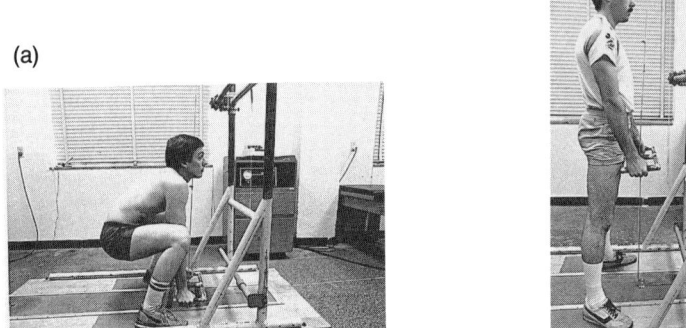

Figure 7.11.   Body posture for measuring (a) isokinetic lifting strength with Cybex, and (b) isokinetic elbow flexion strength with Cybex

The demographic and strength data of subjects are given in Table 7.8.

Table 7.8.   Age, weight, height and isokinetic strength (Nm) data for males (Aghazadeh and Ayoub, 1985)

Age (years)		Weight (kg)		Height (cm)		Lifting strength		Elbow flexion strength	
$\bar{x}$*	S†	$\bar{x}$	S	$\bar{x}$	S	$\bar{x}$	S	$\bar{x}$	S
24·1	3·7	76·7	8·9	175·9	4·3	301·1	75·2	132·3	37·7

*$\bar{x}$ = mean; †S = standard deviation.

7.3.4 Assessment of Isoinertial Strengths

The isoinertial strength testing does not measure maximum voluntary contraction. Instead, it is a method of measuring the maximum weight a person is willing to handle, at his/her own selected speed within a specified range of movements. The definition clearly indicates that this kind of testing is a variation of the psychophysical methodology. Chapters 4 and 5 include psychophysical strength data from a number of studies. In this section, only the variation of the psychophysical methodology, known as isoinertial strength testing, will be discussed.

The isoinertial strength testing has been used by the US Air Force for several years to determine the weight of lifting capabilities of the Air Force personnel. Kroemer (1983, 1985) has recently formalized the procedure. The equipment used was a modified Mach I 'press station' with guide rails extended to 3 m. The original handle assembly was changed to two handles, 46 cm apart, horizontal and parallel to each other, pointing forward (Figure 7.12).

The testing procedure requires the individual to grasp both handles, located 5 cm above the floor, and lift the weight to knuckle or overreach height and then lower the weight. The individual starts with an initial weight of 11·4 kg. If lifted successfully, an additional 11·4 kg weight is added. Increments of 11·4 kg weight are added until the cut-off limit is reached (77·3 kg in knuckle height tests and 45·5 kg in overhead reach height test — to prevent the risk of over-exertion injury) or an attempt fails. If an attempt fails, weight is reduced by 6·8 kg. If this weight is successfully lifted then 4·5 kg is added; otherwise a 2·3 kg weight is subtracted. This process results in quick determination of the weight that a person can lift.

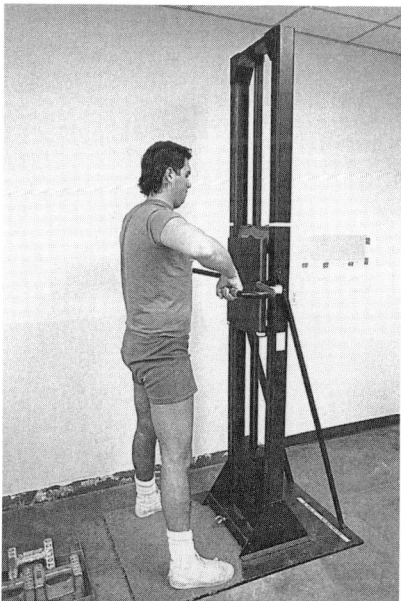

Figure 7.12.   Modified Mach I "Press Station" for measuring isoinertial strengths (Kroemer, 1983, 1985). From Human Factors, Vol. 25. © 1983 by The Human Factors Society, Inc., and reproduced by permission

The above procedure was applied, and weight lifting capabilities of 25 males and 14 females were determined. Table 7.9 gives the age, height, weight and isoinertial strengths for the knuckle and overhead reach heights of males and females who participated in Kroemer's study (1983, 1985).

Only 6 of the 25 males exceeded the 45·5 kg cut-off limit for the overhead reach lifting, while 17 males exceeded the 77·3 kg cut-off limit for the knuckle height lifting. None of the females exceeded the cut-off limits. The test-retest reliability was observed to be high.

Table 7.9. Isoinertial strengths (kg) of males and females (Kroemer, 1983, 1985)

	Age		Height (cm)		Weight (kg)		Strength			
							Overhead reach		Knuckle height	
	$\bar{x}$*	S†	$\bar{x}$	S	$\bar{x}$	S	$\bar{x}$	S	$\bar{x}$	S
Males	21·8	1·6	177·4	6·6	71·4	8·3	34·8	5·2	62·2	7·8
Females	20·0	1·7	165·8	6·9	58·4	9·0	16·3	3·7	49·1	13·7

*$\bar{x}$ = mean; †S = standard deviation.

Jiang (1984) used finer increments in his study. Instead of the 4·5 kg increments used by Kroemer, he used 1·13 kg increments to make the test more precise. Measurements were made on 12 male college students (mean height = 173·2 cm, S.D. = 6·9 cm; mean weight = 74·9 kg, S.D. = 12·6 kg). The various isoinertial strengths measured are given in Table 7.10.

The values obtained by Jiang are much higher than those reported by Kroemer even though the subject population is similar in age, height and weight.

Table 7.10. Isoinertial strengths (kg) of males (Jiang, 1984)

Strength	Mean	Standard deviation
1·82 m max lift	60·0	10·6
Elbow height lift	80·0	11·3
Knuckle to shoulder lift	49·9	9·7
Knuckle height lift	88·7	7·3

7.4 Prediction of Strengths

As mentioned earlier, knowledge of worker strength is essential in designing jobs and workplaces. Strength testing should also be an integral part of any employee selection procedure for proper operator–job match (Kraus, 1967; Chaffin et al., 1977a). Experimental measurement of strength is time consuming and requires trained personnel to conduct the testing. Several of these difficulties, including procurement and maintenance of the equipment, can be eliminated if polynomials which can predict worker strength become available. In order to be useful, such polynomials should utilize input data which are easy to measure. By quickly determining an individual's strength, an appropriate task assignment can be made and the potential of personal injury can be reduced. The following subsections describe prediction models, developed by various researchers, to predict different kinds of strengths.

7.4.1 Prediction of Isometric Strengths

Attempts have been made to develop isometric strength prediction equations which utilize body size parameters as input (Chaffin et al., 1977a; Keyserling et al., 1978; Mital and Ayoub, 1980; Mital and Manivasagan, 1982, 1984). The rationale has been quick and easy measurement of body size parameters. Some of these attempts, however, have not been very successful. The problem in using anthropometric variables to predict isometric strengths lies in the fact that correlations between strength and body size measures are low. According to Laubach and McConville (1969), measurements of body size, typology and composition are not good predictors of isometric strength and, yet, attempts have continued.

Chaffin et al. (1977a) and Keyserling et al. (1978) attempted to predict arm lifting, back lifting, and composite, or leg, lifting strengths of 443 male and 108 females workers using their age, stature, weight and gender. The resulting regression equations are given in Table 7.11.

These equations explain less then 35% variance in the observed strength value and, therefore, are not very reliable.

Table 7.11. *Isometric strength prediction equations (Chaffin et al., 1977a; Keyserling et al., 1978)*

Strength (lbs)	Equation*	Multiple correlation coefficient (R)	Standard prediction error
Arm lifting	$= 56\cdot848 - 32\cdot36(S) + 0\cdot0035(H)(W) - 0\cdot002647(A)(W)$	0·56	25·7
Back lifting	$= 21\cdot736 + 0\cdot01102(H)(W) - 0\cdot006296(A)(W) - 0\cdot24974(S)(W)$	0·56	46·8
Composite or leg lifting	$= 128\cdot07 - 95\cdot125(S) + 0\cdot0111(H)(W) - 0\cdot000143(A)(H)(W)$	0·59	67·5

*A = age (years); H = stature (in.); W = body weight (lbs); S = gender (0 = male, 1 = female).

Mital and Ayoub (1980) used age, sex and anthropometric variables as inputs to predict isometric arm, shoulder, back and composite strengths of industrial males and females. Their attempt was relatively more successful than previous attempts. In order to explain as much variance as possible, predicted values of shoulder strength were used when predicting arm and composite strengths. In turn, the back strength prediction equation utilized predicted arm strength of individuals. Thus, all models were, directly or indirectly, a function of age, sex and anthropometric variables.

The multiple correlation coefficients for strength models varied from 0·68 (back strength) to 0·85 (shoulder strength), a substantial gain over the models given in Table 7.11. The four different models and their performance statistics are summarized in Table 7.12. These models were also used to predict strengths of 5, 50 and 95% of the population using 5th, 50th and 95th percentile values of anthropometric variables, respectively. Table 7.13 summarizes the outcome.

Table 7.12. Models for predicting isometric strengths (lbs) of male and female industrial workers (Mital and Ayoub, 1980)

Strength	Model*	R^2	Mean error (predicted-actual)	Error standard deviation	Standard error of mean
Shoulder	$= 31 \cdot 028 - 33 \cdot 723(\text{Sex}) - 1 \cdot 5(\text{body-weight}) - 0 \cdot 874(\text{knuckle height}) - 2 \cdot 187(\text{iliac crest height}) + 28 \cdot 71(\text{chest depth}) + 1 \cdot 215(\text{chest width}) + 1 \cdot 584(\text{RPI}) + 0 \cdot 014(\text{body-weight})(\text{shoulder height}) - 0 \cdot 753(\text{chest depth})^2$	0·723	−3·874	23·080	1·110
Arm	$= -80 \cdot 648 + 0 \cdot 688(\text{age}) - 0 \cdot 414(\text{shoulder height}) - 2 \cdot 01(\text{abdominal depth}) + 34 \cdot 264(\lg(\text{body-weight})) + 0 \cdot 622(\text{predicted shoulder strength})$	0·580	3·370	15·440	0·743
Back	$= -8 \cdot 274 - (\text{age}) + 1 \cdot 85(\text{height}) - 3 \cdot 088(\text{knuckle height}) + 0 \cdot 051(\text{chest depth})^2 + 1 \cdot 052(\text{predicted arm strength})$	0·463	−2·790	31·810	1·530
Composite	$= -177 \cdot 435 + 2 \cdot 116(\text{iliac crest height}) + 0 \cdot 116(\text{chest depth})^2 - 0 \cdot 062(\text{abdominal depth})^2 + 1 \cdot 704(\text{predicted shoulder strength})$	0·616	0·233	49·450	2·380

*Age (years); body-weight (lbs); body measurements (cm); sex = 0, males and 1, females.
RPI = HT/$3\sqrt{\text{body-weight}}$

Table 7.13. Comparison of measured and predicted (Table 7.12) isometric strengths (lbs)

		Percent Population*					
		5		50		95	
Strength	Sex	Measured	Predicted	Measured	Predicted	Measured	Predicted
Shoulder	Male	67	87	113	111	158	124
	Female	37	37	60	62	83	79
Arm	Male	46	62	81	85	116	97
	Female	30	35	52	56	73	68
Back	Male	103	144	159	159	216	164
	Female	57	100	110	113	162	119
Composite	Male	205	254	294	293	383	319
	Female	77	95	142	150	208	187

*Assuming normal distribution.

The models given in Table 7.12 clearly indicate that if the use of inputs is not restricted to age, height, body-weight and sex, and other anthropometric variables, such as chest depth and knuckle height are included, a relatively higher multiple correlation coefficient can be obtained. For instance, the arm strength model based upon age, sex, height and body-weight (Chaffin et al., 1977a; Keyserling et al., 1978) explains about 31% variance, while the model based upon shoulder height, age, abdominal depth, body-weight and predicted shoulder strength explains nearly twice (58%) as much variance.

The modelling attempts discussed so far have relied solely on the stepwise regression analysis and the majority of the models developed have been simple low order equations. It is very likely that polynomials of higher order than previously attempted need to be examined. Selection of input transformations, interactions, and the degree of non-linearity, however, cause practical difficulties. Mital and Manivasagan (1982, 1984) overcame these difficulties by utilizing the modified basic group method of data handling (GMDH). The technique has some basic and practical advantages:

1. The need to select interaction (cross-products) and higher order inputs (e.g., squares and cubes, of the input variables) is eliminated;
2. Existing software, such as Statistical Analysis System (SAS Institute, North Carolina), can be used to quickly calculate coefficients of terms in the polynomial; and
3. The degree of non-linearity, which best explains the trend, is automatically reached.

The GMDH technique has been described by many researchers through different applications (Ivakhnenko, 1971; Duffy and Franklin, 1975; Inooka and Inoye, 1978; Mital and Manivasagan, 1982; Yoshimura et al., 1982; Mital, 1984c). It is a heuristic self-organization approach and requires that a second-order polynomial of the form:

$$\bar{y} = a_0 + a_1 x_1 + a_2 x_2 + a_3 x_1^2 + a_4 x_2^2 + a_5 x_1 x_2 \dots \quad (1)$$

to be fitted to all combinations of inputs (x's), taken two at a time, to predict the response, or output, y, whose estimate is $\bar{y}$. If there are n different inputs there will be $n(n-1)/2 (= n!/(n-2)!)$ different second-order polynomials ($\bar{y}$ values). Using a 'self-selection threshold', the combinations ($\bar{y}$ values) which best describe the behaviour of the output are allowed to pass to the succeeding layers. Criterion, such as mean square error (MSE) or correlation, may be used to determine the best combinations. These combinations, which are passed to the succeeding layer, in turn, become inputs and again a second-order polynomial, such as described above, is fitted to all these inputs taken two at a time. A new 'self-selection threshold' is selected and the above procedure is repeated until the algorithm starts degenerating. This is indicated by the correlation criterion (R^2 value stops increasing significantly and may even start decreasing) or the mean square criterion (MSE starts increasing).

For the n different inputs (input variables), the $[n(n-1)]/2$ different second-order polynomials are:

$$\bar{y}_i = a_{0i} a_{1i} x_j + a_{2i} x_k + a_{3i} x_j^2 + a_{4i} x_k^2 + a_{5i} x_j x_k \ldots \qquad (2)$$

$$\text{where } i = 1, 2, \ldots, \frac{n(n-1)}{2}; \ k = 1, 2, \ldots, n; j = 1, 2, \ldots, k-1$$

The coefficients of second-order polynomials given by equation (2) are obtained from the Gauss normal equations. Briefly, given N data points, the matrix takes the form:

$$A = (X^T X)^{-1} X^T Y \qquad (3)$$

where A is the 6×1 vector of the coefficient estimates, X is the $N \times 6$ matrix of measurements, and Y is the $N \times 1$ vector of measurements of the response or output, y. This procedure is repeated for each second-order combination (equation (2)) of the input variables (x's). Coefficients in the succeeding layers are determined similarly. Statistical Analysis System software (SAS Institute, North Carolina), or any other similar software, may be used to determine the coefficients.

The step-by-step procedure is described below:

Step 1. Inputs (independent variables) and their logical transformations (e.g., logarithmic, exponents) are selected.

Step 2. The experimental data (values of inputs and outputs or responses; anthropometric and isometric strength values, respectively, in this case) are randomly divided into two groups: (i) the training group, which is used to develop the polynomials; and (ii) the testing group, which is used to verify polynomials developed on the training group. (This separation of data into two groups is essential to independently evaluate which second-order polynomials are better predictors.) Each group should have approximately equal data points in it.

Step 3. Second-order (quadratic) polynomials, of the form given by equation (1), are fitted to the training group data using SAS statistical software package. Stepwise regression analysis is used for each combination of inputs taken two at a time. (The stepwise regression technique helps in eliminating terms that are collinear with other terms already in the regression equation.) The partial polynomials are verified on the testing set data by calculating the mean square error and performance statistics (e.g., mean error, error standard deviation).

Step 4. Using multiple criteria, mean square error (MSE), performance statistics, and explained variance (R^2), self-imposed thresholds are used to determine the best fitting polynomials to the training set data. These 'best' polynomials are passed on to the next layer as inputs. Some subjectivity and trade-offs are involved since multiple screening criteria are used.

Step 5. A stepwise regression equation is developed using 'best' polynomials, determined in step 4, taken all at the same time. (This allows less complex combinations of more inputs at all levels, without requiring high order polynomials, to obtain combinations of more inputs.) This stepwise regression equation is compared with the polynomials determined in step 4 and with previous layer results (if any) to obtain the best overall polynomial. Performance statistics play an important role in determining the overall 'best' polynomial.

Step 6. Steps 3–5 are repeated on 'best' polynomials until the technique starts to degenerate. This is indicated by significant increases in MSE and performance. statistics and significant decreases in R^2.

Self-imposed thresholds are the cut-off limits of criteria used to determine the best fitting polynomials to the training set data. These limits change from one layer to the next layer so that the 'best' polynomials are passed on to the next layer as inputs. The method for selecting these limits is subjective. The following guideline should be used:

Cut-off limits of MSE and R^2 are selected such that the number of partial polynomials passed to the next layer, when used as inputs, do not yield more than one-third the number of partial polynomials generated in the preceeding layer. For instance, 20

Table 7.14. Arm strength polynomial (Mital and Manivasagan, 1984). Reprinted from *International Journal of Computers and Industrial Engineering*, Vol. 8, © 1984 Pergamon Press PLC

Layer	Polynomial	R^2	MSE
I	$T_1 = 66 \cdot 49 - 10 \cdot 662(\text{sex code})\dagger + 0 \cdot 01(\text{age})^2 - 0 \cdot 506(\text{sex code})(\text{age})$;	0·432	0·360
	$T_2 = 224 \cdot 757 + 33 \cdot 82(\text{sex code}) + 17 \cdot 275(\text{chest width}) - 0 \cdot 235 (\text{chest width})^2 - 1 \cdot 896(\text{sex code})(\text{chest width})$;	0·445	0·396
	$T_3 = 105 \cdot 33 - 33 \cdot 9(\text{sex code}) - 13 \cdot 1(\text{dynamic endurance}) + 1 \cdot 012 (\text{dynamic endurance})^2 + 1 \cdot 826(\text{sex code})(\text{dynamic endurance})$;	0·432	0·328
	$T_4 = 224 \cdot 906 + 36 \cdot 276(\text{sex code}) - 111 \cdot 685(\log(\text{age})) + 20(\log(\text{age}))^2 - 18 \cdot 547(\text{sex code})(\log(\text{age}))$;	0·432	0·354
	$T_5 = 103 \cdot 22 - 36 \cdot 57(\text{sex code}) - 40 \cdot 54(\log(\text{dynamic endurance})) + 11 \cdot 48(\log(\text{dynamic endurance}))^2 + 8 \cdot 58(\text{sex code}) (\log(\text{dynamic endurance}))$;	0·472	0·384
	Self-imposed threshold — $R^2 = 0 \cdot 43$; MSE $= 0 \cdot 40$.		
	Stepwise regression equation $= -7 \cdot 32 - 1 \cdot 82 T_1 + 0 \cdot 497 T_2 - 1 \cdot 04 T_3 + 1 \cdot 99 T_4 + 1 \cdot 45 T_5$;	0·547	0·385
II	$T_6 = 113 \cdot 31 - 2 \cdot 63 T_1 - 0 \cdot 008 T_1^2 - 0 \cdot 0067 T_5^2 + 0 \cdot 025 T_1 T_5$;	0·551	0·351
	$T_7 = 63 \cdot 42 + 0 \cdot 694 T_2 - 17 \cdot 78 T_5 - 0 \cdot 016 T_5^2$;	0·546	0·355
	$T_8 = 94 \cdot 24 - 2 T_4 - 0 \cdot 011 T_4^2 - 0 \cdot 0114 T_4 T_5$;	0·545	0·343
	Self-imposed threshold — $R^2 = 0 \cdot 54$; MSE $= 0 \cdot 39$.		
	Stepwise regression equation $= -4 \cdot 527 + 1 \cdot 185 T_6$;	0·548	0·355
III	$T_9 = 11 \cdot 81 + 2 T_7 - 1 \cdot 340 T_8 - 0 \cdot 021 T_8^2 - 0 \cdot 019 T_7 T_8$;	0·584	0·318

†sex code = 0, for males; 1 for females.

Table 7.15. *Back strength polynomial (Mital, 1984c)*

Layer	Polynomial	R^2	MSE
I	$T_1 = 151.108 - 107.38(\text{sex code})\dagger + 0.459(\text{sex code})(\text{body-weight})$;	0.367	1.884
	$T_2 = 151.108 - 292.6(\text{sex code}) + 1.888(\text{sex code})(\text{shoulder height})$;	0.323	1.981
	$T_3 = 151.108 - 223.655(\text{sex code}) + 1.885(\text{sex code})(\text{iliac crest height})$	0.321	1.900
	$T_4 = 151.15 - 0.945(\text{sex code})(\text{knee height})$	0.309	2.077
	$T_5 = 49.07 - 134.49(\text{sex code}) + 10.180(\text{chest depth}) - 0.248(\text{chest depth})^2 + 4.809(\text{sex code})(\text{chest depth})$;	0.360	2.075
	$T_6 = 151.108 - 143.215(\text{sex code}) + 3.7(\text{sex code})(\text{chest width})$;	0.336	1.962
	$T_7 = 170.02 - 43(\text{sex code}) - 10.93(\text{dynamic endurance}) + 1.076(\text{dynamic endurance})^2$;	0.323	2.023

Self-imposed threshold — $R^2 = 0.305$; MSE = 2.1

Stepwise regression equation = $-2.953 + 0.793T_1 + 0.652T_2 - 1.337T_4 + 0.751T_5 - 0.664T_6 + 0.8255T_7$;

| II | $\boxed{T_9 = -336.13 - 5.696T_1 + 12.279T_2 - 0.014T_1^2 - 0.1T_2^2 - 0.093T_1T_2;}$ | 0.405 | 3.569 |
| | | 0.429 | 2.133 |

Self-imposed threshold — $R^2 = 0.40$; MSE = 2.15.

†Sex code = 1 for males; 2 for females.

Table 7.16. Composite strength polynomial (Mital and Manivasagan, 1984). Reprinted from *International Journal of Computers and Industrial Engineering*, Vol. 8, © 1984 Pergamon Press PLC

Layer	Polynomials	R^2	MSE
I	$T_1 = 162.65 + 244.08(\text{sex code})\dagger - 0.005(\text{height})^2 - 1.855(\text{sex code})(\text{height}) + 1.638(\text{height})$;	0.537	4.782
	$T_2 = 544.46 - 32.12(\text{shoulder height}) + 95.39(\text{chest width}) + 0.139(\text{shoulder height})^2 - 1.57(\text{chest width})^2$;	0.520	6.929
	$T_3 = -888.78 - 63.43(\text{sex code}) + 46.17(\log(\text{shoulder height})) - 17.25(\log(\text{dynamic endurance}))$;	0.542	3.857
	$T_4 = 3619.54 - 381.057(\text{sex code}) + 0.356(\text{iliac crest height})^2 + 3.13(\text{sex code})(\text{iliac crest height}) - 69.93(\text{iliac crest height})$;	0.520	4.056
	$T_5 = 67.69 - 83.35(\text{sex code}) + 6.58(\log(\text{body-weight}))$;	0.504	3.803
	$T_6 = -465.25 + 977.7(\text{sex code}) + 37.67(\log(\text{knuckle height})) - 246.93(\text{sex code})(\log(\text{knuckle height}))$;	0.507	4.050
	$T_7 = 120.42 - 83.02(\text{sex code}) - 0.001(\text{body-weight})^2 - 1.02(\text{body-weight})$;	0.505	3.930
	$T_8 = 82.82 + 125.03(\text{sex code}) + 0.027(\text{knuckle height})^2 - 2.83(\text{sex code})(\text{knuckle height})$;	0.506	4.083
	$T_9 = 10.49 - 64.81(\text{sex code}) + 0.01(\text{shoulder height})^2$;	0.528	4.168
	Self-imposed threshold — $R^2 = 0.50$; MSE = 7.00		
	Stepwise regression equation = $-365.48 + 244.08(\text{sex code}) + 46.17 T_1^2 - 33.67 T_1 T_3 - 82.82 T_5^2$;	0.587	4.186
II	$T_{10} = -109.85 - 2.67 T_4 + 1.83 T_7 - 0.008 T_2^2 - 0.0145 T_4 T_7$;	0.587	4.680
	$T_{11} = -65.633 + 1.586 T_7 - 0.01 T_7^2 - 0.007 T_9^2 - 0.016 T_7 T_9$;	0.582	4.598
	$T_{12} = 83.5 + 0.0025 T_2 T_7$;	0.599	4.566
	$T_{13} = -1011.32 + 2.79 T_1 + 3.137 T_2 + 0.007 T^2 - 0.011 T_1 T_2$;	0.588	4.360
	Self-imposed threshold — $R^2 = 0.55$; MSE = 4.7		
	Stepwise regression equation = $-667.76 + 0.124 T_6 - 0.222 T_{10}^2 + 0.011 T_{10} T_{11}$	0.592	4.209
III	$T_{14} = 0.929 + 0.3156 T_{10} + 0.6583 T_{12}$;	0.590	4.490
	$T_{15} = -195.885 + 4.3036 T_{11} - 1.3077 T_{12} - 0.0162 T_{12}^2 - 0.021 T_{11} T_{12}$;	0.610	4.090
	Self-imposed threshold — $R^2 = 0.59$; MSE = 4.5		
	Stepwise regression equation = $0.848 + 0.989(T_{15})$.	0.617	4.010

†Sex code = 0, for males; 1, for females.

Table 7.17. Shoulder strength polynomial (Mital and Manivasagan, 1984). Reprinted from *International Journal of Computers and Industrial Engineering*, Vol. 8, © 1984 Pergamon Press PLC

Layer	Polynomials	R^2	MSE
I	$T_1 = 82.247 - 42.8(\text{sex code})\dagger + 0.1557(\text{body-weight})$;	0.552	0.622
	$T_2 = 52.246 - 105.2(\text{sex code}) + 0.0097(\text{knuckle height})^2 - 2.039(\text{sex code})(\text{knuckle height})$	0.580	0.621
	$T_3 = 43.935 - 40.907(\text{sex code}) + 2.142(\text{chest width})$;	0.533	0.635
	$T_4 = 128.25 - 44.19(\text{sex code}) - 12.39(\text{dynamic endurance}) + 1.524.(\text{dynamic endurance})^2 - 2.07(\text{sex code})(\text{dynamic endurance})$;	0.544	0.628
	$T_5 = 41.8 - 42.59(\text{sex code}) + 2.55(\log(\text{body-weight}))^2$;	0.552	0.622
	$T_6 = 125.42 - 49.17(\text{sex code}) - 37.13(\log(\text{dynamic endurance})) + 15.723(\log(\text{dynamic endurance}))^2$;	0.524	0.641
	$T_7 = 10.33 - 39.8(\text{sex code}) - 14.43(\text{shoulder height}) + 0.053(\text{shoulder height})^2$;	0.729	0.634
	$T_8 = 1712.65 - 131.12(\text{sex code}) - 32.197(\text{iliac crest height}) + 0.16(\text{iliac crest height})^2 + 8.9(\text{sex code})(\text{iliac crest height})$;	0.690	0.624
	Self-imposed threshold — $R^2 = 0.52$; MSE = 0.645.		
	Stepwise regression equation = $876.54 - 5.36T_1 + 0.327T_2 - 0.508T_4 + 5.675T_5 + 1.39T_6 - 1.018T_7 - 0.098T_8$;	0.698	2.862
II	$T_9 = 37.361 + 0.029T_1^2 + 0.028T_6^2 - 0.052T_1T_6$;	0.671	0.874
	$T_{10} = 106.03 - 1.72T_2^2 + 0.028T_6^2 - 0.052.T_2T_6$;	0.676	1.121
	$T_{11} = 37.91 + 0.006T_3T_6$;	0.661	0.740
	$T_{12} = 37.00 + 0.031T_5^2 + 0.029T_6^2 - 0.055T_5T_6$;	0.672	1.163
	Self-imposed threshold — $R^2 = 0.66$; MSE = 1.2.		
	Stepwise regression equation =		
III	$T_{13} = 173.53 - 2.27T_{10} + 0.012T_{10}^2 - 0.023T_{11}^2 - 0.02T_{10}T_{11}$;	0.689	0.776
	$T_{14} = 28.12 + 0.253T_{12} + 0.0052T_{10}T_{12}$;	0.684	0.755
	Self-imposed threshold — $R^2 = 0.68$; MSE = 0.78.		
	Stepwise regression equation = $93.124 - 0.041T_{13} - 0.745T_{14}$;	0.685	0.756
IV	$\boxed{T_{15} = 66.96 + 0.004T_{10}T_{11}}$;	0.650	0.699

*Simplified equation. †Sex code = 0, for males; 1, for females.

input variables in layer I will yield 190 (= 20 × 19/2) partial polynomials; only those many of these 190 polynomials should be passed to layer II as inputs that, in turn, will generate approximately 64 (= 190/3) partial polynomials (12 layer I partial polynomials in this case). Layer III should generate approximately 21 partial polynomials.

The use of stepwise regression analysis simplifies selection of partial polynomials that are to be passed to the next layer. Many input combinations are found statistically not significant (at the 15% level of significance if SAS software package is used) and, thus, are eliminated from consideration. Grouping of R^2 and MSE values also help in selecting self-imposed thresholds.

Using the GMDH technique, as described above, Mital and Manivasagan (1982, 1984) developed non-linear polynomials to predict arm, back, composite and shoulder strengths. The experimental data base used was the same as used by Mital and Ayoub (1980). Tables 7.14 to 7.17 give the prediction equation for arm, back, composite and shoulder strengths, respectively. The final equation is boxed for quick reference. The performance statistics for the 'best' polynomial are given in Table 7.18. The final models given in Tables 7.14–7.17 show a significant improvement in the prediction accuracy compared to the models in Table 7.12. The GMDH models also demonstrate that even though the measurements of body size independently are not good predictors of human isometric strengths, when used in combination, they yield much more satisfactory results than obtained through simple stepwise regression analysis (Chaffin et al., 1977a; Mital and Ayoub, 1980; Mital, 1984c). In all equations (Tables 7.14–7.17), strength is in pounds, age in years, dynamic endurance in minutes, body weight in pounds and anthropometric measurements in centimetres. If the dynamic endurance is not known, it can be determined from the arm strength equation given in Table 7.12 and the procedure described in section 7.3.2.

Table 7.18. Performance statistics for the best isometric strength (kg) polynomials (Tables 7.14–7.17)

Polynomial	Mean error*	R^2	Standard deviation of error*	Standard error of mean*
Arm strength	−0·157	0·584	6·350	0·256
Back strength	2·541	0·429	15·327	0·601
Composite strength	−0·768	0·617	23·178	0·910
Shoulder strength	1·095	0·650	9·618	0·378

*Calculated on the test data set.

7.4.2 Prediction of Isokinetic Strengths

Attempts to predict isokinetic strengths have been few. As a matter of fact, only one study was identified (Mital et al., 1985a). This work by Mital et al. led to the development of GMDH prediction equations for the floor to shoulder height and knuckle to shoulder height isokinetic strength. The data used in developing these equations were collected on 30 male and 30 female college students. The various measurements made are given in Table 7.19. The Super-2 Mini Gym (Figure 7.13) was used to determine dynamic strengths. The speed of handle movement was maintained at 75 cm/s.

Table 7.19. *Anthropometric (cm) and strength (kg) measurements of the subject population (Mital et al., 1985a)*

Variable	Males (n = 30)		Females (n = 30)	
	Mean	Standard deviation	Mean	Standard deviation
Age (years)	22·80	2·68	21·23	1·52
Body weight (kg)	78·63	8·32	58·98	10·40
Stature	178·84	5·84	161·53	7·26
Acromial height	148·32	6·72	134·57	7·46
Standing iliac crest height	104·51	5·97	95·09	7·64
Knuckle height	77·49	3·51	70·69	5·17
Knee height	50·22	2·88	45·55	3·20
Forearm grip distance	37·83	3·77	32·92	2·48
Chest width	33·71	2·98	29·06	2·44
Chest circumference	97·35	5·91	88·11	5·53
Abdominal circumference	85·80	7·20	72·14	7·04
Biceps circumference	32·98	2·43	26·87	2·85
Static arm strength	40·17	8·92	18·85	9·64
Static stooped back strength	94·28	26·29	39·94	14·60
Static shoulder strength	50·58	12·29	20·82	6·96
Static composite strength	135·64	26·36	60·85	23·24
Dynamic knuckle to shoulder strength	56·83	19·98	19·95	10·61
Dynamic floor to shoulder strength	114·29	25·22	50·29	14·16

The GMDH technique (see section 7.4.1) was applied and the final equations to predict isokinetic strengths for the knuckle to shoulder and floor to shoulder heights were developed. Tables 7.20 and 7.21 show these equations.

The equations in Tables 7.20 and 7.21 again indicate that human strengths are non-linear in nature and such modelling would be lacking if the conventional regression technique is used. This is proved again by comparing the 'best' dynamic strength models given in Tables 7.20 and 7.21 with respective strength models developed by using stepwise regression analysis. Besides the 13 anthropometric variables listed in Table 7.19, interactive terms appearing in partial polynomials given in Tables 7.20 and 7.21 and some of the relevant interactive terms found useful in previous work (Mital and Ayoub, 1980) were considered in the regression analysis. Performance statistics were generated using the best possible regression equations. A comparison between GMDH models and stepwise regression models is shown in Table 7.22. In both cases, the performance of regression models lagged far behind the GMDH models. The best stepwise regression models for the two dynamic strengths were:

Knuckle to shoulder dynamic strength = $-70.63 + 0.0037$(iliac crest height)^2 + 0.037 (chest width)^2 + 0.037(biceps circumference)^2;

Floor to shoulder dynamic strength = $179.79 + 3.52$(biceps circumference) + 0.017 (stature)^2 − 0.017(acromial height)^2.

The performance statistics for these equations are given in Table 7.22.

Table 7.20. *Dynamic strength polynomial for the knuckle to shoulder height (Mital et al., 1985a)*

Layer	No. of equations	No. chosen	Polynomial	R^2	MSE
I	78	9	$T_1 = 21 \cdot 866 + 0 \cdot 009(\text{Weight})^2 - 0 \cdot 287(\text{sex code})\dagger(\text{weight})$;	0·733	286·70
			$T_2 = 21 \cdot 422 + 0 \cdot 064(\text{Chest width})^2 - 0 \cdot 633(\text{sex code})(\text{chest width})$;	0·764	258·06
			$T_3 = -5 \cdot 5776 + 0 \cdot 009(\text{Chest circumference})^2 - 0 \cdot 256(\text{sex code})(\text{chest circumference})$;	0·744	284·61
			$T_4 = 6 \cdot 633 + 0 \cdot 064(\text{biceps circumference})^2 - 0 \cdot 621(\text{sex code})(\text{biceps circumference})$;	0·698	207·26
			$T_5 = -75 \cdot 678 + 0 \cdot 021(\text{stature})(\text{chest width})$;	0·704	303·67
			$T_6 = -69 \cdot 673 + 0 \cdot 021(\text{stature})(\text{biceps circumference})$;	0·691	243·16
			$T_7 = -67 \cdot 581 + 0 \cdot 025(\text{acromial height})(\text{biceps circumference})$;	0·688	254·79
			$T_8 = -52 \cdot 499 - 0 \cdot 022(\text{forearm grip distance}) - 0 \cdot 111(\text{forearm grip distance})(\text{biceps circumference})$;	0·697	240·96
			$T_9 = -47 \cdot 540 - 0 \cdot 089(\text{chest width})(\text{biceps circumference})$;	0·723	263·77
			Self-imposed threshold — $R^2 = 0 \cdot 650$; MSE = 310·0.		
			Stepwise regression equation = $-0 \cdot 085 + 0 \cdot 999 T_2$;	0·764	256·31
II	36	6	$T_{10} = 14 \cdot 096 + 0 \cdot 013 T_2 T_4$;	0·789	212·33
			$T_{11} = 13 \cdot 749 + 0 \cdot 013 T_2 T_6$;	0·798	220·65
			$T_{12} = 13 \cdot 642 + 0 \cdot 013 T_2 T_7$;	0·787	215·05
26			$T_{13} = 14 \cdot 490 + 0 \cdot 013 T_2 T_8$;	0·788	213·51
			$T_{14} = 13 \cdot 277 + 0 \cdot 014 T_4 T_5$;	0·796	213·39
			$T_{15} = 14 \cdot 044 + 0 \cdot 013 T_5 T_8$;	0·770	242·87
			Self-imposed threshold — $R^2 = 0 \cdot 77$; MSE = 250·0.		
			Stepwise regression equation = $0 \cdot 786 + 4 \cdot 794 T_{11}^2 - 3 \cdot 803 T_{12}$;	0·816	306·17
III	15	4	$T_{16} = -0 \cdot 318 + 1 \cdot 023 T_{11}$;	0·798	215·05
			$T_{17} = -0 \cdot 411 + 1 \cdot 029 T_{10}$;	0·789	208·46
			$T_{18} = -0 \cdot 328 + 1 \cdot 024 T_{12}$;	0·787	209·50
			$T_{19} = 0 \cdot 515 + 1 \cdot 013 T_{13}$;	0·788	211·29
			Self-imposed threshold — $R^2 = 0 \cdot 787$; MSE = 216·0		
			Stepwise regression equation = $1 \cdot 058 + 4 \cdot 686 T_{15}$;	0·816	306·33
IV	6	1	$\boxed{T_{20} = -0 \cdot 00008 + 1 \cdot 002 T_{17}}$;	0·789	208·43
			Self-imposed threshold — $R^2 = 0 \cdot 788$; MSE = 209·0.		

†Sex code = 1 for males; 2 for females.

Table 7.21. Dynamic strength polynomial for the floor to shoulder height (Mital et al., 1985a)

Layer	No. of equations	No. chosen	Polynomial	R^2	MSE
I	78	9	$T_1 = 129.392 - 19.509(\text{sex code})^2$†;	0.647	363.61
			$T_2 = 59.647 + 0.819(\text{weight}) - 14.567(\text{sex code})^2$;	0.694	386.27
			$T_3 = 25.479 - 13.405(\text{sex code})^2 + 0.003(\text{stature})^2$;	0.674	408.52
			$T_4 = 50.624 - 15.399(\text{sex code})^2 + 0.029(\text{knee height})^2$;	0.699	496.87
			$T_5 = 88.494 - 32.648(\text{sex code})^2 + 1.601(\text{sex code})(\text{chest width})$;	0.681	368.49
			$T_6 = 27.312 + 2.961(\text{biceps circumference}) - 14.035(\text{sex code})^2$;	0.689	302.26
			$T_7 = -83.048 + 0.038(\text{acromial height})(\text{biceps circumference})$;	0.609	495.69
			$T_8 = -86.558 + 0.114(\text{knee height})(\text{biceps circumference})$;	0.649	572.52
			$T_9 = -53.080 - 0.038(\text{forearm grip distance})^2 + 0.171(\text{forearm grip distance})(\text{biceps circumference})$;	0.591	474.99

Self-imposed threshold — $R^2 = 0.58$; MSE = 595.
Stepwise regression equation = $-5.695 + 0.656T_4 + 0.413T_8$;   0.699   477.51

II	36	6	$T_{10} = 0.025 + T_2$;	0.694	386.00
			$T_{11} = -0.006 + T_5$;	0.681	368.54
			$T_{12} = -0.003 + T_6$;	0.689	301.19
			$\boxed{T_{13} = 31.980 + 0.007T_1T_7}$;	0.687	285.14
			$T_{14} = 31.039 + 0.007T_1T_8$;	0.728	372.34
			$T_{15} = 31.902 + 0.007T_1T_9$;	0.673	308.47

Self-imposed threshold — $R^2 = 0.67$; MSE = 390.
Stepwise regression equation = $0.231 + 0.993T_{14}$;   0.728   377.35

III	15	3	$T_{16} = -0.00005 + 1.0003T_{10}$;	0.694	288.15
			$T_{17} = 1.701 + 0.987T_{12}$;	0.689	302.71
			$T_{18} = 0.472 + 0.985T_{13}$; MSE = 303.00.	0.687	295.15

Self-imposed threshold — $R^2 = 0.687$; MSE = 303.00.
Stepwise regression equation = $-0.000001 + 0.99T_{16}$.   0.694   386.72

†Sex code = 1 for males; 2 for females.

Table 7.22. *Performance statistics* for the 'best' dynamic strength polynomials and its comparison with regression equations (Mital et al., 1985a)*

Polynomial	Technique	Mean error (actual-predicted)	R^2	Standard deviation of error	Standard error of mean	Error range	MSE
Knuckle to shoulder dynamic strength	GMDH	−1·32	0·79	14·62	2·67	−37·39 to 33·19	208·46
	Regression	307·36	0·77	88·17	16·10	128·10 to 465·59	101,983·30
Floor to shoulder dynamic strength	GMDH	−2·63	0·69	16·96	3·10	−57·59 to 22·70	285·14
	Regression	−17·43	0·72	33·72	6·16	−162·51 to 10·69	1,403·11

*Rounded to the nearest two decimal places.

7.4.3 Prediction of Isoinertial Strengths

The prediction of psychophysical strength, for intermittent and repetitive activities, has been discussed at length in Chapter 3. In this section, only those prediction models which are based on Kroemer's methodology (1982, 1985) are presented.

Kroemer utilized 25 male and 14 female students in his efforts to develop regression equations to predict isoinertial strengths for the knuckle and overhead inch height. The methodology for determining isoinertial strengths has been described in section 7.3.4. Table 7.23 shows the demographic, anthropometric and isometric strength data of the sample population. Isoinertial strengths of the subjects are given in Table 7.9.

Table 7.23. *Demographic, anthropometric, and isometric strength data of the sample utilized in developing isoinertial strength prediction equations (Kroemer, 1983)*

Parameter	Male (n = 25)		Female (n = 14)	
	Mean	Standard deviation	Mean	Standard deviation
Age (years)	21·80	1·61	20·00	1·75
Weight (kg)	71·39	8·26	58·39	9·05
Height (cm)	177·42	6·59	165·79	6·87
Knuckle height (cm)	77·81	4·20	75·17	5·04
Overhead reach (cm)	203·70	20·13	194·22	8·81
Arm strength (kg)	29·73	8·75	15·46	5·09
Leg or composite strength (kg)	78·65	28·25	40·64	15·81
Torso or back strength (kg)	72·15	25·80	38·48	12·50

Using the stepwise regression analysis, the final prediction models were developed:

Overhead reach strength (kg) = $63·225 - 16·11(S) + 0·309(H) - 0·322(W) - 0·0507(A) - 0·0038(L) + 0·077(B)$; $R^2 = 0·9$.

Knuckle height = 72·35 + 1·055(W) − 0·62(H) + 0·507 (A) − 0·131(L) + 0·151(B);
strength (kg) $R^2 = 0·61$.

Where S = sex code; 0 for males and 1 for females; W = body-weight (kg); H = stature (cm); A = average isometric arm strength (kg); L = average isometric leg or composite strength (kg); B = average isometric back or torso strength (kg).

The above prediction equations clearly indicate that average isometric arm, composite and back strengths must be utilized in predicting overhead reach and knuckle height isoinertial strengths. If these isometric strengths cannot be measured, the prediction equations given in Table 7.12 may be used for their estimation.

7.5 Additivity of Strengths

Usually, in industrial settings, a person can manage to handle objects alone. Occasionally, however, situations demand that two, or maybe three, individuals work together to complete the assignment. This happens when the objects are excessively heavy and bulky and mechanical aids cannot be used because either they are not available or workplace geometry and load location do not permit their use.

The situation, described above, is not an uncommon occurrence in manufacturing and non-manufacturing industries, such as construction, health and agriculture.

As has been shown by a number of researchers (Poulsen and Jorgensen, 1971; Chaffin et al., 1977a; Kamon et al., 1982; Aghazadeh and Ayoub, 1985; Mital et al., 1986c; Yates et al., 1980), muscular strength is an important determinant of MMH capabilities of workers. It is logical to expect, therefore, that muscular strength would be equally, if not more, important in determining work limits for multiple worker tasks, and yet only sketchy information is available regarding the magnitude of human strength capabilities in team work. It has been speculated that in the case of simultaneous pushing or pulling by two or three individuals, push or pull force recommendations for one person should be doubled and tripled, respectively (Kroemer, 1974). It is, however, very unlikely that two or three people in a team would exert their peak force simultaneously (Davies, 1972).

Karwowski et al. (1985) and Karwowski and Mital (1986) tested the validity of additivity assumption for teams of two and three males. Six healthy male students participated in the investigation. Prior to team strength measurements, individual isometric and isokinetic strengths and anthropometric characteristics were recorded. Table 7.24 shows some of the anthropometric data and all strength data. The individual strengths were measured in accordance with the procedure described in sections 7.3.1 and 7.3.3. A Super-2 Mini Gym was used for the measurement of isokinetic strengths.

Special handles were designed and constructed so that strengths of teams of two and three persons could be measured. Figure 7.13 shows the handles used.

An H-bar handle, weighing 3·5 kg, was used in testing the two-man team isometric back strength and both isokinetic strengths (Table 7.24). For measuring two-man composite and arm strengths, the long handle shown in Figure 7.3 was modified. It was padded on the sides for the measurement of composite strength.

Table 7.24. *Anthropometric (cm) and strength (N) data of the subjects participating in team work study (Karwowski et al., 1985; Karwowski and Mital, 1986)*

Characteristic	Mean	Standard deviation	Range
Age (years)	24·0	1·7	21–26
Weight (kg)	79·7	19·4	58·9–115·6
Height	180·0	6·7	171·6–189·0
Knee height	55·7	3·2	51·3–61·4
Chest circumference	102·3	9·3	88·9–118·1
Abdominal circumference	89·9	19·6	66·7–124·5
Grip strength	459·0	66·7	372·8–554·3
Isometric arm strength	438·5	21·6	402–461
Isometric back strength	1006·5	171·6	745–1266
Isometric composite strength	1231·2	154·0	961–1481
Dynamic lift strength	861·3	96·1	706–1020
Dynamic back extension strength	683·7	76·5	559–785

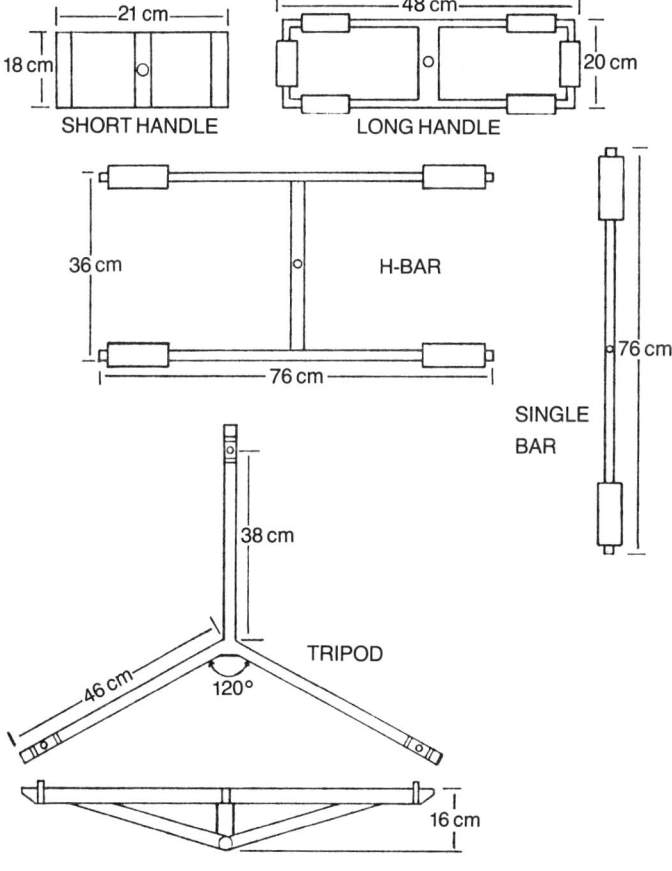

Figure 7.13. *Handles used to measure strengths of 2-man and 3-man teams (Karwowski et al., 1985; Karwowski and Mital, 1986)*

In order to measure strengths of teams of three males, a tripod handle, weighing 8·85 kg, was constructed (Figure 7.13). Special mountings were provided so that each subject could use the short handle, long handle, or the single bar, just as in the case of individual measurements.

The strength measurement procedure was similar in each case. A verbal countdown was used to ensure that each subject synchronized his peak exertion with others in the team. The countdown was: -2, -1, start, 1, 2, 3, 4, 5, 6, 7, stop. The first 2 s were the get-ready time; the next 4 s were the build-up time, and the last 3 s were the duration of the steady state.

To ensure that each member of the team was recruiting the same muscles as applied in the individual measures (Grieve, 1984; Karvonen, 1985), and to eliminate potential disadvantages due to differences in stature among the team members, 0·65 cm thick wooden boards were used to accommodate shorter subjects. As many as six boards were used in order to provide similar statures and postures for all subjects.

Two trials (replications) were performed for each isometric and isokinetic strength. In case the readings differed by more than 10%, a third trial was made.

Tables 7.25 and 7.26 show the actual strengths and corresponding sums of individual strengths for two- and three-man teams, respectively.

With the exception of isometric arm strength for the team of three men, the actual team strengths were significantly lower than the corresponding sums of individual's strengths. As shown in Table 7.27, the differences were greater for isokinetic strengths than for isometric strengths. Nevertheless, in all cases, the 'additivity of strength' assumption proved to be incorrect.

Table 7.25. Isometric and isokinetic strengths (N) for a team of two males (Karwowski et al., 1985; Karwowski and Mital, 1986)

Strength	Mean	Standard deviation	Range
Isometric arm strength			
Actual	814	123	608–1017
Sum	878	25	825–919
Difference (%)	7·3	13·3	$-12·8$–29·7
Isometric back strength			
Actual	1884	228	1389–2354
Sum	2011	205	1620–2379
Difference	5·5	13·7	$-20·5$–33
Isometric composite strength			
Actual	2312	286	1829–2934
Sum	2458	187	2126–2752
Difference	5·7	16·3	$-23·7$–22·5
Isokinetic lift strength			
Actual	1146	103	963–1398
Sum	1723	114	1509–1931
Difference	33·4	5·2	18·3–41·4
Isokinetic back extension strength			
Actual	929	74	755–1059
Sum	1364	92	1216–1536
Difference	31·8	5·5	20–43

Table 7.26. Isometric and isokinetic strengths (N) for a team of three males (Karwowski et al., 1985; Karwowski and Mital, 1986)

Strength	Mean	Standard deviation	Range
Isometric arm strength			
Actual	1384	167	1025–1691
Sum	1316	27	1265–1367
Difference (%)	−5	11·5	−24·5–22
Isometric back strength			
Actual	2678	314	2024–3298
Sum	3017	216	2637–3396
Difference	10·7	12·6	−13–30·3
Isometric composite strength			
Actual	3269	381	2637–4329
Sum	3687	197	3347–4029
Difference	11·4	8·4	−7·5–27·8
Isokinetic lift strength			
Actual	1559	133	1283–1838
Sum	2584	120	2366–2801
Difference	39·7	3·5	33–46·5
Isokinetic back extension strength			
Actual	1137	81	1007–1345
Sum	2045	98	1869–2224
Difference	44·4	3·5	37·2–51·9

Table 7.27. Actual strength in teamwork as a percent of the sum of individual strengths (Karwowski and Mital, 1986)

Strength	Team	Mean	Standard deviation	Range
Isometic arm strength	2	92·6	13·3	70–113
	3	105	11·5	78–125
Isometric back strength	2	94·5	13·6	67–120
	3	88·6	8·4	69–113
Isometric composite strength	2	94·3	11·6	77–124
	3	88·5	8·4	72–107
Isokinetic lift strength	2	66·6	5·2	58–82
	3	60·3	3·5	53–67
Isokinetic back extension strength	2	68·0	5·5	57–80
	3	55·6	3·5	48–62

It appears that, in general, the actual strength values for teams of two or three men are significantly lower than the corresponding sums of individual strengths. On average, isometric strengths of a team of two males is approximately 94% of the sum of their individual strengths. As the number of workers in the team increases to three, the team isometric strengths decline to about 90% of the total strength.

Isokinetic strengths of teams of two males are, on the average, only 68% of the total of individual's isokinetic strengths. When a third member is added to the team, the actual strengths exerted by the team accounts only for approximately 58% of the summed up strengths. Since the isokinetic strength decreases as the number of workers increases from one to two to three, and isokinetic strengths are more strongly correlated with the manual lifting capabilities than to isometric strength (section 7.6). It is logical to expect, therefore, that the maximum lifting capability of teams of two or three men will also decline proportionately.

7.5 Relationship between Strengths and Acceptable Weights of Lift

As pointed out in Chapters 3 and 6, isometric and dynamic strengths have been frequently used in order to predict MMH capabilities of individuals, lifting capability in particular. The efforts, until the end of the last decade, concentrated mainly on the use of isometric strengths to determine what an individual or a certain population percentile would accept to do. Even in the early 1980s, researchers have, from time to time, proposed models to predict the manual lifting capabilities of workers as defined by the maximum acceptable weight of lift (Garg et al., 1980; Mital and Ayoub, 1980). Many such efforts still continue (Mital, 1985c, d; Mital and Ayoub, 1986; Mital, 1986b).

The pervasive use of isometric strengths to predict dynamic lifting capabilities does not, however, suggest that the use of isometric strengths as predictors is highly valid. One of the major weaknesses is the low correlations between isometric strengths and the maximum acceptable weight of lift (Kamon et al., 1982; Aghazadeh and Ayoub, 1985; Mital et al; 1986a, b, c; Mital, 1986b). The low correlations result from the fact that the actual manual lifting act is dynamic in nature, while the isometric strengths are not. All isometric strengths fail to account for inertial forces resulting from the movement of the external object and body segments. When these movements are taken into consideration, as in dynamic strengths, higher correlations result (Aghazadeh and Ayoub, 1985; Mital et al., 1986a, b, c). Additionally, fewer independent variables are required to predict the maximum acceptable weight of lift.

Mital (1985e), Mital and Karwowski (1985) and Mital et al. (1986c) collected isometric strength, simulated job dynamic strength (SJDS — maximal isokinetic muscular exertion, which is the total force including inertia, measured while the body assumes the same configuration as during the actual task), and maximum lifting capability (MLC — maximum acceptable weight of lift for single lifts) data on 19 males and

Table 7.28. *Isometric strength (kg) and maximum lifting capability (MLC in kg) data (Mital, 1985e; Mital and Karwowski, 1985; Mital et al., 1986c)*

Variable	Males (n = 19)[*]			Females (n = 6)		
	$\bar{x}$[†]	S[‡]	Range	$\bar{x}$	S	Range
Isometric arm strength	40·4	6·7	30·0–55·7	21·8	2·2	19·2–24·2
Isometric back strength	91·6	44·5	25·3–204·1	60·2	23·6	35·3–102·0
Isometric composite strength	116·3	39·4	44·9–197·3	71·3	21·1	39·7–105·5
Isometric shoulder strength	52·6	8·5	30·3–66·0	32·6	16·6	17·5–64·3
MLC—						
Floor to 81 cm	46·5	12·3	30·9–79·7	23·6	7·6	15·2–37·7
81 cm to 152 cm	31·2	6·3	15·5–43·1	17·5	3·3	14·4–23·5
At 81 cm height	26·9	6·5	10·4–37·6	13·4	4·4	10·7–22·1
At 152 cm height	44·8	9·9	28·1–66·3	24·0	7·7	10·4–31·4

[*]n = sample size; [†]$\bar{x}$ = mean; [‡]S = standard deviation.

6 females. SJDS and MLC values were determined for the floor to 81 cm and 81–152 cm heights and at 81 cm and 152 cm. The age, height, weight and SJDS data of the subject population are given in Table 7.7. Table 7.28 provides distribution of their isometric strengths and maximum lifting capabilities (MLC).

The results of the correlation analysis indicated that, as expected, SJDS and MLC were much more strongly correlated than MLC and various isometric strengths. Table 7.29 shows the results of the correlation analysis.

Thus, there are strong indications that isokinetic strengths, not isometric strengths, are superior predictors of lifting capabilities. The use of lifting capability prediction equations utilizing isometric strengths, therefore, should be avoided.

Table 7.29. Correlations between strengths and MLC (Mital, 1985e; Mital and Karwowski, 1985; Mital et al., 1986c)

Strength	MLC			
	Floor to 81 cm	81 cm–152 cm	At 81 cm	At 152 cm
SJDS				
Floor to 81 cm	0·517	—	—	—
81 cm to 152 cm	—	0·672	—	—
At 81 cm height	—	—	0·533	—
At 152 cm height	—	—	—	0·574
Isometric				
Arm	0·356 (maximum value)			
Back	0·295 (maximum value)			
Composite	0·380 (maximum value)			
Shoulder	0·383 (maximum value)			

Of the four isometric strengths, the maximum correlation (0·383) was obtained between the isometric shoulder strength and MLCs. This value is even lower than the correlation between respective SJDSs and MLCs which range from 0·52–0·67.

The correlations between SJDS and isometric strengths ranged from 0·14, between isometric back strength and SJDS at 81 cm height, to 0·60, between isometric shoulder strength and SJDS at 81 cm height. Only 3 of the 16 correlation values were above 0·5. Most correlations were around 0·30.

While Aghazadeh and Ayoub (1985) did not report correlations between MLC and isokinetic strength, and MLC and isometric strengths, they did provide lifting capacity prediction models based on dynamic and static strengths. Using the stepwise regression analysis, they developed the following equations:

(A) Predicted load (kg) = 16·88 − 0·004 (lift height) − 1·14 (frequency) + 0·11 (dynamic strength); $R^2 = 0·775$;

(B) Predicted load (kg) = − 8·65 − 0·004 (lift height) − 1·14 (frequency) + 0·0235 (isometric shoulder strength) + 0·0163 (isometric leg strength); $R^2 = 0·775$.

where lift height = 51 cm for knuckle to shoulder height; 127 cm for floor to shoulder height; frequency = 2 or 6 lifts/min; dynamic strength = dynamic strength from knuckle to shoulder height (Nm) as measured by CYBEX; shoulder and leg strengths = isometric strengths (Ayoub et al., 1977).

The validation of model (A) yielded an average error of 2·2 kg, while that of model (B) yielded 4.2 kg. Model (A) also utilizes one less independent variable. Mital (1985e) and Mital *et al.* (1986a, b, c) have also arrived at the conclusion that dynamic (isokinetic) strengths, especially SJDS, are more reliable predictors of the maximum acceptable weights of lift, and the resulting prediction equations need relatively fewer different inputs while guaranteeing greater prediction accuracy.

While the relationships between strengths and maximum acceptable weights of lift have been firmly established, as indicated by the use of strengths in various lifting capability prediction models, questions are now being raised about their reliability. For instance, Mital (1986a) showed that even though two individuals may have equal isometric strengths, their lifting capabilities could be quite different if there is a wide difference in their experience. Experienced workers in Mital's work were observed to accept significantly heavier loads, for the same task conditions than their equally strong but inexperienced colleagues (less than 6 months experience in handling materials). This raises the possibility of severely overpredicting or underpredicting a person's lifting capabilities. While underprediction of lifting capability is simply inefficient, overprediction can result in dire consequences. Prediction models developed from data collected on industrial workers, therefore, should not be used to predict lifting capabilities of inexperienced individuals.

7.7 Repetitive Dynamic Strengths and Manual Lifting Capabilities

As pointed out in the previous section, dynamic strength of individuals are more highly correlated to their lifting capabilities than their static strengths and, therefore, are superior and more reliable predictors of how much weight individuals can safely lift. The majority of studies supporting this observation, however, have addressed only infrequent lifting. Mital *et al.* (1986b) suggest that repetitive dynamic strength (RDS), instead of peak dynamic strength, should be used to screen workers for frequently performed manual lifting tasks. In their work, Mital *et al.* hypothesized that maximum dynamic strength (MDS) will not be as strongly correlated with lifting capabilities for frequent lifting tasks as those dynamic strengths which take into consideration the effect of repetition or frequency (RDSs).

To test this hypothesis, 21 physically healthy males exerted repeatedly on a dynamic strength simulator (Figure 7.13) at frequencies of one per day, once per mintue, three times a minute, and six times a minute. All exertions were carried out across two height levels: floor to 81 cm height and 81 cm to 152 cm height. In addition to RDSs at

Table 7.30. *Overall correlation coefficients between PAW and strengths (Mital et al., 1986b) by permission of the publishers, Butterworth & Co. (Publishers) Ltd.* © *1986*

Strength	Correlation	Level of significance
Isometric		
Arm	0·103	18·37
Shoulder	0·193	1·23
Back	0·079	30·99
MDS	0·293	0·01
RDS	0·560	0·01

Table 7.31. Correlation coefficients between PAW and dynamic strengths for two lifting heights (Mital et al., 1986b) by permission of the publishers, Butterworth & Co. (Publishers) Ltd. © 1986

Lifting height	Strength	Correlation	Level of significance
Floor to 81 cm	MDS	0·107	32·99
	RDS	0·561	0·01
81 cm to 152 cm	MDS	0·158	15·07
	RDS	0·463	0·01

Table 7.32. Correlation coefficients between PAW and dynamic strengths for various lifting frequencies (Mital et al., 1986b) by permission of the publishers, Butterworth & Co. (Publishers) Ltd. © 1986

Lifting height	Strength	Correlation	Level of significance
Once/day*	MDS	0·749	0·01
	RDS	0·749	0·01
Once/minute	MDS	0·328	3·39
	RDS	0·390	1·07
Three/minute	MDS	—	
	RDS	—	
Six/minute	MDS	—	
	RDS	—	

*For lifting frequency of once per day, MDS and RDS values were the same.
—No appreciable correlation.

these lifting frequencies and height levels, isometric strengths (arm, shoulder and back) and psychophysically acceptable weights (PAWs) for various frequency–lifting height combinations were also determined.

The results indicated that both maximum repetitive dynamic strengths were more highly correlated with PAW than isometric strength (Table 7.30).

Between MDSs and RDSs, RDSs were more highly correlated with PAWs than MDSs (Tables 7.31 and 7.32). The effect of repetition (frequency) on dynamic strengths is shown in Figures 7.14a and b.

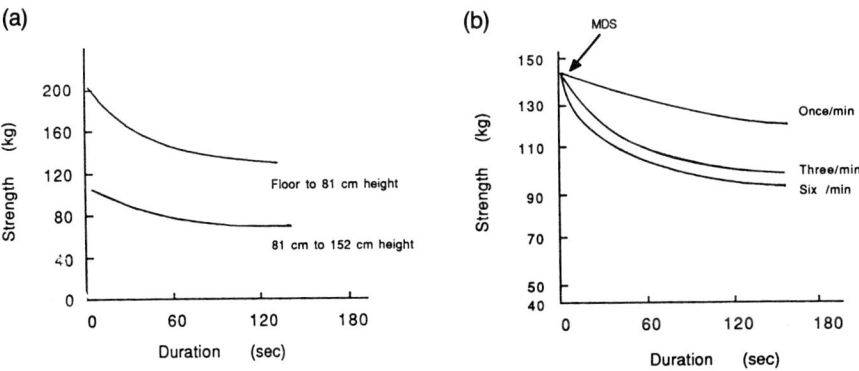

Figure 7.14. Relationship between RDS and duration of frequent exertions for (a) two heights of lifting, (b) various lifting frequencies (Mital et al., 1986b) by permission of the publishers, Butterworth & Co. (Publishers) Ltd. © 1986

The correlations in Table 7.32 also suggest that for manual lifting tasks performed more frequently than once per minute, individual's strengths become unsatisfactory means of assessing his or her psychophysical lifting capacity. It appears that somewhere between one lift per minute and three lifts per minute, the orientation of lifting tasks changes and muscular strength no longer remains the limiting factor. Perhaps physiological responses, such as steady state metabolic energy expenditure rate and heart rate, which are well known to vary with work-rate for high frequency tasks, should be utilized to assess lifting capacity in such cases.

Chapter 8
Training and Manual Handling

8.1 Introduction

A major area of controversy regarding MMH concerns the effectiveness of training programmes in reducing MMH-related injuries. It has been asserted by several researchers that training in 'safe' lifting, as it is being administered presently, does not effectively control MMH-related injuries, specifically low-back injuries (e.g., Snook *et al.*, 1978). However, even the critics of such training programmes do not advocate that they be dismissed out of hand. Rather, training programmes should represent one portion of an overall MMH programme consisting of the previously discussed job design/redesign, work practices, (on occasion) employee selection, and training. In this chapter, the various types of training programmes practices used today, and their relative effectiveness, will be presented.

8.2 The Concept of Safe Lifting

Prior to any discussion of training programmes, the concept of 'safe' lifting needs to be examined. Virtually all training programmes are based on the assumption that there are fundamentally correct (safe) and incorrect (unsafe) methods to perform MMH tasks. There are four general rules that underlie the concept of safe lifting:

1. Maintain a straight back when lifting, using the leg muscles to lower the body and lift the load.
2. When lifting, keep the load as close to the body as possible.
3. Lift the load with a smooth body motion (avoid jerking).
4. During asymmetric lifting or carrying, turn with the feet rather than twist the trunk (NIOSH, 1981).

The above rules are designed to minimize the compressive, shear and torsional forces acting on the intervertebral discs. However, the dogmatic application of these rules, without consideration of the specific work environment, can result in problems. The straight back/bent leg (or squat) lifting method does produce smaller compressive forces than bent/straight leg (stoop) lifting, provided that the load is compact enough to be lifted from between the legs. Bulky containers that cannot be lifted close to the body can produce greater load moments if a squat rather than stoop lifting method is employed. Squat lifting also assumes that the quadriceps femoris muscles are adequate to perform the lift. Finally, a higher physiological cost is associated with squat lifting

Table 8.1. Comparison of lifting methods

Method	Back orientation	Back curvature	Propulsion force	Trunk muscles used (theoretical)	Trunk Muscles used (Actual)	Spinal force	Consistency of force
Mechanical	As vertical as possible	Straight	Legs	Back	All	Compression and shear (minimize shear)	Constant
Hip-Flex	As vertical as possible (not critical)	Straight	Legs	Back	All	Compression and shear (minimize shear)	Constant
Kinetic	As vertical as possible (not critical)	Straight	Legs	Back	All	Compression and shear (minimize shear)	Variable
Dynamic	As vertical as possible (varies with circumstances)	Varies within limits of safe movement	Legs, arms and back (varies with circumstances)	All	All	Tension, compression, shear or neutral	Variable
Natural	As dictated by personal choice and habit	Varies within limits of safe movement	Legs, arms and back (varies with circumstances)	All	All	Tension, compression shear or neutral	Variable

relative to stoop lifting, which must be considered if repetitive load lifting is required by the job (Chaffin and Baker, 1970).

There are several lifting techniques reported in the literature by Jones (1985). He states that each lifting method has its merits and is suitable for some situations. Table 8.1 summarizes these methods. Figure 8.1 shows the basic postures assumed using each of these methods.

One lifting method which has received at least guarded acceptance is the kinetic lifting technique (Figure 8.1). The kinetic lifting technique involves planning the trajectory of the object and adding kinetic energy to a horizontally moving load or vertically moving body prior to lifting the load vertically.

1. The mechanical or traditional leg lift is only used for occasional or intermittent lifting of weights under 50 lbs. It should also be used for loads about 12 in×12 in or smaller. The schematic diagram would appear as shown at right.

2. The hip-flex lift is similar to that of the mechanical lift with the exception of the trunk being flexed at the hip.

3. The kinetic lift is, according to Brown, very infrequently used (only about 20% of industrial lifting is by this method). It does, however, possess components of lift number 4.

4. The dynamic lift is somewhat similar to number 3 only with a bent back. The method is used for approximately 30% of industrial lifting.

5. The natural lift is used for approximately 50% of the industrial lifting.

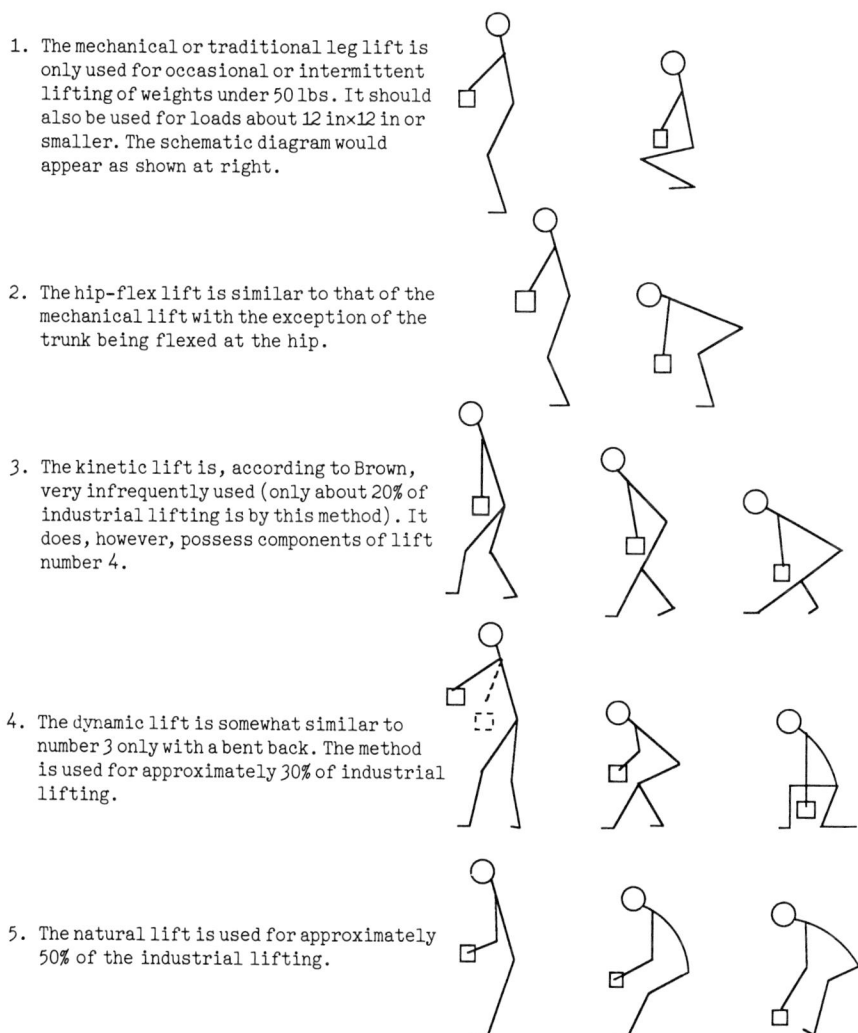

Figure 8.1. *Schematic diagrams of the most popular lifting classifications*

Kinetic lifting is presently recommended by the International Labour Office (1967). However, as Jones (1985) notes, while kinetic lifting appears to fit situations in which the load movement has both horizontal and vertical components, its benefits when the lift is entirely vertical are debatable. In addition, Leskinen *et al.* (1983a) reported that a leg lift produced smaller peak compressive forces, and a back lift produced less stress in terms of compression × time interval, than either a load kinetic lift or a trunk kinetic lift.

It should be emphasized that most training programmes presently in use understand that rules for safe lifting must be dictated by the specific parameters of the job. Possibly the most useful rule regarding safe lifting is that there is no single, correct way to lift. Lifting, like any job, can be done several ways, and because of this on site tailoring of a training programme is essential.

8.3 Training Programmes: Methods and Approaches

In general, training programmes in MMH have as their objective one or both of the following:

1. Prevention of MMH-related injuries; and
2. Rehabilitation of MMH-related injuries.

For each of these the methodology used to achieve the objective has generally emphasized one or more of the following:

1. Training in safe lifting procedures;
2. Training to increase overall strength and fitness; and
3. Methods to reduce existing back pain.

Training individuals to lift safely has been discussed. The other two methodologies will be detailed in conjunction with programmes utilizing them.

8.3.1. Training Programmes within Industrial Organizations

Training programmes within industrial settings generally emphasize prevention of MMH-related injuries through training in safe lifting procedures and, to a lesser extent, training to increase worker capacities. In the establishment of such a programme, four major areas need to be addressed:

1. What are the specific aims of the programme?
2. Who will administer the programme?
3. What will be the content of the programme?
4. How will the effectiveness of the programme be determined?

Regarding the first issue, the aims of a training programme must be outlined in as much detail as possible in order to facilitate the implementation of the programme. Without 'operationalized' aims, it becomes virtually impossible to determine the optimal administrator of the training programme, the most productive programme content and viable measures of programme effectiveness. The *Work Practices Guide for Manual Lifting* (NIOSH, 1981) suggests three aims for a training programme in MMH:

1. Make the trainee aware of the dangers of careless and unskilled MMH;
2. Demonstrate means of avoiding unnecessary stress; and
3. Teach workers to be aware of what they can handle safely.

The above aims are generally sound. However it is the position of the authors that the third aim should be encompassed by proper job design and employee placement procedures, rather than permitting the worker to make such decisions.

With the aims of the training programme established, the issue of who is to administer the programme must be addressed. Glover (1976) has asserted that an organization must retain a permanent instructor(s) in order for long-term beneficial effects to be derived from a training programme. For large firms, a physiotherapist is recommended by Glover; for smaller firms, a part-time instructor who has undergone outside training. The logic behind a permanent 'in-house' instructor is that he would be best suited to tailor the training programme to the specifics of the jobs being performed at that organization. However, as Selby (1983) notes, all too often training programmes fall on the shoulders of line supervisors or other in-house personnel with inadequate experience in areas such as ergonomics, occupational safety and injury prevention. Whoever is selected to administer an MMH training programme, that individual(s) must have a working knowledge of the scientific principles relevant to MMH (e.g., biomechanics, ergonomics, anatomy) and have practical experience with the MMH situations and problems specific to the company.

In terms of training programme content, imparting an understanding of anatomy (in particular a working knowledge of the structure and function of the spine) is an essential first lesson. Along with this, the basic physics of MMH should also be taught. The NIOSH (1981) guide suggests the following in this area:

1. The principle of levers;
2. The difference between the force needed to resist gravitational forces on a load and the forces required to lift that load;
3. The forces associated with changes in direction of motion;
4. Momentum and kinetic energy; and
5. Newton's third law of motion.

This knowledge serves as the basis for understanding disc pressure and how body positions affect pressure. Training in basic body mechanics should not be limited to lifting. It should be stressed to workers that body mechanics are involved in sitting, standing, bending, reaching, stooping, turning, etc., and that all of these motions are important in reducing the risk of MMH injuries.

It is at this point in a training programme that procedures for safe lifting are introduced. To summarize the previous discussion of this concept, rules for safe lifting must be dictated by the parameters of the job. As such, this portion of the training programme should occur at the job location or, at the very least, in an environment closely simulating the relevant job conditions. In addition to the general rules of safe lifting presented previously, the NIOSH (1981) guide suggests some physical factors associated with work environments which workers need to recognize:

1. Make sure load is free to move (i.e., not stuck);
2. Availability of lifting aids;
3. Presence of or provisions for proper handles to grasp;
4. Need for protective clothing; and
5. Make sure the area for MMH is clean, dry, non-slip and clear of obstructions.

In addition to training in safe lifting procedures, several programmes establish exercise regimens designed to increase the overall strength and fitness of workers. Several such programmes have been described by Snook and White (1984). Those programmes described by Snook and White emphasized fitness/strength factors such as endurance, abdominal muscle strength, flexibility, exercise blood pressure and recovery heart rate. The studies reported indicated a reduction in injury rates as a function of strength and fitness training. Two additional strength/fitness training programmes warrant mention. R.J. Hawley (pers. comm.) implemented a training programme consisting of flexibility exercises prior to the beginning of the work shift and following the mid-shift meal break. Congleton (1983) reported a training programme consisting of flexibility exercises prior to the beginning of the work shift reduction in back injuries was associated with the implementation of each programme. However, despite these findings, Snook and White note that, due to epidemiologic problems and lack of control groups, the efficacy of strength and fitness training as a technique for back injury prevention remains debatable.

Selby (1983) provides some additional content areas for an MMH training programme. She notes the importance for these techniques to be transferable to the home situation (e.g., doing yard work, exercising) so that home injuries are not brought to work by the employee. A similar viewpoint is expressed by Glover (1976). Workers should also be given information regarding first aid treatment techniques (e.g., massage, stretching, aspirin) they can utilize when back pain occurs. Finally, it should be noted that a training programme is not a one-off occurrence. Refresher courses should be scheduled at regular intervals. Moreover, the content of the training programme should be updated in parallel with any changes made in the work environment.

8.3.2. Training/Rehabilitation Programmes Outside the Industrial Organization

Programmes conducted outside the industrial environment generally encompass the programme aims and contents discussed above. The major distinction is that programmes involving rehabilitation of existing back injuries are usually performed outside the industrial organization. Such programmes are often grouped under the generic heading 'back schools', although emphasis, format and curriculum can vary widely from programme to programme. The discussion of back schools/rehabilitation programmes that follows is based largely on the information compiled by Bergquist-Ullman and Larsson (1977).

The back school described by Bergquist-Ullman and Larsson consists of four 45-min sessions conducted over 2 weeks under the supervision of a physiotherapist. The session contents are outlined below.

Session 1. Presentation of course contents; different aspects of back disorders (how, why, when, and to whom do back disorders occur); anatomy and back function; methods of treatment; use of the Semi-Fowler position.

Session 2. Mechanical strain due to position and movement; influence of centre of gravity on back strain; function of muscles and their influence on the back; performance of isometric abdominal muscle exercises.

Session 3. Practical application of previous knowledge (proper sitting and standing postures, proper lifting techniques – based on working conditions); performance of exercises for strengthening leg muscles; training to remain active during back pain.

Session 4. Physical training in water; encouragement to participate in physical activities; examination to assess learning; patients provided with written summary of back school contents.

In addition, suggestions are provided by the physiotherapist to adjust the work site to improve working conditions or to transfer the patient to a more suitable job. The therapist conducts a follow-up visit to the work site to assess the extent to which the instructions are being followed.

As noted, programme content varies considerably from one back school to another. Some programmes incorporate intensive physical therapy administered by physiotherapists trained in manual therapy. The programme described was developed by Cyriax (1959), Kaltenborn (1975) and Lewit (1977). The initial step in such therapy is diagnosis via examination of the spine while standing, sitting, lying prone, supine and in the lateral position. The spine is examined through a series of standard tests for evidence of scoliosis, lateral deviations, hyper- and hypomobility. Based on the diagnosis, a specific treatment regimen is prescribed. Back injury diagnosis and their related therapy regimens are summarized in Table 8.2.

Table 8.2. Diagnosed back problems and their related treatment (Bergquist-Ullman and Larsson, 1977).

Diagnosis	Treatment
Hypomobility	Hypomobile segment is moved passively (articulated) while adjacent mobile segments are fixed
Hypermobility	Stabilization exercises (patient applies force with back against manual pressure) combined with strict back regimen
Blocked iliosacral joint	Gapping (with sacrum and ilium on opposite side of blockage fixed, patient performs maximal trunk rotation in direction opposite to blockage while exhaling)
'Rotated' pelvis	Articulation wherein therapist presses backwards on anterior superior iliae spine and pulls on ischial tuberosity while patient lays on side opposite of rotated ilium
Contracted postural muscles	Stretching exercises for involved muscle groups involving isometric contractions and passive flexion
Weak muscles	Static and dynamic exercise

A final issue regarding rehabilitation is the reintroduction of the worker into the industrial organization. Snook and White (1984) reported data from McGill (1968) indicating that 50% of workers off work for at least 6 months due to back injuries never return to productive employment, with the percentage falling near zero for workers off work for more than 2 years. These data indicate the need for methods whereby back-injured workers can return to the industrial setting as quickly as possible following the injury. Selby (1983) notes that 4 days of rehabilitation are required to achieve full mobility for each day that a worker is immobile. Selby recommends light duty for injured workers as a means to facilitate the rehabilitation process, and to provide a physiological environment conducive to recovery. While the recommendation is sound, it should be noted that often management and the worker are not receptive to light-duty work. Management sometimes perceives light-duty work as 'make-work', and the worker is sometimes resistant due to a perceived drop in work status and/or a drop in pay. Also, before the institution of light-duty work, it must be empirically determined that the 'light' work represents no hazard to a worker with diminished physical capacity.

8.4 Effectiveness of Training Programmes

A conclusive statement regarding whether or not training programmes represent an effective means for preventing back injuries is difficult to make because evidence exists supporting both positions. In support of training programmes, Glover (1976) cites a preliminary study indicating a 65% reduction of lost working hours due to back pain, in a factory where all employees were required to participate in an intensive lifting course. Snook and White (1984) report several studies reporting reductions in back injuries as a function of training programmes emphasizing safe lifting techniques (Glover, 1976; Miller, 1977) and programmes designed to increase overall strength and fitness (Rowe, 1969; Cady et al., 1979; Imrie, 1983). The back school described by Bergquist-Ullman and Larsson (1977) was compared to physical therapy alone and placebo treatments, and was found to be superior in terms of days required to return to work. The back school was superior to the placebo treatments but not significantly different from physical therapy alone in terms of days required to achieve relief from back pain. According to Asfour et al. (1984a), training can significantly increase an individual's strength and endurance. Evidence presented further suggests that no more than 10 training sessions, 2 h each, are needed.

Despite findings such as these, doubts remain regarding the efficacy of training programmes. One reason for these continued doubts is that studies regarding training programme effectiveness tend to be poorly controlled. Lack of control groups is a major problem (the Bergquist-Ullman and Larsson, 1977, study being an exception), and in most instances back injury control measures in addition to the training programme exist in the industrial organization. A second reason for doubting training programme effectiveness occurs when comparing training with other back injury control methods. Snook et al. (1978) compared the occurrence of back injuries in industries employing training programmes in safe lifting, selection procedures (medical histories, physical examinations and X-rays), and injuries attributable to job design practices. No significant differences were found in the expected number of injuries for companies with or without training programmes. No significant differences were found in the expected number of injuries for companies with or without selection programmes. A significant difference was found between expected injuries when comparing jobs which less than 75% of the working population could perform without exertion, and jobs which 75% or more of the population could perform without exertion. Based on these findings, Snook *et al.* (1978) concluded that ergonomic job design represents the most appropriate means of reducing back injuries in industry. It should be noted that the quality of the various training programmes was not evaluated, and selection programmes involving techniques such as strength testing were not included.

8.5 Summary

The effectiveness of training programmes in reducing MMH-related injuries remains debatable despite some promising reports. More research is required to identify those components of a training programme which facilitate its effectiveness. This can only be accomplished through well controlled training programme evaluations which so far have been painfully absent. However, despite criticism, enough evidence is available in support of training programme effectiveness to warrant its further employment, provided these programmes are conducted in conjunction with ergonomic job design and employee selection procedures.

Chapter 9
Determination of Rest Allowances

9.1 The Need for Rest Allowances

The ability to perform physical tasks depends on muscular contractions. These muscular contractions become weak as the metabolic energy resources are depleted. Muscles also get tired and ache when significant amounts of lactic acid are accumulated within the muscular system. Whenever these conditions occur, a state of fatigue, known as physiological muscle fatigue, results and the level of work output starts decreasing. The quality of output also starts to deteriorate. In order to avoid these detrimental effects, rest allowances are provided. During rest periods, workers recover from the fatigue that is generated as a result of producing work.

9.2 Methods of Determining Rest Allowances

The necessary rest allowances are determined with the help of either a time study or a physiological evaluation. The time study method is faster, but less sophisticated. It frequently results in different standards for the same job in different industries. The reason for variation in industrial time standards, for the same job, results from different levels of 'normal' within a firm, and the subjectivity and inaccuracy involved in the determination of performance rating. Rest allowances which are an arbitrary percentage of erroneous normal times, therefore, are either inadequate or unnecessary. Moreover, the same rest allowance is applicable to each worker since the individual differences are ignored. The physiological evaluation method of determining rest allowances, on the other hand, is an accurate and reliable evaluation method since all recommendations are based on objective data. It also accounts for individual's capabilities by taking his/her aerobic capacity into consideration. Until recently, the physiological method was regarded as cumbersome and slow, but the development of portable and battery operated oxygen consumption computers and heart rate meters have simplified the procedural difficulties. A complete physiological evaluation of the job–worker system can now be made fairly quickly. Also, in many situations rest allowances can be determined only by a physiological evaluation. For instance, when jobs are performed in hot and humid climates.

Two different kinds of physiological evaluation methods have been recommended in the literature: (i) strain based, and (ii) metabolic energy expenditure based. In the strain based methodology, working heart rate (heart rate above resting level) is used for evaluating strain. On this basis, Rohmert (1973a, b) has proposed rest allowances for static muscular work (Figure 9.1) and dynamic muscular work (Figure 9.2). These rest allowances, however, were not specifically developed for MMH activities.

The metabolic energy expenditure rate methodology generally allows for rest allowances when, and if, the metabolic energy demands of the job, for 8 h of work, exceeds about 33% (one-third) of the aerobic capacity of a person (Spitzer, 1952; Lehman, 1962; Murrell, 1964; Monod, 1967; American Industrial Hygiene Association, 1971; Grandjean, 1982; Krager and Hancock, 1982). Without rest, higher percentages, up to 50% can be maintained only for a short time (an hour or less) (Astrand and Rodahl, 1977). There are some exceptions, such as long distance runners, who can maintain substantially higher percentages of metabolic energy expenditure rate for up to 2–3 h without a break. However, in their cases the individuals are highly trained and are usually born with greater physical capabilities. For most industrial workers, the job metabolic energy demand limit of 33% is widely accepted.

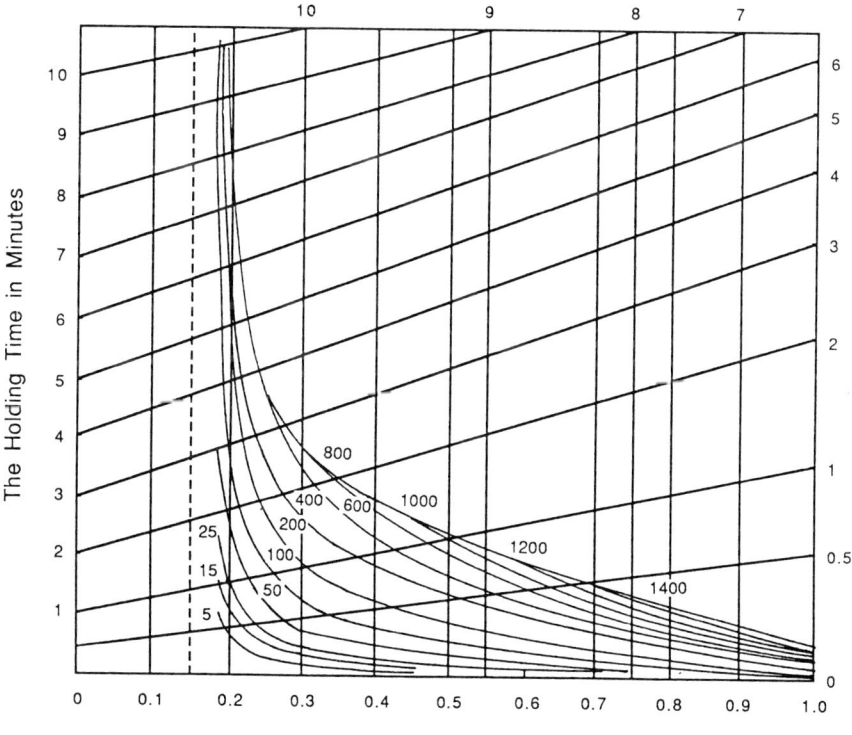

Figure 9.1. Percentage rest allowances for various combinations of holding forces and time (Rohmert, 1973a)

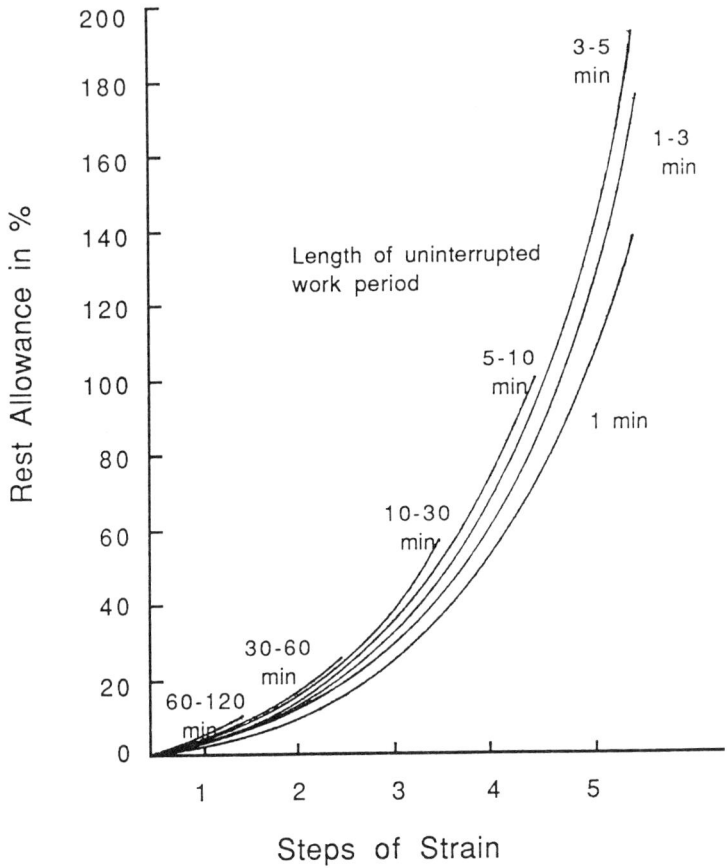

Figure 9.2. Rest allowances in dynamic muscular work

9.2.1 Metabolic Energy Expenditure Rate Models for Determining Rest Allowances

Two different models have been proposed in the past. Muller (1953) reported the rest–pause schedule for some tasks such as carrying load, pulling carriage, ascending stairs, walking and bicycling. He used 4 kcal/min as the net amount of energy expenditure which one could perform per day (or 5 kcal/min including basal metabolic rate). He developed a straight line relationship between the net energy expenditure (kcal/min) and percentage of recovery time to working time. If a task requires 8 kcal/min and working time is 10 min, a worker needs recovery time 100% of working time or 10 min of rest.

Spitzer (1952) proposed resting time as a percentage of working time:

$$R = \frac{(M-1) \times 100}{4}$$

where R = resting time as a % of working time; M = net energy cost (kcal/min) = Total energy cost − resting energy cost.

The same example is in endurance time is solved below.

$$M = (1·25 - 0·25) \text{ l/min} \times 5·05 \text{ kcal/l}$$
$$= 5·05 \text{ kcal/min}$$
$$R = \frac{(5·05 - 1) \times 100}{4}$$
$$= 26·25\% \text{ of working time.}$$

In an 8h-day the rest period is:

$$-\frac{26·25 \times 8}{126·25} = 1·66 \text{ h} = 99·8 \text{ min.}$$

In an 8 h-day the work period is:

$$\frac{100}{126·25} \times 8 - 6·35 \text{ h} = 380·20 \text{ min.}$$

Murrell (1964) proposed resting time as a percentage of total work time:

$$R = \frac{T \times (M - 4)}{M - 1·5}$$

where R = resting time (min); T = total working time (min); M = net energy cost (kcal/min).

A modification of this model is made for an 8 h day.

$$R = 480 \times \frac{(M - 4)}{2M - 5·5}$$

Solving the same example

$$R = \frac{480 \times (5·05 - 4)}{(2 \times 5·05 - 5·5)}$$
$$= 109·5 \text{ min.}$$

9.3 Limitations of the Metabolic Energy Expenditure Rate Method

Appropriateness of using 33% of aerobic capacity as the cut-off limit, beyond which rest allowances are to be provided, is questionable since it does not take into consideration the total energy available for producing work. Also, what this percentage should be for 10- or 12-h shifts is presently not known.

A variation of the physiological methodology, described above, is the use of acceptable levels of metabolic energy expenditure rate (which does not lead to excessive fatigue), based on total energy consumption, for determining rest allowances. This approach is also widely used (American Industrial Hygiene Association, 1971; Grandjean, 1982). There is, however, some conflict as to what is the total energy consumption per day. While one source suggests a value of 4200 kcal (American Industrial Hygiene Association, 1971) the other proposes a 4800 kcal (Grandjean, 1982). Surprisingly, both sources recommend an occupational energy expenditure rate limit of 5 kcal/min as the limit of acceptable level of job demand. Obviously, both sources differ in their assessment of basal and leisure metabolism. Regardless of which one is correct, for most industrial workers these limits are unrealistically high (National Research Council, 1980). Recent nutritional studies also indicate that the assumption that humans have plenty of food reserves to meet the demand may not be correct (R.C. Bozian, 1983). Given excess food supply and adequate oxygen consumption, still only a certain amount of calories wil be consumed. Additional calories (excess food) will increase the weight of the individual by adding to the body fat. Inadequate supply of food, on the other hand, will first result in the reduction of fat and then reduction in the muscle mass.

The above discussion raises the obvious question. Which approach is appropriate for physiological evaluation and estimation of the duration of rest breaks? Both approaches have been recommended in the published literature (American Industrial Hygiene Association, 1971; Krager and Hancock, 1982).

9.4 A Comprehensive Metabolic Energy Model for Determining Rest Allowances

Mital and Shell (1984) sought the answer to the above paradox and attempted to refine the physiological methodology. The procedure developed is based on physiological evaluation of job demands and worker capabilities and provides rest allowances unique to the individual performing a physical task. The data for developing the procedure were collected experimentally. The experimental details are given in the following subsections.

9.4.1 Data Collection and Results

Ten industrial workers, five males and five females, ranging in age from 24–55 years, were recruited from industries located in the greater Cincinnati area. The subjects were experienced (at least 6 months) in industrial palletizing and stacking tasks. Table 9.1 shows the subject population descriptors. The aerobic capacity of the subjects was estimated (Chaffin, 1972). It was assumed that subjects reflected the capabilities of the average industrial work-force rather than large-sized highly active professionals such as lumberjacks or construction workers.

Table 9.1. *Subject descriptors (Mital and Shell, 1984)*. Reproduced with permission of Institute of Industrial Engineers

	Males ($n = 5$)		Females ($n = 5$)	
	Mean	Range	Mean	Range
Age (years)	32	24–45	36·2	24–55
Height (cm)	166·2	155·7–171·5	163·4	154·1–176·7
Body-weight (lbs)	165·25		160·07	120·6–190·8
Experience (years)	7	1–20	5	3–10
Aerobic capacity* (kcal/min)	16·9	14·5–18·5	11·0	9·7–11·6

*Estimated from Chaffin (1972)

The experiment simulated palletizing/stacking tasks which were performed by these subjects under controlled laboratory conditions. Each subject was asked to stack containers of different sizes from one height to another height. Containers varied in size from 30·48–60·96 cm while the height ranged from approximately 81–152 cm. Four different paces were included in the experiment: containers were stacked at a rate of either 1, 4, 8 or 12/min. Each subject performed only one combination of container size, height and pace (out of 36 possible), selected at random. He or she was allowed to control one element of the work-rate (weight) and was asked to set it up such that the final work-rate would be the maximum rate that could be sustained continually, with breaks for lunch, supper and other personal needs, for 8 or 12 h. It took each subject approximately 20–45 min to set the maximim acceptable work-rate.

On the second day, the subjects were asked to perform with this maximum acceptable work-rate for 8 h, and eventually for 12 h. Subjects were given a 15 min break after 2 h, a lunch break of 30 min after 4 h, another 15 min break after 6 h, a 30 min break after 8 h, and finally another 15 min break after 10 h. No two subjects performed on any given day. Heart rate and metabolic energy expenditure rate of the workers were monitored continually.

A special palletizing equipment was constructed to simulate the task in the laboratory. An MRM–1 Oxygen Consumption Computer and a Quinton Heart Rate Meter were used to record physiological responses of the subjects.

The results of the experiment indicated that the subjects initially (during the 45-min estimation period on the first day) selected a work-rate, that they thought they could sustain for 8 and 12 h, corresponding to approximately one-third of their aerobic capacity. As has been indicated in the literature (American Industrial Hygiene Association, 1971; Astrand and Rodahl, 1977; Mital et al., 1982), this should be the case if individuals work at their peak but at levels which avoid undue fatigue. However, on the second day, as the day progressed, subjects made reductions in the maximum acceptable work-rate (Figure 9.3).

These adjustments in the work-rate continued all day long. The metabolic energy expenditure rate of the subjects also declined continually with them. Figure 9.4 shows the change in metabolic energy expenditure rate with time both for males and females. Heart rates of the individuals, on the other hand, did not change significantly with time. Figure 9.5 shows the variation in heart rate of males and females with time. The decline in heart rate due to reduced workload was compensated by increasing fatigue, thereby causing almost no change in heart rate with time. The decrease in metabolic energy expenditure rate, on the other hand, was not expected.

Since subjects had selected a work-rate which elicited an energy expenditure rate equivalent to approximately one-third of their aerobic capacity, it was expected that the selected work-rate could be sustained throughout the working day. It did not happen. Work-rates corresponding to one-third of aerobic capacity could not be sustained. The actual work-rates sustained for 8 h corresponded to approximately 28–29% of subject's aerobic capacity; for 12 h this number was 23–24%. Since the aerobic capacity approach could not explain the observed pattern, an explanation based on total energy available to produce work was sought.

Using the basis of total available energy, and the fact that it has limits, a compre-

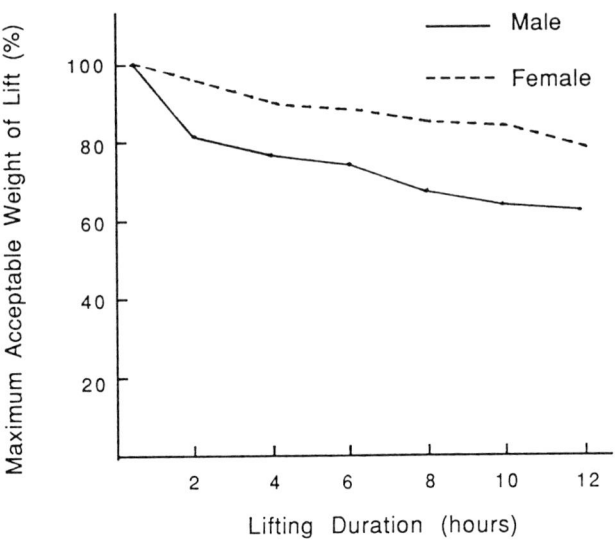

Figure 9.3. *Decline in weight with time (Mital and Shell, 1984)*

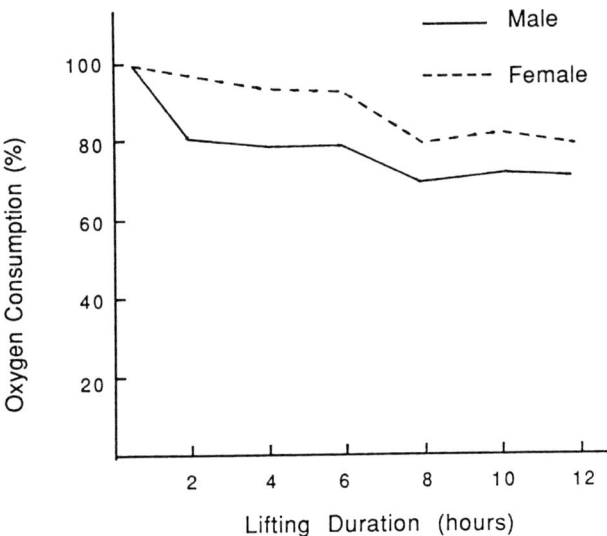

Figure 9.4. *Decline in oxygen consumption with time (Mital and Shell, 1984)*

hensive metabolic energy model structure was developed to determine physiological fatigue allowances unique to the worker. The subsequent sections describe the modelling concept of its validation.

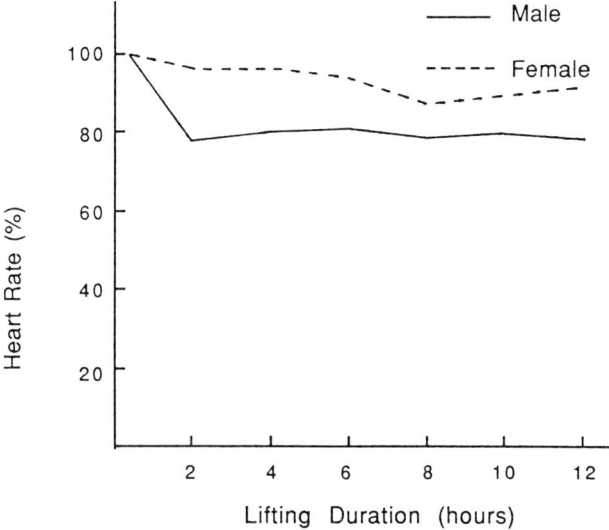

Figure 9.5. Change in heart rate with time (Mital and Shell, 1984).
Figures 9.3–9.6 reproduced with permission of Institute of Industrial Engineers

9.5 Model Development

9.5.1 Concept

This proposed model also utilizes the total energy concept as suggested earlier (American Industrial Hygiene Association, 1971; Grandjean, 1982). There are, however, major differences between the proposed procedure and earlier procedures. The total daily energy requirement of an individual is based on recent data (National Research Council, 1980) and is adjusted for age. The energy requirement is also adjusted for energy required for food ingestion. The calculation of basal and leisure metabolism is based on the hours during which these activities are performed rather than standard assumption that each lasts for 8 h. The work (shift) duration is also taken into consideration.

9.5.2 Assumption

It was assumed that there is a limit to the amount of metabolic energy available to produce work. If the job requires more energy expenditure from the individual, excess energy cannot be supplied by additional food intake. Recent nutritional studies indicate (R.C. Bozian, 1983) that excess food in such cases only adds to the body-weight and is not burned to meet greater job demands. Inadequate food supply, on the other hand, leads to a reduction in the body fat followed by a reduction in the body muscle mass.

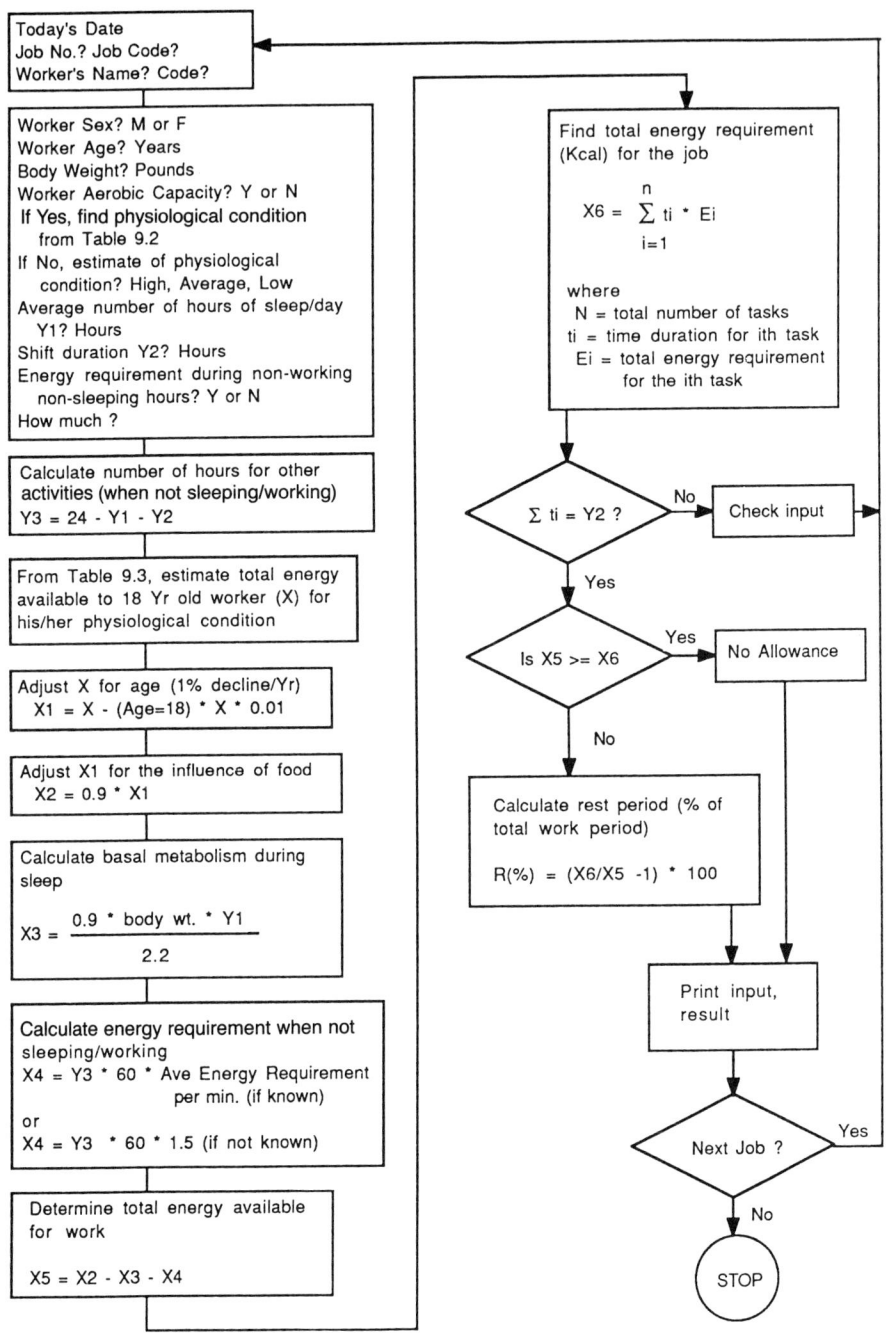

Figure 9.6. The model structure (Mital and Shell, 1984)

9.5.3 Model Structure

Figure 9.6 shows the model structure. The inputs required are:

(i) Worker gender (M or F);
(ii) Worker age (years);
(iii) Body-weight (pounds);
(iv) Number of hours of sleep per day (hours);
(v) Shift duration (hours);
(iv) Number of tasks performed during the shift;
(vii) Time duration of each task (hours);
(viii) Metabolic energy requirement for each task (kcal/h); and
(ix) An estimate of worker's physiological condition (high, average, or low).

To increase the accuracy of the procedure two additional pieces of information are needed. This optional information is:

(i) Worker aerobic capacity (ml/min/kg); and
(ii) Average energy requirement/min when not working or sleeping (kcal).

In the absence of these, the recommended rest allowances will be conservative.

As shown in Figure 9.6, if the aerobic capacity of the worker is known then his/her physiological condition is calculated from Table 9.2. This table gives a relationship between worker sex, age, aerobic capacity and physiological condition and is reproduced here from the original reference (American Heart Association, 1972). Five different categories (low, fair, average, good and high) are used to describe the worker's physiological condition. In case the aerobic capacity of the worker is not known, a qualitative estimate (low, average or high) should be provided. Once the general physiological condition of the worker is known, the total amount of energy available to an 18-year-old, per day, is determined from Table 9.3. Next, this total amount is adjusted for age. A decline of 1% per year of age, above 18 years, is used as

Table 9.2 The range of expected maximal aerobic capacity, by sex and age (American Heart Association, 1972).

Age (years)	Maximal oxygen uptake (ml/kg/min)				
	Low	Fair	Average	Good	High
Men					
20–29	25	25–33	34–42	43–52	53*
30–39	23	23–30	31–33	39–49	49*
40–49	20	20–26	27–35	36–44	45*
50–59	18	18–24	25–33	34–42	43*
60–69	16	16–22	23–30	31–40	41*
Women					
20–29	24	24–30	31–37	38–48	49*
30–39	20	20–27	28–33	34–44	45*
40–49	17	17–23	24–30	31–41	42*
50–59	15	15–20	21–27	28–37	38*
60–69	13	13–17	18–23	24–34	35*

*Or higher.

the adjustment factor. This decline was determined from the nutritional requirement data that have been reported (National Research Council, 1980). Further adjustments are made to this new value for the energy required in digesting the food. Approximately 10% of the total energy is expended in this process (Darden, 1976).

Table 9.3. Total energy available per day (kcal) to an 18-year-old (Recommended Dietary Allowance, 1980; Bozian, 1983).

Sex	Aerobic capacity	Physical condition				
		Low	Fair	Average	Good	High
Male	Known	2100	2575	3050	3225	4000
	Not known	2100	—	3050	—	4000
Female	Known	2100	2325	2550	2775	3000
	Not known	2100	—	2550	—	3000

Once the adjustments have been made for age and the influence of food, basal metabolism and energy expenditure during the leisure period are calculated. A value of 0·9 kcal/kg of body-weight/h of sleep (corrected for metabolic savings in sleep) is used for determining the basal metabolism. For determining the energy requirement during leisure time either the actual value (in kcal/min) is used or a default value of 1·5 kcal/min is used (this value represents energy expenditure during a typical leisure time sedentary task).

The basal and leisure metabolism are subtracted from the adjusted total energy to determine the amount available for producing work. Next the total energy requirement for the job is determined as follows:

$$\text{Total energy requirement for the job (kcal)} = \sum_{i=1}^{N} t_i E_i$$

where t_i = time (in hours) for the ith task and E_i = its energy requirement in kcal. The last step is the determination of the rest duration. The rest period, as a percent of the total work period (shift duration), is given by:

$$\text{Rest (\%)} = \left[\left(\frac{\text{total energy requirement for the job}}{\text{adjusted, total energy available for work}} \right) - 1 \right] \times 100$$

The above procedure is applicable to any shift duration and any combination of physical tasks.

9.5.4 Model Validation

The model proposed in the previous section was validated by comparing its predicted performance with the actual data obtained in the experiment described earlier. As shown in Figure 9.4, the metabolic energy expenditure rate of the objects declined continuously with time. On the average, oxygen uptake of males declined by approximately 2·6%/h. It meant that males could sustain an energy expenditure rate of 3·96 kcal/min for 8h: for 12h, the metabolic rate was 3·44 kcal/min. For females the decline was 1·9%/h and the corresponding numbers were 2·97 and 2·70 kcal/min. If, in fact, subjects performed at these levels of physical demand, they would not require any rest allowances. The following example is given.

In this example, the total energy available for producing work was calculated. The procedure described in Figure 9.6 was used for this purpose. If the metabolic energy expenditure rate that could be sustained for the working duration (3·96 and 3·44 kcal/min for males for 8 and 12 h, respectively, and 2·97 and 2·70 kcal/min for females and 8 and 12 h, respectively; for shift durations between 8 and 12 h, linear interpolations between the two values may be used) was maintained, there would be no need for rest allowances. In the following example, the subject operated at a metabolic rate that could be sustained for the entire working duration without rest.

9.5.5 Example

The subject involved was a male with the following particulars:

Age = 29 years;
Body-weight = 145 lbs;
Aerobic capacity = 55 ml/kg/min;
Hours of sleep/day = 8;
Shift duration (hours) = 8 (case 1) and 12 (case 2);
Energy requirement during leisure time (kcal/min) = 1·7;
Energy required for the job (kcal/h) = 237·6 (= 60 min × 3·96 kcal/min) for case 1, = 206 (= 60 min × 3·44 kcal/min) for case 2.

Two different cases were examined: case 1: 8-h shift; case 2: 12-h shift. Table 9.4 shows the program output for the two cases. In both cases, no rest allowances are needed. The total energy requirement for the job is quite close to the total energy

Table 9.4. Program output (Mital and Shell, 1984; 1985a, b)

Date:	7/3/84	
Job number:	Example	
Job code:	Case 1	
Worker name:		Mr. X
Sex:		M
Age (years):		29
Weight in pounds:		145
Aerobic capacity in ml/kg/min:	55	
Physiological condition computed from the table:		5
Hours of sleep per day:	8	
Shift duration (hours):		8
Energy available for 18-year-old from the table (X):		4000
Energy required when not sleeping and not working:		816
Total energy available for work:	1913·45	
Toral number of jobs:	2	
Time duration of job:	Energy required for job	
7	237·6	
1	102	
Total energy requirement for the job (kcal):		1765·2
Rest period as percent of total work period:		0

Program output

Table 9.4. Continued

Date:	7/3/84
Job number:	Example
Job code:	Case 2
Worker name:	Mr. X
Sex:	M
Age (years):	29
Weight in pounds:	145
Aerobic capacity in ml/kg/min:	55
Physiological condition computed from the table:	5
Hours of sleep per day:	8
Shift duration (hours):	12
Energy available for 18-year-old from the table (X):	4000
Energy required when not sleeping and not working:	408
Total energy available for work:	2321·45
Total number of jobs:	2
Time duration of jobs:	Energy required for job
10·25	206
1·75	102
Total energy requirement for the job (kcal):	2290
Rest period as percent of total work period:	0

available for producing work. The differences in the total energy available and the total energy requirement for the job are primarily due to:

(i) Energy expenditure during leisure time was estimated, not measured;
(ii) A linear decline in oxygen uptake was assumed. The actual decline is shown in Figure 9.4;
(iii) Hours of sleep were estimated.

It should be kept in mind that the above analysis is based on the experimental procedure which allows subjects to work only 7 out of 8 h and 10·25 out of 12 h.

To date, the model has also been validated on a number of subjects. In each case, it proves to be quite reliable.

9.6 Computer Program

In order to use the methodology described in sections 9.4 and 9.5, a user-friendly computer program listed in the Appendix was developed by Mital and Shell (1985b). This program allows industrial engineers and ergonomists to determine the appropriate duration of rest periods unique to the individual and the job. Thus, it ensures that individual productivity will be maximized and any over-exertion or undue physical fatigue will be prevented.

The program requires several inputs related to the job and the worker performing it: sex; age (in years); body-weight (in pounds); aerobic capacity and qualitative physiological condition (in ml/kg/min or in qualitative terms such as low, average, high); hours of sleep (per 24 h day); shift duration (in hours); task duration (in minutes) and total metabolic energy requirement for each task (in kcal) for the duration of the task.

The program is written in BASIC and was developed for a Vector Graphic System

5005 microcomputer. The date on which the program is run is the first piece of information required after the program is loaded and run. Once the program is executed, the screen displays:
  DATE:
The user responds with the date. Next, the job number, job code and worker name are entered.
The program then requests: SEX (M—MALE, F—FEMALE):
User response, in this example, is: M
The program then flashes: Age (BETWEEN 20 AND 69):
User response is: 35
Next, the program requests: WEIGHT IN POUNDS:
User response is: 150

The worker's physiological condition is determined next. If the aerobic capacity (in ml/kg/min) is known, it is entered; otherwise, it is estimated from qualitative responses.
The program requests: IS AEROBIC CAPACITY KNOWN? (Y/N):
User response is: Y
The program responds: ENTER AEROBIC CAPACITY IN ML/KG/MIN (1–100):
User response is: 44
The program flashes: COMPUTED PHYSIOLOGICAL CONDITION: 4
It then asks: HOURS OF SLEEP PER DAY (Y1):
User response is: 7
The program then requests: SHIFT DURATION IN HOURS (Y2):
User response is: 8
The program responds with: ENERGY AVAILABLE TO 18 YEAR OLD: 3525.

Next, the energy requirement during non-working and non-sleeping hours is determined.
The program asks: IS ENERGY (KCAL/MIN) WHEN NOT SLEEPING/WORKING KNOWN?
User response is: N
The program responds: X5 = 1393·63
It then requests: NUMBER OF JOBS N:
User response is: 3

The duration (in hours) and energy requirement for each job (in kcal/hour) are entered next.
The program requests: ENTER TIME DURATION IN HOURS FOR ITH TASK:
User response is: 2
The program asks: ENTER TOTAL ENERGY/HOUR REQUIRED FOR THE ITH TASK:
User response is: 180
The last two requests are repeated for each job. The respective user responses are: 3   200   3   200

After the last piece of information is entered and the return key is pressed. The program responds with: REST PERIOD AS A PERCENT OF WORK PERIOD: 11.3979
It then asks: DO YOU WANT A HARD COPY? (Y/N)
User response is: Y
and the hard copy is printed (see Table 9.5).
Next, the program asks: DO YOU WANT TO RUN THE NEXT JOB? (Y/N)
User response is: N
At this, the program stops.

If the aerobic capacity of the worker is not known and the user response to: IS AEROBIC CAPACITY KNOWN? (Y/N):
is N
then the program responds with: SINCE AEROBIC CAPACITY IS NOT KNOWN, ENTER ESTIMATED PHYSIOLOGICAL CONDITION (1 = LOW, 2 = AVERAGE, 3 = HIGH)
User response is: 3

For the same input data when the aerobic capacity is not known, the program output is shown in Table 9.6. The rest allowances determined in this way are somewhat liberal because of the assumptions made in estimating the worker's physiological condition.

Table 9.5. Program output with known aerobic capacity

Date:	12/3/83		
Job number:	1		
Job code:	Example		
Worker name:		Mr. X	
Sex:		M	
Age (years):		35	
Weight in pounds:		150	
Aerobic capacity:		44	
Physiological condition computed from the table:			4
Hours of sleep per day:	7		
Shift duration (hours):	8		
Energy available for 18 year old from the table (X):		3525	
Energy required when not sleeping and not working:		810	
Total energy available for work:		1393·63	
Total number of jobs:	3		
Time duration of job:		Energy required for job	
2		180	
3		200	
3		200	
Total energy requirement for the job:		1560	
Rest period as percent of total work period:			11·9379

Table 9.6. Program output with unknown aerobic capacity

Date:	12/3/83			
Job number:	1			
Job code:	Example			
Worker name:		Mr. X		
Sex:		M		
Age (years):		35		
Weight in pounds:		150		
Estimated physiological condition:			3	
Hours of sleep per day:		7		
Shift duration (hours):			8	
Energy available for 18 year old from the table (X):				3050
Energy required when not sleeping and not working:				810
Total energy available for work:			$1038 \cdot 8$	
Total number of jobs:		3		
Time duration of job:		Energy required for job		
2		180		
3		200		
3		200		
Total energy requirement for the job (kcal):			1560	
Rest period as percent of total work period:				$50 \cdot 1726$

Appendix

Software Program: Calculating Rest Period

9.6.1. *Source Listing*

```
10  '   PROGRAM TO COMPUTE THE REST PERIOD AS A PERCENT
        OF TOTAL WORK PERIOD BY
20  '   BY ANIL MITAL
30  '   UNIVERSITY OF CINCINNATI, OHIO
40  '
50  '
60  '   THIS PROGRAM COMPUTES THE REST PERIOD AS A PERCENT
        OF TOTAL WORK PERIOD
70  '   TWO LOOKUP TABLES ARE IMPLEMENTED IN THE PROGRAM
        GIVEN THE AGE AND
80  '   AEROBIC CAPACITY OF THE WORKER, IT COMPUTES THE
        PHYSIOLOGICAL CONDITION
90  '   WHICH COULD FALL IN FIVE CATEGORIES.
100'
110'    SECOND TABLE COMPUTES TOTAL AEROBIC ENERGY FOR 18
        YEAR OLD MALE OR FEMALE
120'    FOR 24 HOURS WHEN THE PHYSIOLOGICAL CONDITION IS
        KNOWN
130'
140'
150     DIM T(20), E(20), MX(5), FX(5)
160     DIM WCAP1(5), WCAP2(5), WCAP3(5), WCAP4(5), WCAP5(5)
170     DIM MCAP1(5), MCAP2(5), MCAP3(5), MCAP4(5), MCAP5(5)
180'
190'
200'    THE FOLLOWING IS THE TABLE FOR 18 YEAR OLD MALE (MX)
        AND FEMALE (FX)
210     MX(1) = 2100: MX(2) = 2575: MX(3) = 3050: MX(4) = 3525:
        MX(5) = 4000
220     FX(1) = 2100: FX(2) = 2325: FX(3) = 2550: FX(4) = 2775: FX(5) = 3000
230'
240'
```

250′	TABULAR VALUES FOR WOMEN'S PHYSIOLOGICAL CONDITION DEPENDING UPON AGE
260′	AND AEROBIC CAPACITY
270′	
280	WCAP1(1) = 24: WCAP1(2) = 30: WCAP1(3) = 37: WCAP1(4) = 48: WCAP1(5) = 100
290	WCAP2(1) = 20: WCAP2(2) = 27: WCAP2(3) = 33: WCAP2(4) = 44: WCAP2(5) = 100
300	WCAP3(1) = 17: WCAP3(2) = 23: WCAP3(3) = 30: WCAP3(4) = 41: WCAP3(5) = 100
310	WCAP4(1) = 15: WCAP4(2) = 20: WCAP4(3) = 27: WCAP4(4) = 37: WCAP4(5) = 100
320	WCAP5(1) = 13: WCAP5(2) = 17: WCAP5(3) = 23: WCAP5(4) = 34: WCAP5(5) = 100
330′	
340′	
350′	TABULAR VALUES FOR MEN'S PHYSIOLOGICAL CONDITION DEPENDING UPON AGE
360′	AND AEROBIC CAPACITY
370′	
380	MCAP1(1) = 25: MCAP1(2) = 33: MCAP1(3) = 42: MCAP1(4) = 52: MCAP1(5) = 100
390	MCAP2(1) = 23: MCAP2(2) = 30: MCAP2(3) = 33: MCAP2(4) = 49: MCAP2(5) = 100
400	MCAP3(1) = 20: MCAP3(2) = 26: MCAP3(3) = 35: MCAP3(4) = 44: MCAP3(5) = 100
410	MCAP4(1) = 18: MCAP4(2) = 24: MCAP4(3) = 33: MCAP4(4) = 42: MCAP4(5) = 100
420	MCAP5(1) = 16: MCAP5(2) = 22: MCAP5(3) = 30: MCAP5(4) = 40: MCAP5(5) = 100
430′	
440′	
450	AGE1 = 29: AGE2 = 39: AGE3 = 49: AGE4 = 59: AGE5 = 69
460′	
470′	
480	INPUT "DATE : ", DATE$
490	INPUT "JOB NUMBER: ", JOB$
500′	
510	INPUT "JOB CODE", JOB CODE$
520	INPUT "WORKER NAME : ", W. NAME$
530	INPUT "SEX (M-MALE, F-FEMALE) : ", SEX$
540	IF SEX$ = "M" OR SEX$ = "F" THEN GOTO 570
550	GOTO 530
560′	
570	INPUT "AGE (BETWEEN 20 AND 69): ", AGE

```
580    IF AGE <19 OR AGE>69 THEN GOTO 570
590'
600    INPUT "WEIGHT IN POUNDS: ", WEIGHT
610    IF WEIGHT <1 THEN GOTO 600
620'
630'
640    INPUT "IS AEROBIC CAPACITY KNOWN? (Y/N): ", AC
       YESNO$
650    IF AC YESNO$ = "N" THEN GOTO 1260
660    INPUT "ENTER AEROBIC CAPACITY IN ML/KG/MIN (1-100):",
       ACAPACITY
670'
680'
690'   SINCE AEROBIC CAPACITY IS KNOWN, PHYSIOLOGICAL
       CONDITION IS COMPUTED
700'   FROM THE TABLE
710'   IF SEX$ = "M" GOTO 1000
720'
730'
740'   THIS SECTION COMPUTES PHYSIOLOGICAL CONDITION OF
       WOMEN
750'
760'   IF AGE>AGE1 THEN GOTO 800
770    FOR I=1 to 5
780    IF ACAPACITY<WCAP1(I) THEN GOTO 1210
790    NEXT I
800    IF AGE>AGE2 THEN GOTO 840
810    FOR I=1 TO 5
820    IF ACAPACITY<WCAP2(I) THEN GOTO 1210
830    NEXT I
840    IF AGE>AGE3 THEN GOTO 880
850    FOR I=1 TO 5
860    IF ACAPACITY<WCAP3(I) THEN GOTO 1210
870    NEXT I
880    IF AGE>AGE4 THEN GOTO 920
890    FOR I=1 TO 5
900    IF ACAPACITY<WCAP4(I) THEN GOTO 1210
910    NEXT I
920    FOR I=1 TO 5
930    IF ACAPACITY<WCAP5(I) THEN GOTO 1210
940    NEXT I
950'
960'
970'
980'   THIS SECTION COMPUTES PHYSIOLOGICAL CONDITION OF
       MEN
```

```
990'
1000  IF AGE > AGE1 THEN GOTO 1040
1010  FOR I = 1 TO 5
1020  IF ACAPACITY < MCAP1(I) THEN GOTO 1210
1030  NEXT I
1040  IF AGE > AGE2 THEN GOTO 1080
1050  FOR I = 1 TO 5
1060  IF ACAPACITY < MCAP2(I) THEN GOTO 1210
1070  NEXT I
1080  IF AGE > AGE3 THEN GOTO 1120
1090  FOR I = 1 TO 5
1100  IF ACAPACITY < MCAP3(I) THEN GOTO 1210
1110  NEXT I
1120  IF AGE > AGE4 THEN GOTO 1160
1130  FOR I = 1 TO 5
1140  IF ACAPACITY < MCAP4(I) THEN GOTO 1210
1150  NEXT I
1160  FOR I = 1 TO 5
1170  IF ACAPACITY < MCAP5(I) THEN GOTO 1210
1180  NEXT I
1190'
1200'
1210  PHYS. COND = I
1220  PRINT "COMPUTED PHYSIOLOGICAL CONDITION: ", PHYS. COND
1230  GOTO 1310
1240'
1250'
1260  PRINT "SINCE AEROBIC CAPACITY IS NOT KNOWN, ENTER ESTIMATED"
1270  INPUT "PHYSIOLOGICAL CONDITION (1 = LOW, 2 = AVERAGE, 3 = HIGH)", PHYS. COND
1280  IF PHYS. COND < 1 OR PHYS COND > 3 THEN 1270
1290'
1300'
1310  INPUT "HOURS OF SLEEP PER DAY (Y1): ", Y1
1320  INPUT "SHIFT DURATION IN HOURS (Y2): ", Y2
1330  LET Y3 = 24 - Y1 - Y2
1340'
1350'
1360' ENERGY AVAILABLE FOR 18 YEAR OLD IS COMPUTED FROM THE TABLE
1370  IF SEX$ = "F" THEN GOTO 1400
1380  X = MX (PHYS. COND)
1390  GOTO 1410
```

```
1400  X = FX (PHYS. COND)
1410  PRINT "ENERGY AVAILABLE TO 18 YEAR OLD: ", X
1420'
1430'
1440' ADJUSTING FOR AGE – 1% DECLINE
1450  LET X1 = X – (Age-18) × X × .01
1460'
1470'
1480' ADJUST X1 FOR THE INFLUENCE OF FOOD
1490  LET X2 = .9 × X1
1500'
1510'
1520' CALCULATE BASAL METABOLISM DURING SLEEP
1530  LET X3 = (.9 × WEIGHT × Y1)/2.2
1540'
1550'
1560' CALCULATE X4, THE ENERGY REQUIREMENT WHEN NOT
      SLEEPING OR WHEN NOT
1570' AT WORK
1580  INPUT "IS ENERGY (KCAL/MIN) WHEN NOT
      SLEEPING/WORKING KNOWN?", ER. YESNO$
1590  IF ER. YESNO$ = "Y" THEN INPUT "ENERGY REQUIRED: ",
      ER
1600  IF ER. YESNO$ < > "Y" THEN GOTO 1630
1610  LET X4 = Y3 × 60 × ER
1620  GOTO 1670
1630  LET X4 = Y3 × 60 × 1. 5
1640'
1650'
1660' DETERMINE TOTAL ENERGY AVAILABLE FOR WORK (X5)
1670  LET X5 = X2 – X3 – X4
1680  PRINT "X5 = ", X5
1690'
1700'
1710  INPUT "NUMBER OF JOBS N: ", N
1720  LET X6 = 0
1730  TT = 0
1740  FOR I = 1 TO N
1750  INPUT "ENTER TIME DURATION IN HOURS FOR ITH TASK:
      ", T(I)
1760  INPUT "ENTER TOTAL ENERGY REQUIRED/HOUR FOR THE
      ITH TASK: ", E(1)
1770  LET X6 = X6 + T(I) × E(I)
1780  LET TT = TT + T(I)
1790  NEXT I
```

```
1800'
1810'
1820   IF TT = Y2 THEN GOTO 1860
1830   PRINT "TOTAL TIME DURATION IS NOT EQUAL TO Y2 (SHIFT
       DURATION) CHECK INPUT"
1840   GOTO 640
1850'
1860   IF X5 > X6 THEN R = O ELSE R = (X6/X5 – 1) × 100
1870   PRINT "REST PERIOD AS A PERCENT OF WORK PERIOD: ", R
1880'
1890   INPUT "DO YOU WANT A HARD COPY? (Y/N):.",
       HARD.YESNO$
1900   IF HARD YESNO$ < > "Y" GOTO 2280
1910'
1920   LPRINT
1930   LPRINT "  PROGRAM OUTPUT"
1940   LPRINT
1941   LPRINT
1950   LPRINT "DATE: ", DATE$
1951   LPRINT
1960   LPRINT "JOB NUMBER: ", JOB$
1961   LPRINT
1970   LPRINT "JOB CODE: ", JOB.CODE$
1971   LPRINT
1980   LPRINT "WORKER NAME: ", W.NAME$
1981   LPRINT
1990   LPRINT "SEX : ",SEX$
1991   LPRINT
2000   LPRINT "AGE (YEARS) :", AGE
2001   LPRINT
2010   LPRINT "WEIGHT IN POUNDS : ", WEIGHT
2011   LPRINT
2020'
2030   IF AC. YESNO$ = "N" GOTO 2070
2040   LPRINT "AEROBIC CAPACITY IN ML/KG/MIN: ", ACAPACITY
2050   LPRINT "PHYSIOLOGICAL CONDITION COMPUTED FROM
       THE TABLE: ", PHYS. COND
2060   GOTO 2080
2070   LPRINT "ESTIMATED PHYSIOLOGICAL CONDITION: ", PHYS.
       COND
2080'
2081   LPRINT
2090   LPRINT "HOURS OF SLEEP PER DAY: ", Y1
2100   LPRINT "SHIFT DURATION (HOURS) : ", Y2
2110'
```

```
2111  LPRINT
2120  PRINT "ENERGY AVAILABLE FOR 18 YEAR OLD FROM THE
      TABLE (X): ", X
2130  LPRINT "ENERGY REQUIRED WHEN NOT SLEEPING AND
      NOT WORKING: ", X4
2140  LPRINT "TOTAL ENERGY AVAILABLE FOR WORK: ", X5
2150  LPRINT
2151  LPRINT
2160  LPRINT "TOTAL NUMBER OF JOBS: ", N
2170  LPRINT
2180  LPRINT "TIME DURATION OF JOB     ENERGY REQUIRED
      FOR JOB"
2190  FOR I = 1 TO N
2200  LPRINT T(I)," ", E(I)
2210  NEXT I
2220'
2221  LPRINT
2230  LPRINT
2240  LPRINT "TOTAL ENERGY REQUIRED FOR THE JOB (KCAL): ",
      X6
2250  LPRINT "REST PERIOD AS PERCENT OF TOTAL WORK
      PERIOD: ", R
2260'
2270'
2280  INPUT "DO YOU WANT TO RUN THE NEXT JOB? (Y/N) ",
      JOB.YESNO$
2290  IF JOB.YESNO$ = "Y" GOTO 480
2300  END
```

References

Aberg, U. (1961). Physiological and mechanical studies of material handling. *Ergonomics*, **4**, 282.

Adams, W.C. (1967). Influence of age, sex and body weight on the energy expenditure of bicycle riding. *Journal of Applied Physiology*, **22**, 539–545.

Aghazadeh, F. (1974). Lifting capacity as a function of operator and task variables. Unpublished M.S. thesis, Texas Tech University, Lubbock, Texas.

Aghazadeh, F. (1982). Simulated dynamic lifting strength models for manual lifting. Ph.D. Dissertation, Texas Tech University, Lubbock, Texas.

Aghazadeh, F. and Ayoub, M.M. (1985). A comparison of dynamic and static strength models for prediction of lifting capacity. *Ergonomics*, **28**, 1409–1417.

Alles, G.A. and Feigan, G.A. (1942). The influence of bengardine on work-decrement and patellar reflex. *American Journal of Physiology*, **136**, 392.

Alston, W., Carlson, K.E., Feldman, D.J., Grimm, Z. and Gerontinos, E. (1966). A quantitative study of muscle factors in the chronic low back syndrome. *Journal of American Geriatric Society*, **14**, 1041–1047.

Alvik, I. (1949). Tuberculosis of the spine II. The mobility of the lumbar spine after tuberculosis spondylitis. *Acta Chirurgica Scandinavia*, Suppl. 141.

American College of Radiology (1973). *Conference on Low Back-Back X-rays in Pre-employment Physical Examinations.* (NIOSH Contract HSM–99–72–153).

American Heart Association (1972). *Exercise Testing and Training of Apparently Healthy Individuals: A Handbook for Physicians.* Published by the American Heart Association.

American Industrial Hygiene Association (1971). Ergonomics guide to assessment of metabolic and cardiac costs of physical work. *American Industrial Hygiene Association Journal*, **32**, 560–564.

Anderson, C.K., Chaffin, D.B., Herrin, G.D. and Matthews, L. (1985). A biomechanical model of the lumbosacral joint during lifting activities. *Journal of Biomechanics*, **18**, 571–588.

Andersson, G.B.J. and Ortengren, R. (1984). Assessment of back load in assembly line work using electromyography. *Ergonomics*, **27**, 1157–1169.

Andersson, G.B.J., Ortengren, R. and Herberts, P. (1977). Quantitative electrographic studies of back muscle activity related to posture loading. Symposium on the Lumbar Spring-II. *Orthopedic Clinics of North America*, **8**, 85–96.

Andersson, G.B.J., Ortengren, R. and Nachemson, A. (1976). Quantitative studies of back loads in lifting. *Spine*, **1**, 178–185.

Andersson, G.B.J., Ortengren, R. and Nachemson, A. (1977). Intradiskal pressure, intraabdominal pressure and myoelectric back muscle activity related to posture and loading. *Clinical Orthopaedics*, **129**, 156–164.

Aquilano, N.J. (1968). A physiological evaluation of time standards for strenuous work as set by stopwatch time study and two predetermined motion time data systems. *Journal of Industrial Engineering*, **19**, 425–432.

Armstrong, J.R. (1965). *Lumbar Disc Lesions*, (Baltimore, MA: Williams and Williams), pp. 42–45.

Asfour, S.S. (1980). Energy cost prediction models for manual lifting and lowering tasks. Ph.D. Dissertation, Texas Tech University, Lubbock, Texas.

Asfour, S.S. and Ayoub, M.M. (1980). Effects of training on manual lifting. *Paper presented at the American Industrial Hygiene Association Conference, Houston, Texas*.

Asfour, S.S., Ayoub, M.M. and Mital, A. (1984a). Effects of an endurance and strength training programme on lifting capability of males. *Ergonomics*, **27**, 435–442.

Asfour, S.S., Genaidy, A.M., Khalil, T.M. and Greco, E.C. (1984b). Physiological and psychophysical determination of lifting capacity for low frequency lifting tasks. In *Trends in Ergonomics/Human Factors I*, edited by A. Mital (Amsterdam: North Holland), pp. 149–153.

Asfour, S.S., Genaidy, A.M., Khalil, T.M. and Greco, E.C. (1985). Lifting capacity norms based on strength and endurance of workers. In *Trends in Ergonomics/Human Factors II*, edited by R. Eberts and C.G. Eberts (Amsterdam: North Holland), pp. 609–615.

Asmussen, E. and Heeboll-Nielson, K. (1961). Isometric muscle strength of adult men and women. *Testing and Observation Institute of the Danish National Association for Infantile Paralysis, Communication*, **11**, 1–43.

Asmussen, E. and Heeboll-Nielson, K. (1962). Isometric muscle strength in relation to age in men and women. *Ergonomics*, **5**, 167–169.

Asmussen, E., Hansen, O. and Lammert, O. (1965). The relation between isometric and dynamic muscle strength in man. *Testing and Observation Institute of the Danish National Association for Infantile Paralysis, Communication*, **20**.

Asmussen, E., Heeboll-Nielson, K. and Melbech, S. (1959). Methods of evaluation of muscle strength. *Testing and Observation Institute of the Danish National Association for Infantile Paralysis, Communication*, **5**, 3–13.

Astrand, I. (1967a). Aerobic work capacity. *Circulation Research*, **20, 21**, suppl. I, 211–217.

Astrand, I. (1967b). Degree of strain during building work as related to individual aerobic work capacity. *Ergonomics*, **10**, 293–303.

Astrand, I., Astrand, P.O., Hallback, I. and Kilbom, A.S.A. (1973). Reduction in maximal oxygen uptake with age. *Journal of Applied Physiology*, **35**, 649–654.

Astrand, I., Guharay, A. and Wahren, J. (1968). Circulatory responses to arm exercise with different arm positions. *Journal of Applied Physiology*, **25**.

Astrand, P.O. and Rodahl, K. (1977). *Textbook of Work Physiology*, 2nd edition (New York: McGraw-Hill).

Avellini, B.A., Kamon, E. and Krajewski, J.T. (1980). Physiological responses of physically fit men and women to acclimation to humid heat. *Journal of Applied Physiology*, **49**, 254–261.

Ayoub, M.A. (1982a). Control of manual lifting hazards: II. Job redesign. *Journal of Occupational Medicine*, **24**, 668–676.

Ayoub, M.A. (1982b). Control of manual lifting hazards: III. Preemployment screening. *Journal of Occupational Medicine*, **24**, 751–761.

Ayoub, M.M. (1977). Lifting capacity of workers. *Journal of Human Ergology*, **6**, 187–192.

Ayoub, M.M., Bethea, N.J., Asfour, S.S., Calisto, G. and Grasley, C. (1979). *Review of the Strength and Capacity Data for Manual Materials Handling Activities*. (Technical Report, N63126-77-M-1719). (Point Mugu, California: Pacific Missile Test Center).

Ayoub, M.M., Bethea, N.J., Bobo, M., Burford, C.L., Caddel, K., Intaranont, K., Morrissey, S. and Selan, J.L. (1981a). *Mining in Low Coal, Vol. 1, Biomechanics and work physiology* (Final Report US Bureau of Mines Contract No. H03087022).

Ayoub, M.M., Bethea, N.J., Deivanayagum, S., Asfour, S.S., Bakken, G.M., Liles, D., Mital, A. and Sherif, M. (1978). *Determination and Modeling of Lifting Capacity*. (Final Report, HEW (NIOSH), Grant No. 5R01-OH-00545-02).

Ayoub, M.M. and Chen, H.C. (1986). Dynamic model for sagittal lifting. Presented at *American Industrial Hygiene Association Conference, Dallas, Texas*.

Ayoub, M.M., Deivanayagam, S. and Bakken, G.M. (1977). *A Preliminary Manual for Anthropometric and Strength Measurements*. (Technical Report). (Lubbock, Texas: Department of Industrial Engineering, Texas Tech University).

Ayoub, M.M., Denardo, J.D., Smith, J.L., Bethea, N.J., Lambert, B.K., Alley, L.R., and Duran, B.S. (1982a). Establishing physical criteria for assigning personnel to Air Force jobs. Final Report. (Air Force Office of Scientific Research, Contract No. F49620-79-C-0006.)

Ayoub, M.M. Dryden, R.D., McDaniel, J.W., Knipfer, R.E., and Aghazadeh, F. (1976). Modeling of lifting capacity as a function of operator and task variables. Paper 13, presented at *International Symposium: Safety in Manual Materials Handling*, Buffalo, New York.

Ayoub, M.M. and El-Bassoussi, M.M. (1976a). Dynamic biomechanical mdel for sagittal plane lifting activities. In C.G. Drury (Ed.), *Safety in Manual Materials Handling*, (Cincinnati, Ohio: DHEW (NIOSH) Publication No. 78-185, 88–95).

Ayoub, M.M., and El-Bassoussi, M.M. (1976b). Dynamic biomechanical model for sagittal plane lifting activities. *Proceedings of the 6th Congress of the International Ergonomics Association*, 355–359.

Ayoub, M.M. and El-Bassoussi, M.M. (1978). Dynamic biomechanical model for sagittal plane lifting activities. In C.G. Drury (Ed.) *Safety in Manual Materials Handling* (Cincinnati, Ohio: DHEW (NIOSH) Publication no. 78-105), pp. 88–95.

Ayoub, M.M., Gidcumb, C.F., Hafez, H., Intaranont, K., Jiang, B.C. and Selan, J.L. (1983a). *A Design Guide for Manual Lifting Tasks*. Prepared for OSHA.

Ayoub, M.M., Gidcumb, C.F., Reeder, M.J., Beshir, M.Y., Hafez, H.A., Aghazadeh, F. and Bethea, N.J. (1981b). *Development of an Atlas of Strengths and Establishment of an Appropriate Model Structure*. (Final Report, Institute for Biotechnology, Texas Tech University).

Ayoub, M.M., Gidcumb, F.F., Reeder, M.J., Hafzz, H.A., Beshir, M.Y., Aghazadeh, F., and Bethea, N.J. (1982b). *Development of a Female Atlas of Strengths* (Final Report Institute of Biotechnology, Texas Tech University).

Ayoub, M.M. and McDaniel, J.W. (1974). Effects of operator stance on pushing and pulling tasks. *Transactions of the American Institute of Industrial Engineers*, 6, 185–195.

Ayoub, M.M., Mital, A., Asfour, S.S. and Bethea, N.J. (1980a). Review, evaluation, and comparison of models for predicting lifting capacity. *Human Factors*, 22, 257–269.

Ayoub, M.M., Mital, A., Bakken, G.M., Asfour, S.S. and Bethea, N.J. (1980b). Development of strength and capacity norms for manual materials handling activities: the state of the art. *Human Factors*, 22, 271–283.

Ayoub, M.M. and Selan, J.L. (1983). Job design for manual materials handling. *Proceedings of the U.S. Bureau of Mines Technology Transfer Seminar* (Pittsburg, PA).

Ayoub, M.M., Selan, J.L. and Jiang, B.C. (1987). Manual materials handling. In *Handbook in Human Factors* edited by G. Salvendy (New York: John Wiley), pp. 790–818.

Ayoub, M.M., Selan, J.L., Karwowski, W. and Rao, H.P.R. (1983b). Lifting capacity determination in back injuries. *Proceedings of Bureau of Mines Technology Transfer Symposia, United States Bureau of Mines, Information Circular,* 8948, 54–63.

Ayoub, M.M., Selan, J.L. and Liles, D.H. (1983c). An ergonomics approach for the design of manual materials handling tasks. *Human Factors*, 25, 507–516.

Bakke, S.N. (1931). Roentgenologische beobachtungen uber die bewegungen der wirbelsan. *Acta Radiologica*, suppl. 13.

Bakken, G.M. (1983). Lifting capacity determination as a function of task variables. Ph.D. Dissertation, Texas Tech University, Lubbock, Texas.
Ball, J. (1978). New knowledge of intervertebral disc disease. *Journal of Clinical Pathology*, **31**, 200–204.
Bartelink, D.L. (1957). The role of abdominal pressure in relieving the pressure on the lumbar intervertebral discs. *Journal of Bone and Joint Surgery*, **39B**, 718–725.
Becker, W.F. (1955). Prevention of back injuries through pre-placement examinations. *Industrial Medicine and Surgery*, **24**, 486–490.
Belding, H.S., Hatch, T.F., Hertig, B.A. and Riedesel, M.L. (1961). Recent developments in understanding of effects of exposure to heat. In *Proceedings of the 13th International Congress on Occupational Health* (New York: Book Craftsmen Associated).
Bendix, T. and Eid, S.E. (1983). The distance between the load and the body with three bimanual lifting techniques. *Applied Ergonomics*, **14**, 185–192.
Bergquist-Ullman, M. and Larsson, U. (1977). Acute low back pain in industry: a controlled prospective study with special reference to therapy and confounding factors. *Acta Orthopaedica Scandinavica*, **170**, suppl. 1–117.
Berkson, M., Nachemson, A. and Schultz, A. (1979). Mechanical properties of human lumbar spine motion segments: Part II: responses in compression and shear; influence of gross morphology. *Journal of Biomechanical Engineering*, **101**, 55–57.
Billings, C.E., Bason, R., Mathews, D.K. and Fox, E.L. (1971). Cost of submaximal and maximal work during chronic exposure at 3800 m. *Journal of Applied Physiology*, **30**, 406–408.
Bink, B. (1962). The physical working capacity in relation to working time and age. *Ergonomics*, **5**, 25–28.
Blow, R.J. and Jackson, J.M. (1971). Rehabilitation of registered dock workers. *Proceedings of the Royal Society of Medicine*, **64**, 753–760.
Bobo, M., Bethea, N.J., Ayoub, M.M. and Intaranont, K. (1983). Energy expenditure and aerobic fitness of male low seam coal miners. *Human Factors*, **25**, 43–48.
Bonjer, F.H. (1962). Actual energy expenditure in relation to the physical work capacity. *Ergonomics*, **5**, 29–31.
Booynes, J. and Keating, W.R. (1957). The expenditure of energy by men and women walking. *Journal of Physiology*, **138**, 165–171.
Boudrifa, H. and Davis, B.T. (1984). The effect of backrest inclination, lumbar support and thoracic support on the intra-abdominal pressure while lifting. *Ergonomics*, **27**, 379–387.
Boussenne, M., Corlett, E.N. and Pheasant, S.T. (1982). The relation between discomfort and postural loading at the joints. *Ergonomics*, **25**, 313–322.
Bozian, R.C., (1983). Personal communication with A. Mital.
Brooke, J.D. (1967). Extraversion, physical performance and pain perception in physical education students. *Research in Physical Education*, **1**, 25–30.
Brouha, L. (1967). *Physiology in Industry* (New York: Pergamon Press).
Brown, J.R. (1971). *Lifting as an Industrial Hazard* (Ontario, Canada: Labour Safety Council).
Brown, J.R. (1974). Lifting as an industrial hazard. *American Industrial Hygiene Association Journal*, **34**, 294.
Brown, J.R. (1976). *Manual Lifting and Related Fields: An Annotated Bibliography*. (Ontario, Canada: Ministry of Labour).
Brown, J.R. (1977). *Low Back Pain: Its Etiology and Prevention* (Ontario, Canada: Labour Safety Council).
Brown, T., Hansen, R.S. and Yorra, A.J. (1957). Some mechanical tests on the lumbrosacral spine with particular reference to the intervertebral discs. *Journal of Bone and Joint Surgery*, **39A**, 1135–1164.

Cady, L.D., Biscoff, D.P., O'Connell, E.R., Thomas, P.C. and Allen, J.H. (1979). Strength and fitness and subsequent back injuries in firefighters. *Journal of Occupational Medicine*, **21**, 269–272.

Caillet, R. (1981). *Low Back Pain Syndrome* (Philadelphia: F.A. Francis).

Caiozzo, V.J., Davis, J.A., Ellis, J.F., Vandagriff, R., Prietto, C.A. and McMaster, W.C. (1982). A comparison of gas exchange indices used to detect the anaerobic threshold. *Journal of Applied Physiology: Respiratory, Environmental and Exercise Physiology* **53**, 1184–1189.

Caldwell, L.S. (1964). Body position and the strength and endurance of manual pull. *Human Factors*, **6**, 479–483.

Caldwell, L.S., Chaffin, D.B., Dukes-Dobos, F.N., Kroemer, K.H.E., Laubach, L.L., Snook, S.N. and Wasserman, D.E. (1974). A proposed standard procedure for static muscle strength testing. *American Industrial Hygiene Association Journal*, **35**, 201–206.

Caplan, C.S. (1971). Fatigue. In *Encyclopedia of Sport Sciences and Medicine* (New York: Macmillan).

Chaffin, D.B. (1969). A computerized biomechanical model-development of and use in studying gross body action. *Journal of Biomechanics*, **2**, 429–441.

Chaffin, D.B. (1972). *Some Effects of Physical Exertion*. (Technical report) (Ann Arbor: Department of Industrial Engineering, University of Michigan).

Chaffin, D.B. (1974). Human strength capability and low-back pain. *Journal of Occupational Medicine*, **16**, 248–254.

Chaffin, D.B. (1975a). Biomechanics of manual materials handling and low back pain. In *Occupational Medicine: Principles and Practical Applications* edited by C. Zenz (Chicago: Yearbook Medical Publishers).

Chaffin, D.B. (1975b). Ergonomics guide for assessment of human static strength. *American Industrial Hygiene Association Journal*, **36**, 505–511.

Chaffin, D.B. (1977). A method for evaluating the biomechanical stress resulting from manual materials handling jobs. *American Industrial Hygiene Association Journal*, **38**, 662–675.

Chaffin, D.B. and Andersson, G.B.J. (1984). *Occupational Biomechanics* (New York: John Wiley).

Chaffin, D.B. Andres, R.O., and Garg, A. (1983). Volitional postures during maximal push/pull exertions in the sagittal plane. *Human Factors*, **25**, 541–550.

Chaffin, D.B. and Ayoub, M.M. (1975). The problem of manual materials handling. *Industrial Engineering*, **7**, 24–29.

Chaffin, D.B. and Baker, W.H. (1970). A biomechanical model for analysis of symmetric sagittal plane lifting. *Transactions of the American Institute of Industrial Engineers*, **2**, 16–27.

Chaffin, D.B. and Moulis, E.J. (1969). An empirical investigation of low back strains and vertebrae geometry. *Journal of Biomechanics*, **2**, 89–96.

Chaffin, D.B. and Park, K. (1973). A longitudinal study of low back pain as associated with occupational weight lifting factors. *American Industrial Hygiene Association Journal*, **34**, 513–525.

Chaffin, D.B., Andres, R.O. and Garg, A. (1983). Volitional postures during maximal push/pull exertions in the sagittal plane. *Human Factors*, **25**, 541–550.

Chaffin, D.B., Herrin, G.D. and Keyserling, W.M. (1978). Pre-employment strength testing. *Journal of Occupational Medicine*, **20**, 403–408.

Chaffin, D.B., Herrin, G.D., Keyserling, W.M. and Foulke, J.A. (1977a). *Pre-employment Strength Testing in Selecting Workers for Materials Handling Jobs* (Cincinnati: National Institute for Occupational Safety and Health). Publication CDC-99-74-62.

Chaffin, D.B., Herrin, G.D., Keyserling, W.M. and Garg, A. (1977b). A method for evaluating the biomechanical stresses resulting from manual materials handling jobs. *American Industrial Hygiene Association Journal*, **38**, 661–675.

Chapman, A.E. and Troup, J.D.G. (1969). Electromyographic study of the effect of training on the lumbar erectores spinae. *Journal of Anatomy*, **105**, 186–187.

Ciriello, V.M. and Snook, S.H. (1978). The effects of size, distance, height and frequency on manual handling performance. In *Proceedings of the 22nd Annual Meeting of the Human Factors Society*, pp. 318–322.

Ciriello, V.M. and Snook, S.H. (1983). A study of size, distance, height, and frequency effects on manual handling tasks. *Human Factors*, **25**, 473–483.

Clausen, J., Trap-Jensen, J. and Lassen, N. (1970). The effects of training on the heart rate during arm and leg exercise. *Scandinavian Journal of Clinical Laboratory Investigation*, **26**, 295–301.

Congleton, J.J. (1983). Design and evaluation of a neutral posture chair. Ph.D. Dissertation, Texas Tech University, Lubbock, Texas.

Consolazio, C.F., Matoush, L.R.O., Nelson, R.A., Torres, J.B. and Isqui, G.J. (1963). Environmental temperature and energy expenditure. *Journal of Applied Physiology*, **18**, 65–68.

Contini, R. and Drillis, R. (1966). Kinematic and kinetic techniques in biomechanics. In *Advances in Bioengineering and Instrumentation* (Vol. 1) edited by F. Alt (New York: Plenum Press).

Corlett, E.N. and Bishop, P. (1976). A technique for assessing postural discomfort. *Ergonomics*, **19**, 175–182.

Corlett, E.N., Madeley, S.J. and Manenica, J. (1979). Posture targetting: a technique for recording work postures. *Ergonomics*, **22**, 357–366.

Coury, B.G. and Drury, C.G. (1982). Optimum handle positions in a box-holding task. *Ergonomics*, **25**, 645–662.

Cox, M., Shephard, R.J. and Corey, P. (1981). Influence of an employee fitness program upon fitness, productivity and absenteeism. *Ergonomics*, **24**, 795–806.

Cumming, G.R. (1967). Current levels of fitness. *Canadian Medical Association Journal*, **96**, 868–882.

Cunningham, D.A. and Hill, J.S. (1975). Effect of training on cardiovascular response to exercise in women. *Journal of Applied Physiology*, **39**, 891–895.

Cyriax, J. (1959). *Textbook of Orthopaedic Medicine*, Vol. 2. Cited in Bergquist-Ullman, M., and Larsson, U., (1977).

Darden, E. (1976). *Nutrition and Athletic Performance* (Pasadena: The Athletic Press).

Datta, S.R. and Ramanathan, N.L. (1971). Ergonomics of seven modes of carrying loads on the horizontal plane. *Ergonomics*, **14**, 269–278.

Datta, S.R., Chatterjie, B.B. and Roy, B.N. (1978). The energy cost of rickshaw pulling. *Ergonomics*, **21**, 373–381.

David, G.C. (1985). Intra-abdominal pressure measurements and load capacities for females. *Ergonomics*, **28**, 345–358.

David, J. (1968). Electromyographic and cinematographic analysis of spinal extension under stress. *Physiological Society*, 4P–4P.

Davies, B.T. (1972). Moving loads manually. *Applied Ergonomics*, **3**, 190–194.

Davies, B.T. (1978). Training in manual handling and lifting. In *Safety in Manual Materials Handling* (NIOSH Report No. 78–185).

Davis, J.A., Frank, M.H., Whipp, B.J. and Wasserman, K. (1979). Anaerobic threshold alternations caused by endurance training in middle-aged men. *Journal of Applied Physiology: Respiratory Environmental and Exercise Physiology*, **46**, 1039–1046.

Davis, J.A., Vodak, P., Wilmore, J.H., Vodak, J. and Kurtz, P. (1976). Anaerobic threshold and maximal aerobic power for three modes of exercise. *Journal of Applied Physiology*, **41**, 544–550.

Davis, P.R. (1956). Variations of the intra-abdominal pressure during weight lifting in various postures. *Journal of Anatomy*, **90**, 601(P).

Davis, P.R. (1959a). Posture of the trunk during the lifting of weights. *British Medical Journal*, **1**, 87–89.

Davis, P.R. (1959b). The causation of hernia by weight-lifting. *Lancet*, **2**, 155–157.
Davis, P.R. (1969). Trunk mechanics and intra-truncal pressure. *Journal of Anatomy*, **1–5**, 185–186.
Davis, P.R. and Stubbs, D.A. (1977). Safe levels of manual forces for young males. *Applied Ergonomics*, **8**, 141–150.
Davis, P.R. and Stubbs, D.A. (1980). *Force Limits in Manual Work* (Guilford: IPC Science and Technology Press).
Dehlin, O., Berg, S., Anderson, G.B.J. and Grimby, G. (1981). Effect of physical training and ergonomic counselling on the psychological perception of work and on the subjective assessment of low-back insufficiency. *Scandinavian Journal of Rehabilitation Medicine*, **13**, 1–9.
Dehlin, O., Berg, S., Hedenrud, B., Anderson, G. and Grimby, G. (1978). Muscle training, psychological perception of work and low back symptoms in nursing aides. *Scandinavian Journal of Rehabilitation Medicine*, **10**, 201–209.
Deivanayagam, S. and Ayoub, M.M. (1979). Prediction of endurance time for alternating workload tasks. *Ergonomics*, **22**, 279–290.
deVries, H.A. (1980). *Physiology of Exercise for Physical Education and Athletics*, 3rd edition (Dubuque, Iowa: Wm. C. Brown).
Dillane, J.B., Fry, J. and Kalton, G. (1966). Acute back syndrome — a study from general practice. *British Medical Journal*, **268**, 82–84.
Doelen, J.V. (1981). *Female Capacity for Physical Work* (Ontario, Canada: Ministry of Labour).
Doelen, J.V. and Wright, G.R. (1979). Fitness and occupational injuries: a review. (Safety Studies Service Report), (Ontario, Canada: Ministry of Labour).
Drury, C.G. (1980). Handles for manual materials handling. *Applied Ergonomics*, **11**, 35–42.
Drury, C.G. (1985). Influence of restricted space on manual materials handling. *Ergonomics*, **28**, 167–175.
Drury, C.G., Begbie, K., Ulate, C. and Deeb, J.M. (1985). Experiments on wrist deviation in manual materials handling. *Ergonomics*, **28**, 577–589.
Drury, C.G., Law, C.H. and Pawenski, C.S. (1982). A survey of industrial box handling. *Human Factors*, **24**, 553–565.
Drury, C.G. and Pizatella, T. (1983). Hand placement in manual materials handling. *Human Factors*, **22**, 551–562.
Dryden, R.D. (1973). A predictive model for the maximum permissible weight of lift from knuckle to shoulder height. Ph.D. Dissertation, Texas Tech University, Lubbock, Texas.
Duffner, L.R., Hamilton, L.H. and Schmitz, M.A. (1982). Effect of whole-body vibration on respiration in human subjects. *Journal of Applied Physiology*, **17**, 913–916.
Duffy, J.J. and Franklin, M.A. (1975). A learning identification algorithm and its application to an environmental system. *IEEE Transactions on Systems, Man and Cybernetics*, **SME-5**, 226–240.
Dukes–Dobos, F.N. (1981). Hazards of heat exposure: a review. *Scandinavian Journal of Work Environment and Health*, **7**, 73–83.
Durnin, J.V.G.A. and Namyslowski, L. (1958). Industrial variations in energy expenditure of standard activities. *Journal of Physiology*, **143**, 573–578.
Durnin, J.V.G.A. and Passmore, R. (1967). *Energy, Work and Leisure* (London: Heinemann Educational).
Edholm, O.G. (1967). *The Biology of Work* (New York: McGraw-Hill).
Ekblom, B., Astrand, P.O., Salin, B. and Wallstrom, B. (1968). Effect of training on circulatory response in exercise. *Journal of Applied Physiology*, **24**, 518–527.
El-Bassoussi, M.M. (1974). A biomechanical dynamic model for lifting in the sagittal plane. Ph.D. Dissertation, Texas Tech University, Lubbock, Texas.
Emanuel, I., Chaffee, J. and Wing, J. (1956). *A Study of Human Weight Lifting Capabilities for*

Loading Ammunition into the F-86 Aircraft (Technical Report No. WADC–TR056–367, U.S. Air Force).

Evans, F.G. and Lissner, H.R. (1959). Biomechanical studies of the lumbar spine and pelvis. *Journal of Bone and Joint Surgery*, **41A**, 278–290.

Evans, O.M., Zerbib, Y., Faria, M.M. and Monod, H. (1983). Physiological responses to load holding and load carriage. *Ergonomics* **26**, 161–171.

Fernandez, J.E. (1986). Physiological lifting capacity over extended periods. Unpublished doctoral dissertation, Texas Tech University, Lubbock, Texas.

Fleishman, E.A., Kremer, E.J. and Supar, G.W. (1961). *The Dimensions of Physical Fitness: A Factor Analysis of Strength Tests* (Technical Report 2, Dept. of Industrial Administration and Dept. of Psychology), (New Haven, Connecticut: Yale University).

Flint, M.M., Drinkwater, B.L. and Horvath, S.M. (1974). Effects of training on woman's response to submaximal exercise. *Medicine and Science in Sports*, **6**, 89–94.

Foreman, T.K., Baxter, C.E. and Troup, J.D.G. (1984). Ratings of acceptable load and maximal isometric lifting strengths: the effects of repetition. *Ergonomics*, **27**, 1283–1288.

Fox, R.R. (1982). A psychophysical study of bi-manual lifting. MSIE Thesis, Texas Tech University, Lubbock, Texas.

Fox, E., Bartels, R., Billings, C., Mathews, D., Bason, R. and Mathews, D. (1973). Intensity and distance of interval training programs and changes in aerobic power. *Medicine and Science in Sports*, **5**, 18–22.

Fox, E., Bartels, R., Billings, C., O'Brien, R., Bason, R. and Mathews, D. (1975). Frequency and duration of interval training programs and changes in aerobic power. *Journal of Applied Physiology*, **38**, 481–484.

Fox, E.L. and Mathews, D.K. (1981). *The Physiological Basis of Physical Education and Athletics*, 3rd edition (Philadelphia: Saunders College Publishing).

Frankel, V.H. and Nordin, M. (1980). *Basic Biomechanics of the Skeletal System* (Philadelphia: Lea and Febiger).

Frederick, W.S. (1959). Handling at the workplace II. Human energy in manual lifting. *Modern Materials Handling*, **14**, 74–76.

Frick, M., Elovainio, R. and Somer, T. (1963). Effects of physical training on circulation at rest and during exercise. *American Journal of Cardiology*, **12**, 142–147.

Frick, M., Elovainio, R. and Somer, T. (1967). The mechanism of bradycardia evoked by physical training. *Cardiologia*, **51**, 46–54.

Frievalds, A., Chaffin, D.B., Garg, A. and Lee, K.S. (1984). A dynamic biomechanical evaluation of lifting maximum acceptable loads. *Journal of Biomechanics*, **17**, 251–262.

Frye, A.J. and Kamon, E. (1981). Responses to dry heat of men and women with similar aerobic capacities. *Journal of Applied Physiology*, **50**, 65–70.

Gamberale, F., Strindberg, L. and Wahlberg, I. (1975). Female work capacity during the menstrual cycle: physiological and psychological reactions. *Scandinavian Journal of Work Environment and Health*, **1**, 120–127.

Garg, A. (1976). A metabolic rate prediction model for manual materials handling jobs. Ph.D. Dissertation, The University of Michigan, University Microfilms.

Garg, A. (1979). Methods for estimating physical fatigue. *Proceedings of the American Institute of Industrial Engineers Conference*, 68–75, San Francisco, California.

Garg, A. (1980). An evaluation of physical fatigue and stresses in warehouse operations. *American Industrial Hygiene Association Abstracts*, Paper No. 31.

Garg, A. (1983). Physiological responses to one-handed lift in the horizontal plane by female workers. *American Industrial Hygiene Association Journal*, **44**, 190–200.

Garg, A. and Ayoub, M.M. (1980). What criteria exist for determining how much load can be lifted safely? *Human Factors*, **22**, 475–486.

Garg, A. and Chaffin, D.B. (1975). A biomechanical computerized simulation of human strength. *Transactions of the American Institute of Industrial Engineers*, 7, 1–15.
Garg, A. and Herrin, G.D. (1979), Stoop or squat: a biomechanical and metabolic evaluation. *Transactions of the American Institute of Industrial Engineers*, 11, 293–302.
Garg, A. and Saxena, U. (1979). Effects of lifting frequency and techniques on physical fatigue with special reference to psychophysical methodology and metabolic rate. *American Industrial Hygiene Association Journal*, 40, 894–904.
Garg, A. and Saxena, U. (1980). Container characteristics and maximum acceptable weight of lift. *Human Factors*, 22, 487–495.
Garg, A. and Saxena, U. (1981). *Factors for Establishing Permissible Limits for One-handed Lifts by Women* (NIOSH Report No. 210–80–0084).
Garg, A. and Saxena, U. (1982). Maximum frequency acceptable to female workers for one-handed lifts in the horizontal plane. *Ergonomics*, 25, 839–853.
Garg, A., Chaffin, D.B. and Herrin, G.D. (1978). Prediction of metabolic rates for manual materials handling jobs. *American Industrial Hygiene Association Journal*, 39, 661–674.
Giles, C.G., Sabey, B.E. and Cardew, K.H.F. (1964). *Development and Performance of the Portable Skid-resistance Tester* (Technical Report). (Road Research Laboratory, UK: Department of Scientific and Industrial Research).
Girandola, R.N. and Katch, V. (1973). Effects of nine weeks of physical training on aerobic capacity and body composition in college men. *Archives of Physical Medicine and Rehabilitation*, 54, 521–524.
Givoni, B. and Goldman, R.F. (1972). Predicting rectal temperature response to work, environment and clothing. *Journal of Applied Physiology*, 32, 812–822.
Givoni, B. and Goldman, R.F. (1973). Predicting heart rate response to work, environment and clothing. *Journal of Applied Physiology*, 34, 201–204.
Glorig, A. (1971). Non-auditory affects of noise exposure. *Sound and Vibration*, 5, 28.
Glover, J.R. (1976). Prevention of back pain. In *The Lumbar Spine and Back Pain*, edited by M. Jayson (New York: Grune and Stratton), pp. 47–54.
Goldman, R.F. and Iampietro, P.F. (1962). Energy cost of load carriage. *Journal of Applied Physiology*, 17, 675–676.
Gordon, M.J., Gaslin, B.R., Graham, T. and Hoare, J. (1983). Comparison between load carriage and grade walking on a treadmill. *Ergonomics*, 26, 289–298.
Grandjean, E. (1982). *Fitting the Task to the Man* (London: Taylor and Francis).
Grasley, C., Ayoub, M.M. and Bethea, N.J. (1978). Male-female differences in variables affecting performance. *Proceedings of the Human Factors Society Meeting*, pp. 416–420.
Greater London Council (Architect's Department) (1971). Slip resistance of floors, stairs and pavings. *Development and Materials Bulletin*, 43 (2nd series) Item 5.
Grieve, D.W. (1979). Environment constraints on the static exertion of force: PSD analysis in task design. *Ergonomics*, 22, 1165–1175.
Grieve, D.W. (1984). The influence of posture on power output generated in single pulling movements. *Applied Ergonomics*, 15, 115–117.
Griffin, A.B., Traup, J.D.G. and Lloyd, D.C.E.F. (1984). Tests of lifting and handling capacity — their repeatability and relationship to back symptoms. *Ergonomics*, 27, 305–320.
Grillner, S.J., Nilson, J. and Thorstensson, A. (1978). Intra-abdominal pressure changes during natural movement in man. *Acta Physiologica Scandinavica*, 103, 275–283.
Gyntelberg, F. (1974). One year incidence of low back pain among male residents of Copenhagen ages 40–59. *Danish Medical Bulletin*, 21, 30–36.
Habes, D., Carlson, W. and Badger, D. (1985). Muscle fatigue associated with repetitive arm lifts: effects of height, weight and reach. *Ergonomics*, 28, 471–488.

Hafez, H.A. (1984). Manual lifting under hot environmental conditions. Ph.D. Dissertation, Texas Tech University, Lubbock, Texas.

Hafez, H.A., Gidcumb, F.F., Reeder, M.J., Beshire, M.Y., and Ayoub, M.M. (1982). Development of a human atlas of strengths. *Proceedings of the Human Factors Society 26th Annual Meeting*, 575–579.

Haisman, M.F., Winsman, F.R. and Goldman, R.F. (1972). Energy cost of pushing loaded hand carts. *Journal of Applied Physiology*, 33, 181–183.

Hamilton, B.J. and Chase, R.B. (1969). A work physiology study of the relative effects of pace and weighted in a carton handling task. *Transactions of the American Institute of Industrial Engineers*, 1, 106–111.

Hansson, T. and Roos, B. (1981). The relation between bone mineral content, experimental compression fractures and disc degeneration in lumbar vertebrae. *Spine*, 6, 147–153.

Hansson, T., Roos, B. and Nachemson, A. (1980). A bone mineral content and ultimate compressive strength of lumbar vertebrae. *Spine*, 5, 46–55.

Happey, F. (1980). Studies of the structure of the human intervertebral disc in relation to its functional and aging process. In *The Joint and Synovial Fluid*, (Vol. 2), edited by L. Sokoloff (New York: Academic Press).

Harber, P., Billet, E., Gutowski, M., SooHoo, K., Lew, M. and Roman, A. (1985). Occupational low-back pain in hospital nurses. *Journal of Occupational Medicine*, 27, 518–524.

Harrison, R. and Malkin, F. (1983). On-site testing of shoe and floor combinations. *Ergonomics*, 26, 101–108.

Hartung, G.H. (1974). Responses of middle-age women to maximal cycling exercise. *American Correctional Therapy Journal*, 28, 103–106.

Hawley, R.J. (1982). Personal communication.

Henschel, A. (1971). The environment and performance. In *Physiology of Work Capacity and Fatigue*, edited by E. Simonson (Springfield: Charles C. Thomas).

Hermansen, L. and Saltin, B. (1969). Oxygen uptake during maximal treadmill and bicycle exercise. *Journal of Applied Physiology*, 26, 31–37.

Herndon, R.F. (1927). Back injuries in industrial employees. *Journal of Bone and Joint Surgery*, 9, 234–269.

Herrin, G.D., Chaffin, D.B. and Mach, R.S. (1974). *Criteria for Research on the Hazards of Manual Materials Handling* (NIOSH Contract Report CDC–99–74–118).

Hertzberg, H.T.E. (1955). Some contributions of applied physical anthropometry to human engineering. *Annals of New York Academy of Sciences*, 63, 616–629.

Hettinger, T. (1961). *Physiology of Strength* (Springfield: Charles C. Thomas).

Hettinger, T., Birkhead, N.C., Horwarth, S.M., Issekutz, B. and Rodahl, K. (1961). Assessment of physical work capacity. *Journal of Applied Physiology*, 16, 153–156.

Hickey, D.S. and Hukins, D.W.L. (1980). Relation between the structure of the annulus fibrosis and the function and failure of the intervertebral disc. *Spine*, 5, 106–116.

Himbury, S. (1967). Kinetic methods of manual handling in industry. *Occupational Safety and Health Series No. 10*, (Geneva: I.L.O.).

Hirsch, C. (1966). Low back pain etiology and pathogenesis. *Applied Therapeutics*, 8, 857–862.

Hood, W.B., Murray, R.H., Urschel, C.W., Bowers, J.A. and Clark, J.G. (1966). Cardiopulmonary effects of whole-body vibration in men. *Journal of Applied Physiology*, 21, 1725–1731.

Hult, L. (1954). Cervical, dorsal and lumbar spinal syndromes. *Acta Orthopedical Scandinavica*, suppl. 17.

Hutton, W.C. and Adams, M.A. (1982). Can the lumbar spine be crushed in heavy lifting? *Spine*, 7, 586–590.

Hutton, W.C., Cyron, B.M. and Stott, J.R. (1979). The compressive strength of lumbar vertebrae. *Journal of Anatomy*, **129**, 753–758.

Ikai, M. and Steinhaus, A.H. (1961). Some factors modifying the expression of human strength. *Journal of Applied Physiology*, **16**, 157–163.

Ilmarinen, J. and Louhevaara, V. (1984). Oxygen consumption and health rate in different modes of postal delivery. *Ergonomics*, **27**, 331–339.

Ilmarinen, J. and Rutenfraz, J. (1980). Occupationally induced stress, strain and peak loads as related to age. *Scandinavian Journal of Work, Environment and Health*, **6**, 274–282.

Imrie, D. (1983). *Goodbye Backache* (Scarborough, Ontario: Prentice-Hall).

Imrie, D. (1983b). Personal communication.

Inooka, H. and Inoye, I. (1978). Application of GMDH algorithm to a manual control system. *IEEE Transactions on Systems, Man and Cybernetics*, **SMC-8**, 819–821.

Intaranont, K. (1983). A study of anaerobic threshold for lifting tasks. Ph.D. Dissertation, Texas Tech University, Lubbock, Texas.

International Labour Office. (1967).

International Labour Organization. (1962). *Maximum Permissible Weight to be Carried by One Worker* (Information sheet no. 3), (Geneva: International Labour Organization).

Ivakhnenko, A.G. (1971). Polynomial theory of complex systems. *IEEE Transactions on Systems, Man and Cybernetics*, **SMC-1**, 364–373.

James, D.I. (1983). Rubbers and plastics in shoes and flooring: the importance of kinetic friction. *Ergonomics*, **26**, 83–99.

Jennekens, F.G.I., Tomlinson, B.E. and Walton, J.N. (1971). Histochemical aspects of five limb muscles in old age: an autopsy study. *Journal of Neurological Science*, **14**, 259–276.

Jensen, R.C. (1985). Events that trigger disabling back pain among nurses. In *Proceedings of the 29th Annual Meeting of the Human Factors Society*, pp. 799–801.

Jensen, R.C. (1986). Work-related back injuries among nursing personnel in New York. In *Proceedings of the 29th Annual Meeting of the Human Factors Society*, pp. 244–248.

Jiang, B.C. (1984). Psychophysical capacity modeling of individual and combined manual materials handling activities. Ph.D. Dissertation, Texas Tech University, Lubbock, Texas.

Jiang, B.C. and Ayoub, M.M. (1987). Modeling of maximum acceptable load of lifting by physical factors. *Ergonomics*, **30**, 529–538.

Jiang, B.C. and Mital, A. (1986). A procedure for designing/evaluating manual materials handling tasks. *International Journal of Production Research*. **24**, 913–925.

Jlmarien, J., Rutenfraz, J. (1980). Occupationally induced stress, strain, and peak loads as related to age. *Scandinavian Journal of Work Environment & Health*, **6**, 274–282.

Johck, L.M. and Van Niekerk, J.M. (1961). A roentgenographic study of the motion of the lumbar spine of the Bantu. *South African Journal of Laboratory and Clinical Medicine*, **7**, 67–71.

Jones, D.F. (1985). Back injury prevention — are programs adequate? *Professional Safety*, February, 18–24.

Jorgenson, K. (1970). Back muscle strength and body weight as limiting factors for work in standing slightly-stooped position. *Scandinavian Journal of Rehabilitation Medicine*, **2**, 149–153.

Kaltenborn, F.M. (1975). *Test segmentimobilis. Columna vertebralis*. Course 1, pp. 44–57.

Kamon, E. (1979). Scheduling cycles of work for hot ambient conditions. *Ergonomics*, **22**, 427–439.

Kamon, E. and Ayoub, M.M. (1976). Ergonomic Guide for Assessment of Physical Work Capacity. *American Industrial Hygiene Association Journal*.

Kamon, E. and Belding, H.S. (1971). The physiological cost of carrying loads in temperate and hot environment. *Human Factors*, **13**, 153–161.

Kamon, E. and Goldfuss, A.J. (1978). In-plant evaluation of the muscle strength of workers. *American Industrial Hygiene Association Journal*, **39**, 801–807.

Kamon, E., Kiser, D. and Pytel, J. (1982). Dynamic and static lifting capacity and muscular strength of steelmill workers. *American Industrial Hygiene Association Journal*, **43**, 853–857.

Kamon, E., Krajewski, J.T. and Avellini, B. (1978). Scheduling cycles of work for carrying under heat stress. In *Proceedings of the 22nd Annual Meeting of the Human Factors Society*, pp. 323–327.

Kamon, E. and Pandolf, K.B. (1972). Maximal aerobic power during laddermill climbing, uphill running, and cycling. *Journal of Applied Physiology*, **32**, 467–473.

Kanehisa, H. and Miyashita, M. (1983). Effect of isometric and isokinetic muscle training on static strength and dynamic power. *European Journal of Applied Physiology*, **50**, 365–371.

Karvonen, M.J. (1974). Work and activity classifications. In *Fitness, Health, and Work Capacity*, edited by L.A. Larson (New York: Macmillan).

Karvonen, M.J. (1985). Effects of temporal patterns of work on lifting and handling capacities. *Ergonomics*, **28**, 177–181.

Karvonen, M.J., Jarvinen, T. and Nummi, J. (1977). Follow-up study on the back problems of nurses. *Instructional Occupational Health*, **14**, 8.

Karvonen, M.J., Manizer, J., Rohmert, W., Lowenthal, I., Undeutsch, K., Kupper, R., Gartner, K.H. and Rutenfranz, J. (1980a). Occupational health studies on airport transport workers. *International Archives of Occupational and Environmental Health*, **47**, 233–244.

Karvonen, M.J. and Ronnholm, N. (1964). Electromyographic and energy expenditure studies of rhythmic and paced lifting work. *Annals of the Academy of Science Fenn, Series A, V Medica*, **106/19**, 3–11.

Karvonen, M.J., Viitasalo, J.T., Komi, P.V., Nummi, J. and Varvinen, T. (1980b). Back and leg complaints in relation to muscle strength in young men. *Scandinavian Journal of Rehabilitation Medicine*, **12**, 53–60.

Karwowski, W. (1982). A fuzzy sets based model on interaction between stresses involved in manual lifting tasks. Ph.D. Dissertation, Texas Tech University, Lubbock, Texas.

Karwowski, W. and Ayoub, M.M. (1984a). Effect of frequency on the maximum acceptable weight of lift. In A. Mital (Ed.), *Trends in Ergonomics/Human Factors I*, North-Holland, Amsterdam.

Karwowski, W. and Ayoub, M.M. (1984b). Fuzzy modeling of stresses in manual lifting tasks. *Ergonomics*, **27**, 641–649.

Karwowski, W. and Mital, A. (1985). Isometric and isokinetic testing of lifting strength of males in team-work. (Louisville, Kentucky: Technical Report. Department of Industrial Engineering, University of Louisville).

Karwowski, W. and Mital, A. (1986). Isometric and isokinetic testing of lifting strength of males in teamwork. *Ergonomics*, **29**, 869–878.

Karwowski, W., Mital, A. and Mulholland, N. (1985). Static strength and maximum lifting capacity for a team of two males. In *Trends in Ergonomics/Human Factors II*, edited by R.E. Eberts and C.G. Eberts (Amsterdam: North Holland).

Karwowski, W. and Yates, J.W. (1986). Reliability of the psychophysical approach to manual lifting of liquids by females. *Ergonomics*, **29**, 237–248.

Kassab, S.J. and Drury, C.G. (1976). The effects of working height on maximal lifting tasks. *International Journal of Production Research*, **4**, 381–386.

Kazarian, L. and Graves, G.A. (1977). Compressive strength characteristics of the human vertebral centrum. *Spine*, **2**, 1–14.

Keim, H.A. (1981). *How to Care for Your Back*. Englewood, New Jersey: Prentice-Hall.

Keyserling, W.M., Herrin, G.D. and Chaffin, D.B. (1978). An analysis of selected work muscle strengths. Paper presented at the *Human Factors Society 22nd Annual Meeting, Detroit*.

Keyserling, W.M., Herrin, G.D. and Chaffin, D.B. (1980a). Isometric strength testing as a means of controlling medical incidents on strenuous jobs. *Journal of Occupational Medicine*, **22**, 332–336.

Keyserling, W.M., Herrin, G.D., Chaffin, D.B., Armstrong, T.J. and Foss, M.L. (1980b). Establishing an industrial strength testing program. *American Industrial Hygiene Association Journal*, **41**, 730–736.

Khalil, T.M., Asfour, S.S. and Moty, E.A. (1984). Case studies in low back pain. In *Proceedings of the 28th Annual Meeting of the Human Factors Society*, pp. 465–470.

Khalil, T.M., Genaidy, A.M., Asfour, S.S. and Vinciguerra, T. (1985). Physiological limits in lifting. *American Industrial Hygiene Association Journal*, **46**, 220–224.

Klein, B.P., Roger, M.A., Jensen, R.C. and Sanderson, L.M. (1984). Assessment of workers' compensation claims for back sprain/strains. *Journal of Occupational Medicine*, **26**, 443–448.

Knipfer, R.E. (1974). Predictive models for the maximum acceptable weight of lift. Unpublished Ph.D. Dissertation, Texas Tech University, Lubbock, Texas.

Komi, P.V. and Karlsson, J. (1979). Physical performance, skeletal muscle enzyme activities, and fibre types in monozygous and dizygous twins of both sexes. *Acta Physiologicae Scandinavica*, suppl. 462.

Konz, S., Dey, S. and Bennett, C. (1973). Forces and torques in lifting. *Human Factors*, **15**, 237–245.

Kosiak, M., Aurelius, J.R. and Hartfield, W.F. (1968). The low back problem — an evaluation. *Journal of Occupational Medicine*, **10**, 508–593.

Koyd, L.F. and Hanson, P.M. (1969). *Age, Physical Ability and Work Potential*. (National Council on Aging Report). (Manpower Administration, Department of Labor).

Krager, D.W. and Hancock, W.M. (1982). *Advanced Work Measurement* (New York: Industrial Press).

Krajewski, J.T., Kamon, E. and Avellini, B. (1979). Scheduling rest for consecutive light and heavy loads under hot ambient conditions. *Ergonomics*, **22**, 975–987.

Kraus, H. (1967). Prevention of low back pain. *Journal of Occupational Medicine*, **9**, 555–559.

Kroemer, K.H.E. (1969, August). *Push Forces Exerted in 65 Common Working Positions* (Report No. AMRL-TR-68-143). (Aerospace Medical Research Laboratory, Wright-Patterson Air Force Base).

Kroemer, K.H.E. (1970). Human strength: terminology, measurement and interpretation of data. *Human Factors*, **12**, 297–313.

Kroemer, K.H.E. (1974). Horizontal push and pull forces. *Applied Ergonomics*, **5**, 94–102.

Kroemer, K.H.E. (1983). An isoinertial technique to assess individual lifting capability. *Human Factors*, **25**, 493–506.

Kroemer, K.H.E. (1985). Testing individual capability to lift material: repeatability of a dynamic test compared with static testing. *Journal of Safety Research*, **16**, 1–7.

Kroemer, K.H. and Marras, W.S. (1981). Evaluation of maximal and submaximal static muscle exertions. *Human Factors*, **23**, 643–653.

Kromodihardjo, S. and Mital, A. (1985). Biomechanical analysis of task variables in manual lifting. In *Proceedings of the 2nd Joint ASCE/ASME Mechanics Conference*, edited by D. Butler, T.K. Hung and R.E. Males, pp. 117–120.

Kromodihardjo, S. and Mital, A. (1986a). The cost/accuracy approach: a method for selecting filming rate for manual materials handling activities. *Applied Ergonomics*, **17**, 218–220.

Kromodihardjo, S. and Mital, A. (1986b). Kinetic analysis of manual lifting activities: Part I — Development of a three dimensional computer model. *International Journal of Industrial Ergonomics*, **1**, 77–90.

Kumar, S. (1980). Physiological responses to weight lifting in different planes. *Ergonomics*, **23**, 987–993.

Kumar, S. (1984). The physiological cost of three different methods of lifting in sagittal and lateral planes. *Ergonomics*, **27**, 425–433.

Kumar, S. and Davis, P.R. (1983). Spinal loading in static and dynamic posture: EMG and intra-abdominal pressure study. *Ergonomics*, **26**, 913–922.

Kumar, S. and Magee, D.J. (1982). Energy cost of lifting in sagittal and lateral planes by different techniques. In *Proceedings of the VIIIth Congress of the International Ergonomics Association*, edited by K. Noro (Tokyo: Intergroup) pp. 644–645.

LaRocca, H. and MacNab, I. (1970). Value of pre-employment radiographic assessment of the lumbar spine. *Radiology*, **39**, 31–36.

Larsson, K., Grimby, G. and Karlsson, J. (1979). Muscle strength and speed of movement in relation to age and muscle morphology. *Journal of Applied Physiology*, **46**, 451–456.

Larsson, L. and Karlsson, J. (1978). Isometric and dynamic endurance as a function of age and skeletal muscle characteristics. *Acta Physiologica Scandinavica*, **104**, 129–136.

Laubach, L.L. (1969). Body composition in relation to muscle strength and range of joint motion. *Journal of Sports Medicine and Physical Fitness*, **9**, 89–97.

Laubach, L.L. (1976). Comparative muscular strength of men and women: a review of the literature. *Aviation, Space and Environmental Medicine*, **47**, 534–542.

Laubach, L.L. and McConville, J.T. (1969). The relationship of strength to body size and typology. *Medicine and Science in Sports*, **1**, 189–194.

Laughery, K.R. and Schmidt, J.K. (1984). Scenario analysis of back injuries in industrial accidents. In *Proceedings of the 28th Annual Meeting of the Human Factors Society*, pp. 471–475.

Legg, S.J. and Myles, W.S. (1981). Maximum acceptable repetitive lifting workloads for an 8-hour workday using psychophysical and subjective rating methods. *Ergonomics*, **24**, 907–916.

Legg, S.L. and Pateman, C.M. (1984). A physiological study of the repetitive lifting capabilities of healthy young males. *Ergonomics*, **27**, 259–272.

Leggo, C. and Mathiasen, H. (1973). Preliminary results of pre-employment back X-ray program for state traffic officers. *Journal of Occupational Medicine*, **15**, 973–974.

Lehman, G. (1962). *Practische Arbeitesphysiologie*, 2 Auflage (Stuttgart: Theime Verlag).

Leskinen, T.P.J., Stalhammer, H.R., Kuorinka, I.A.A. and Troup, J.D.G. (1983a). A dynamic analysis of spinal compression with different lifting techniques. *Ergonomics*, **26**, 595–604.

Leskinen, T.P.J., Stalhammer, H.R., Kuorinka, I.A.A. and Troup, J.D.G. (1983b). The effect of inertia factors on spinal stress when lifting. *Engineering in Medicine*, **12**, 87–89.

Lewit, K. (1977). *Manuelle Medizin im Rahmen der Medizinischen Rehabilitation*, Vol. 2. (Leipzig: Vohann Amorasius Barth). Cited in Bergquist-Ullman and Larsson (1977).

Liles, D.H. (1986). The application of the job severity index to job design for the control of manual materials handling injury. *Ergonomics*, **29**, 65–76.

Liles, D.H. Deivanayagam, S., Ayoub, M.M. and Mahajan, P. (1984). A job severity index for the evaluation and control of lifting injury. *Human Factors*, **26**, 683–694.

Liles, D.H., Mahajan, P. and Ayoub, M.M. (1983). An evaluation of two methods for the injury risk assessment of lifting tasks. In *Proceedings of the 27th Annual Conference of the Human Factors Society*, pp. 279–283.

Lind, A.R. (1977). Guide for lifting in industry: physiological factors. An unpublished report proposed to NIOSH for the preparation of a manual lifting guide.

Lind, A.R. and McNichol, G.W. (1967). Circulatory responses to sustained hand-grip contractions performed during other exercise both rhythmic and static. *Journal of Physiology*, **192**, 597–607.

Lind, A.R., Burse, R., Rochelle, R.H., Rinehart, J.S. and Petrofsky, J.S. (1978). Influence of posture on isometric fatigue. *Journal of Applied Physiology*, **45**, 270–274.

Lloyd, D.C.E.F. and Troup, J.D.G. (1983). Recurrent back pain and its prediction. *Journal of Social and Occupational Medicine*, **33**, 66–74.

Loesser, T.A. (1979). Low back pain: introduction to planary session. In *Advances in Pain Research*, Vol. 3, edited by J.J. Bonica *et al*. (New York: Raven Press), pp. 631–633.

Magora, A. (1970). Investigation of the relation between low back pain and occupation, Part I. Age, sex, community, education and other factors. *Industrial Medicine and Surgery*, **39**, 465–471.

Magora, A. and Taustein, I. (1969). An investigation of the problem of sick leave in the patient suffering from low back pain. *Industrial Medicine and Surgery*, **38**, 398–408.

Mairiaux, P., Davis, P.R., Stubbs, D.A. and Baty, D. (1984). Relation between intra-abdominal pressure and lumbar moments when lifting weights in the erect posture. *Ergonomics*, **27**, 883–894.

Maksud, M.G., Coutts, K.D., Tristani, F.E., Dorchak, J.R., Barvoriak, J.J. and Hamilton, L.H. (1972). The effects of physical conditioning and propranolol on physical work capacity. *Medicine and Science in Sports*, **4**, 225–229.

Malhotra, M.S., Ramaswamy, S.S., Dua, G.J. and Sengupta, J. (1966). Physical work capacity as influenced by age. *Ergonomics*, **9**, 305–316.

Marras, W.S., Joynt, R.L. and King, A.I. (1985). The force–velocity relation and intra-abdominal pressure during lifting activities. *Ergonomics*, **28**, 603–613.

Marras, W.S., King, A.I. and Joynt, R.L. (1984). Measurements of loads on the lumbar spine under isometric and isokinetic conditions. *Spine*, **9**, 176–188.

Martin, J.B. and Chaffin, D.B. (1972). Biomechanical computerized simulation of human strength in sagittal plane activities. *Transactions of the American Institute of Industrial Engineers*, **4**, 19–28.

McArdle, W.D., Katch, F.L. and Katch, V.L. (1981). *Exercise Physiology — Energy, Nutrition, and Human Performance* (Philadelphia: Lea and Febiger).

McConville, J.T. and Hertzberg, H.T.E. (1966). *A Study of One-handed Lifting: Final Report* (Wright-Patterson Air Force Base, OH: Aerospace Medical Research Laboratory, Technical Report AMRI-TR-66-17.)

McConville, J.T. and Hertzberg, H.T.E. (1968). A study of one-handed lifting: Final report. *Ergonomics*, **11**, 297.

McDaniel, J.W. (1972). Prediction of acceptable lift capacity. Unpublished Ph.D. Dissertation, Texas Tech University, Lubbock, Texas.

McDonald, I. (1961). Statistical studies of recorded energy expenditure of man II. Expenditure on walking related to weight, sex, age, height, speed and gradient. *Nutrition Abstract Review*, **31**, 739–762.

McGill, C.M. (1968). Industrial back problems — a control program. *Journal of Occupational Medicine*, **10**, 174–178.

McGill, S.M. and Norman, R.W. (1985). Dynamically and statistically determined low back moments during lifting. *Journal of Biomechanics*, **18**, 877–885.

McGlynn, G.H. (1969). The relationship between maximum strength and endurance of individuals with different levels of strength. *Research Quarterly*, **40**, 529–535.

McNab, R.B.J. Conger, P.R. and Taylor, P.S. (1969). Differences in maximal and submaximal work capacity in men and women. *Journal of Applied Physiology*, **27**, 644–648.

Merriam, W.F., Burwell, R.G., Mulholland, R.C., Pearson, J.C.G. and Webb, J.K. (1983). A study revealing a tall pelvis in subjects with low back pain. *Journal of Bone and Joint Surgery*, **65-B**, 153–156.

Michael, E.D., Hutton, K.E. and Horvath, S.M. (1961). Cardiorespiratory responses during prolonged exercise. *Journal of Applied Physiology*, **16**, 997–999.

Miller, J.A.A., Schultz, A.B., Warwick, D.N. and Spencer, D.L. (1986). Mechanical properties of lumbar spine motion segments under large loads. *Journal of Biomechanics*, **19**, 79–84.

Miller, J.C., Farlow, D.E. and Seltzer, M.L. (1977). Physiological analysis of repetitive lifting. *Aviation, Space, and Environmental Medicine*, **48**, 984–988.

Miller, R.L. (1977). Bend your knees! *National Safety News*, May, 57–58.

Mital, A. (1980). Effects of task variable interactions in lifting & lowering. Unpublished Ph.D. dissertation, Texas Tech University, Lubbock, Texas.

Mital, A. (1981). Dynamic endurance of the upper arm, shoulder and leg muscle groups. Unpublished data, Ergonomics Research Laboratory, University of Cincinnati, Ohio.

Mital, A. (1983a). Generalized model structure for evaluating/designing manual materials handling jobs. *International Journal of Production Research*, **21**, 401–412.

Mital, A. (1983b). Prediction of maximum weights of lift acceptable to male and female industrial workers. *Journal of Occupational Accidents*, **5**, 223–231.

Mital, A. (1983c). The psychophysical approach in manual lifting: a verification study. *Human Factors*, **25**, 485–491.

Mital, A. (1984a). Comprehensive maximum acceptable weight of lift database for regular 8-hour workshifts. *Ergonomics*, **27**, 1127–1138.

Mital, A. (1984b). Maximum weights of lift acceptable to male and female industrial workers for extended workshifts. *Ergonomics*, **27**, 1115–1126.

Mital, A. (1984c). Prediction of human static and dynamic strengths by modified basic GMDH algorithm. *IEEE Transactions on Systems, Man and Cybernetics*, **SMC–14**, 773–776.

Mital, A. (1984d). Prediction of maximum weights of lift acceptable to male and female industrial workers. *Journal of Occupational Accidents*, **5**, 223–231.

Mital, A. (1985a). *A Comparison Between Psychophysical and Physiological Approach Across Low and High Frequency Ranges* (Technical Report). (Ergonomics Research Laboratory, University of Cincinnati, Ohio).

Mital, A. (1985b). Preliminary guidelines for designing one-handed material handling tasks. *Journal of Occupational Accidents*, **7**, 33–40.

Mital, A. (1985c). Models for predicting maximum acceptable weight of lift and heart rate and oxygen uptake of that weight. *Journal of Occupational Accidents*, **7**, 75–82.

Mital, A. (1985d). Modeling lifting capabilities of industrial workers for regular and extended workshifts. In *Ergonomics International, 85*, edited by I.D. Brown, R. Goldsmith, K. Coombes and M.A. Sinclair (London: Taylor and Francis).

Mital, A. (1985e). Use of anthropometry and dynamic strength in developing screening and placement procedures for workers. In *Towards the Factory of the Future*, edited by H.J. Bullinger and H.J. Warnecke (Heidelberg: Springer-Verlag).

Mital, A. (1985f). *Lifting Capacities of Student and Industrial Populations* (NIOSH Report, Grant No. 1-R01-OH-01956-02).

Mital, A. (1986a). Comparison of lifting capabilities of industrial and non-industrial populations. In *Proceedings of the 30th Annual Meeting of the Human Factors Society*, pp. 239–243.

Mital, A. (1986b). Prediction models for psychophysical lifting capabilities and the resulting physiological responses for shifts of varied durations. *Journal of Safety Research*, **17**, 155–163.

Mital, A. (1986c). Subjective estimates of load carriage in confined and open spaces. In *Trends in Ergonomics/Human Factors III*, edited by W. Karwowski (Amsterdam: North Holland), pp. 827–833.

Mital, A., Aghazadeh, F. and Karwowski, W. (1986a). Importance of isometric and isokinetic lifting strengths in estimating maximum lifting capacities. *Journal of Safety Research*, **17**, 65–71.

Mital, A., Aghazadeh, F. and Ramanan, S. (1985a). Use of GMDH to predict dynamic strengths. *International Journal of Computers and Industrial Engineering*, **9**, 371–377.

Mital, A. and Asfour, S.S. (1983). Maximum frequency acceptable to males for one-handed horizontal lifting in the sagittal plane. *Human Factors*, **25**, 563–571.

Mital, A., Asfour, S.S. and Ayoub, M.M. (1982). Physiological approach in manual lifting: work rate recommendations and comparison with psychophysical approach. *Journal of Human Ergology*, **11**, 143–156.

Mital, A. and Ayoub, M.M. (1980). Modeling of isometric strength and lifting capacity. *Human Factors*, **22**, 285–290.

Mital, A. and Ayoub, M.M. (1981a). Effect of task variables and their interactions in lifting and lowering loads. *American Industrial Hygiene Association Journal*, **42**, 134–142.

Mital, A. and Ayoub, M.M. (1981b). Task variable based physiological prediction models for lifting/lowering activities. In *Proceedings of the Annual International Conference, Institute of Industrial Engineers*, pp. 133–136.

Mital, A. and Ayoub, M.M. (1986). Equation for predicting lifting capabilities of people. In *Trends in Ergonomics/Human Factors III*, edited by W. Karwowski (Amsterdam: North Holland), pp. 767–774.

Mital, A., Ayoub, M.M., Asfour, S.S. and Bethea, N.J. (1978). Relationship between lifting capacity and injury in occupations requiring lifting. In *Proceedings of the 22nd Annual Meeting of the Human Factors Society*, pp. 469–473.

Mital, A., Chalaka, A. and Karwowski, W. (1985c). The demands and responses of machine-paced and self-paced materials handling tasks. In *Toward the Factory of the Future*, edited by H.J. Bullinger and H.J. Warnecke (Heidelberg: Springer-Verlag).

Mital, A., Channareeraiah, C., Fard, H.F. and Khaledi, H. (1986b). Reliability of repetitive dynamic strengths as a screening tool for manual lifting tasks. *Clinical Biomechanics*, **1**, 125–129.

Mital, A. and Fard, H.F. (1986). Psychophysical and physiological responses to lifting symmetrical and asymmetrical loads symmetrically and asymmetrically. *Ergonomics*, **29**, 1263–1272.

Mital, A. and Ilango, M. (1983a). Load characteristics and manual carrying capabilities. In *Proceedings of the 27th Annual Meeting of Human Factors Society*, pp. 274–278.

Mital, A. and Ilango, M. (1983b). Subjective estimates of one-handed carrying tasks. *Applied Ergonomics*, **14**, 265–269.

Mital, A. and Karwowski, W. (1985). Use of simulated job dynamic strength (SJDS) in screening workers for manual lifting tasks. In *Proceedings of the 29th Annual Meeting of the Human Factor Society*, pp. 513–516.

Mital, A., Karwowski, W., Mazouz, A.K. and Orsarh, E. (1986c). Prediction of maximum weight of lift in the horizontal and vertical planes using simulated job dynamic strengths. *American Industrial Hygiene Association Journal*, **47**, 288–291.

Mital, A., Karwowski, W., Mazouz, A.K. and Orsarh, E. (1985d). *Prediction of Maximum Weight of Lift in the Horizontal and Vertical Planes Using Simulated Job Dynamic Strengths* (Cincinnati, Ohio: Technical Report, Ergonomics Research Laboratory, University of Cincinnati).

Mital, A. and Kromodihardjo, S. (1984). An evaluation of filming speeds for dynamic biomechanical modelling and recommendation for optimum speed. In *Proceedings of the 28th Annual Meeting of the Human Factors Society*, pp. 992–996.

Mital, A. and Kromodihardjo, S. (1986). Kinetic analysis of manual lifting activities: Part II — biomechanical analysis of task variables. *International Journal of Industrial Ergonomics*, **1**, 91–101.

Mital, A. and Manivasagan, I. (1982). Application of a heuristic technique in polynomial identification. In *Proceedings of the IEEE Systems, Man and Cybernetics Society International Conference*, pp. 347–353.

Mital, A. and Manivasagan, I. (1983a). Maximum acceptable weight of lift as a function of material density, center of gravity location, hand preference, and frequency. *Human Factors*, **25**, 33–42.

Mital, A. and Manivasagan, I. (1983b). Subjective estimates of one-handed carrying tasks. *Applied Ergonomics*, **14**, 265–269.

Mital, A. and Manivasagan, I. (1984). Development of non-linear polynomials in identifying human isometric strength behavior. *International Journal of Computers and Industrial Engineering*, **8**, 1–9.

Mital, A. and Okolie, S.T. (1982). Influence of container shape, partitions, frequency, distance, and height level on the maximum acceptable amount of liquid carried by males. *American Industrial Hygiene Association Journal*, **43**, 813–819.

Mital, A. and Shell, R.L. (1984). Determination of rest allowances for repetitive physical activities that continue for extended hours. In *Proceedings of the Annual International Conference, Institute of Industrial Engineers*, pp. 637–645.

Mital, A. and Shell, R.L. (1985a). A comprehensive metabolic energy model for determining rest allowance for physical tasks. *Journal of Methods Time Measurement*, **XI**, 2–8.

Mital, A. and Shell, R.L. (1985b). Program determines rest allowances for physiological fatigue. In *Softcover Software* edited by G. Whitehouse (Atlanta: Industrial Engineering and Management Press).

Mital, A., Shell, R.L., Mital, C., Sanghavi, N. and Ramanan, S. (1984). *Acceptable Weight of Lift for Extended Workshifts* (NIOSH Report, Grant No. 1-R01-OH-01429-02).

Mital, A. and Vinayagamoorthy, R. (1984). Three-dimensional dynamic strength measuring device: a prototype. *American Industrial Hygiene Association Journal*, **45**, B9–B12.

Monod, H. (1967). La despense enegetique chez l'homme. In *Physiologie du travail*, edited by J. Scherrer (Paris: Mason).

Montgomery, C.H. (1976). Pre-employment back X-rays. *Journal of Occupational Medicine*, **18**, 495.

Montoye, H.J. and Lamphiear, D.E. (1977). Grip and arm strength in males and females, age 10 to 69. *The Research Quarterly*, **48**, 109–120.

Morris, J.M., Lucas, D.B. and Bresler, B. (1961). Role of the trunk in stability of the spine. *Journal of Bone and Joint Surgery*, **43-A**, 327–351.

Morrissey, S.J., Bethea, N.J. and Ayoub, M.M. (1983). Talk demands for shovelling in non-erect postures. *Ergonomics*, **26**, 947–951.

Morrissey, S.J. and Liou, Y.H. (1984). Metabolic costs of load carriage with different container sizes. *Ergonomics*, **27**, 847–853.

Mortimer, R.G. (1974). Foot brake pedal force capability of drivers. *Ergonomics*, **17**, 509–513.

Muller, E.A. (1953). Physiological basis of rest pauses in heavy work. *Quarterly Journal of Experimental Physiology*, **38**, 205–215.

Muller, E.A. (1962). Occupational work capacity. *Ergonomics*, **5**, 445–452.

Murrel, K.F.H. (1964). *Ergonomics: Man in his Working Environment* (London: Chapman and Hall).

Muth, M.B. (1976). An optimization model for the evaluation of manual lifting tasks. Thesis, North Carolina State University, Raleigh, California.

Muth, M.B., Ayoub, M.A. and Gruver, W.A. (1978). A nonlinear programming model for the design and evaluation of lifting tasks. In *Safety in Manual Materials Handling* (DHEW (NIOSH) Publication No. 78-185).

Nachemson, A., Andersson, G.B.J. and Schultz, A.B. (1986). Valsalva maneuver biomechanics: effects of lumbar trunk loads on elevated intra-abdominal pressures. *Spine*, **11**, 476–479.

Nachemson, A., Schultz, A.B. and Berkson, M.H. (1979). Mechanical properties of human lumbar spine motion segments. Influence of age, sex, disc level, and degeneration. *Spine*, **8**, 1–8.

Nag, P.K., Sen, R.N. and Ray, U.S. (1979). Cardio-respiratory performance of porters carrying loads on treadmill. *Ergonomics*, **22**, 897–907.

National Research Council (1980). *Recommended Dietary Allowances*. (Washington, DC: Food and Nutrition Board, National Academy of Sciences, National Research Council).

National Safety Council (1972–84) *Accident Facts* (New York: National Safety Council).

National Safety News (1974). *Shoe Sole Slipperiness Standard Status* (Chicago: National Safety Council).

NIOSH (National Institute of Occupational Safety and Health) (1977). *Pre-employment Strength Testing* (Department of Health and Human Services (NIOSH) Publication No. 77–163).

NIOSH (National Institute of Occupational Safety and Health) (1981). *Work Practices Guide for Manual Lifting* (Department of Health and Human Services (NIOSH) Publication No. 81–122).

Nordgren, B. (1972). Anthropometric measures of muscle strength in young women. *Scandinavian Journal of Rehabilitation Medicine*, **4**, 165–169.

Nordgren, B., Schele, R. and Linroth, K. (1980). Evaluation and prediction of back pain during military field service. *Scandinavian Journal of Rehabilitation Medicine*, **12**, 1–8.

Nummi, J., Jarvinen, T., Stambej, U. and Wickstrom, G. (1978). Diminished dynamic performance capacity of back and abdominal muscles in concrete reinforcement workers. *Scandinavian Journal of Work, Environment and Health*, **4**, 39–46.

Nylind, B., Schele, R and Linroth, K. (1978). Changes in male exercise performance and anthropometric variables between the ages of 19 and 30. *European Journal of Applied Physiology and Occupational Physiology*, **38**, 145–150.

O'Neal, P.V. (1983). *Advanced Engineering Mathematics* (Belmont, CA: Wadsworth).

Ortengren, R. and Andersson, G.B.J. (1977). Electromyographic studies of the trunk muscles with reference to the functional anatomy of the lumbar spine. *Spine*, **2**, 44–52.

Ortengren, R., Andersson, G.B.J., Broman, H., Magnusson, R. and Eterson, I. (1975). Vocational electromyography: studies of localized muscle fatigue at the assembly line. *Ergonomics*, **18**, 157–174.

Ortengren, R., Andersson, G.B.J. and Nachemson, A.L. (1981). Studies of relationships between lumbar disc pressure, myoelectric back muscle activity, and intra-abdominal (intragastric) pressure. *Spine*, **6**, 98–103.

Park, K.S. (1973). A computerized simulation model of postures during manual materials handling, Ph.D. Dissertation, University of Michigan, Ann Arbor, Michigan.

Pateman, C. (1981). Prediction of endurance times for repetitive lifting. B.Sc. (Honours) Dissertation, University of Surrey.

Peacock, B., Westers, T., Walsh, S. and Nicholson, K. (1981). Feedback and maximum voluntary contraction. *Ergonomics*, **24**, 223–228.

Pepler, R.D. (1963). Performance and well-being in heat. In *Temperature, its Measurement and Control in Science and Industry* (Vol. 2), Part 3, edited by C.M. Herzfeld (New York: Reinhold).

Perey, O. (1957). Fracture of the vertebral endplate in the lumbar spine: an experimental biomechanical investigation. *Acta Orthopedicae Scandinavica*, suppl 25.

Perkins, P.J. and Wilson, M.P. (1983). Slip resistance testing of shoes — new developments. *Ergonomics*, **26**, 73–82.

Peterson, O.F. and Staffeldt, E.S. (1982). The relationship between four tests of back muscle strength in untrained subjects. *Scand. Journal of Rehabil. Medicine*, **4**, 175–181.

Petrofsky, J.S. and Lind, A.R. (1975a). Aging, isometric strength and endurance, and cardiovascular responses to static effort. *Journal of Applied Physiology*, **38**, 91–95.

Petrofsky, J.S. and Lind, A.R. (1975b). Isometric strength, endurance, and the blood pressure and heart rate responses during isometric exercise in healthy men and women, with special reference to age and body fat content. *Pflügers Archiv European Journal of Physiology*, **360**, 49–61.

Petrofsky, J.S. and Lind, A.R. (1978a). Comparison of metabolic and ventilatory responses of men to various lifting tasks and bicycle ergometry. *Journal of Applied Physiology: Respiratory, Environmental and Exercise Physiology*, **45**, 60–63.

Petrofsky, J.S. and Lind, A.R. (1978b). Metabolic cardiovascular, and respiratory factors in the development of fatigue in lifting tasks. *Journal of Applied Physiology: Respiratory, Environmental and Exercise Physiology*, **45**, 64–68.

Petrofsky, J.S. and Phillips, C.A. (1981). The influence of body fat on isometric exercise performance. *Ergonomics*, **24**, 215–222.

Petrofsky, J.S., Williams, C., Kamen, G. and Lind, A.R. (1980). The effect of hand grip span on isometric exercise performance. *Ergonomics*, **23**, 1129–1135.

Pheasant, S.T. (1982). Anthropometric estimates for British civilian adults. *Ergonomics*, **25**, 993–1001.

Pheasant, S.T. and Grieve, D.W. (1981). The principal features of maximal exertion in the sagittal plane. *Ergonomics*, **24**, 327–338.

Pimental, N.A. and Pandolf, K.P. (1979). Energy expenditure while standing or walking slowly uphill or downhill with load. *Ergonomics*, **22**, 963–973.

Poulsen, E. (1970). Prediction of maximum loads in lifting from measurement of back muscle strength. *Progressive Physical Therapy*, **1**, 146–149.

Poulsen, E. (1971). Prediction of maximum loads in lifting from measurements of muscle strength. *Testing and Observation Institute of the Danish National Association for Infantile Paralysis, Communication*, **31**.

Poulsen, E. (1981). Back muscle strength and weight limits in lifting burdens. *Spine*, **6**, 73–75.

Poulsen, E. and Jorgensen, K. (1971). Back muscle strength, lifting and stooped working postures. *Applied Ergonomics*, **2**, 133–137.

Powell, P.J., Hale, M., Martin, J. and Simon, M. (1971). *2000 Accidents: A Shop Floor Study of Their Causes Based on 42 Months Continuous Observation* (London: National Institute of Industrial Psychology).

Pytel, J.L. and Kamon, E. (1981). Dynamic strength test as a predictor for maximal and acceptable lifting. *Ergonomics*, **24**, 663–672.

Ready, A.E. and Quinney, H.A. (1982). Alterations in anaerobic threshold as the result of endurance training and detraining. *Medicine and Science in Sports and Exercise*, **14**, 292–296.

Recommended Dietary Allowances (1980). (Washington DC: Food and Nutrition Board, National Academy of Sciences, National Research Council.)

Redfield, J.T. (1971). The low back X-ray as a pre-employment screening tool in the forest products industry. *Journal of Occupational Medicine*, **13**, 219–226.

Reinberg, A., Andlauer, P., Guillet, P., Nicolai, A., Vieux, N. and Laporte, A. (1970). Oral temperature, circadium rhythm amplitude, aging and tolerance to shift work. *Ergonomics*, **13**, 55–64.

Ridd, J.F. (1983). Spatial restraints and intra abdominal pressure. In *Proceedings of Seminar on Prevention of Low Back Pain, Commission of the European Communities*, edited by P.R. Davis and J.D.G. Troup (London: Taylor and Francis).

Roebuck, J.A., Jr., Kroemer, K.H.E. and Thompson, W.G. (1975). *Engineering anthropometry methods* (New York: John Wiley).

Rohmert, W. (1973a). Problems in determining rest allowances — Part 1: use of modern methods to evaluate stress and strain in static muscular work. *Applied Ergonomics*, **4**, 91–95.

Rohmert, W. (1973b). Problems of determination of rest allowances — Part 2: determining rest allowances in different human tasks. *Applied Ergonomics*, **4**, 158–162.

Ronnholm, N., Karvonen, M.J. and Lapinleimu, V.O. (1962). Mechanical efficiency of rhythmic and paced work of lifting. *Journal of Applied Physiology*, **17**, 768–770.

Rowe, M.L. (1969). Low back pain in industry: a position paper. *Journal of Occupational Medicine*, **11**, 161–169.

Rowe, M.L. (1971). Low back disability in industry: updated position. *Journal of Occupational Medicine*, **13**, 476–478.

Rowe, M.L. (1983). *Backache at Work* (Fairport, New York: Perinton Press).

Salter, N. (1955). The effect on muscle strength of maximum isometric and isotonic contractions at different repetition rates. *Journal of Physiology*, **130**, 109–113.

Saltin, B., Blomquist, G., Mitchell, J.H., Johnson, R.L., Wildenthal, K. and Chapman, C.B. (1968). Responses to submaximal and maximal exercise after bed rest and training. *Circulation*, **38**, suppl. 7.

Saltin, B., Hartley, L., Kilbom, A. and Astrand, I. (1969). Physical training in sedentary middle-aged and older men. II. oxygen uptake, heart rate and blood lactate concentrations at submaximal and maximal exercise. *Scandinavian Journal of Clinical Laboratory Investigation*, **24**, 323–334.

Samantha, A. and Chatterjee, B.B. (1981). A physiological study of manual lifting of loads in Indians, *Ergonomics*, **24**, 557–564.

Sanchez, J., Monod, H. and Chaband, F. (1979). Effects of dynamic, static, and combines work in heart rate and oxygen consumption. *Ergonomics*, **22**, 935–943.

Schein, A.J. (1968). Back pain and associated nerve root irritation in the New York City Fire Department. *Clinical Observations*, **59**, 119–124.

Schultz, A.B. (1986). Loads on the human lumbar spine. *Mechanical Engineering*, **36**, 36–42.

Schultz, A.B. and Andersson, G.B.J. (1981). Analysis of loads on the lumbar spine. *Spine*, **6**, 76–82.

Schultz, A., Andersson, G.B.J., Ortengren, R., Haderspeck, K. and Nachemson, A. (1982). Loads on the lumbar spine. Validation of a biomechanical analysis by measurements of intradiscal pressures and myoelectric signals. *Journal of Bone and Joint Surgery*, **64-A**, 713–720.

Schultz, A.B., Warwick, D.N., Berkson, M.H. and Nachemson, A.L. (1979). Mechanical properties of human lumbar spine motion segments: Part I: responses in flexion, extension lateral bending, and torsion. *Journal of Biomechanical Engineering*, **101**, 46–52.

Scott, V. and Gibsbers, K. (1981). Pain perception in competitive swimmers. *British Medical Journal*, **283**, 91–93.

Selan, J.L. (1986). Effect of psychosocial variables on the psychophysical maximum acceptable weight of load. Ph.D. Dissertation, Texas Tech University, Lubbock, Texas.

Selby, N.C. (1983). Training procedures to reduce low back injury. In *Proceedings: Bureau of Mines Technical Transfer Symposium*, pp. 81–87.

Sengupta, A.K., Sarkar, D.N., Mukhopadhyay, S. and Goswami, C.D. (1979). Relationship between pulse rate and energy expenditure during graded work at different temperatures. *Ergonomics*, **22**, 1207–1215.

Shapiro, Y., Randolf, K.B., Avellini, B.A., Pimental, N.A. and Goldman, R.F. (1980). Physiological responses of men and women to humid and dry heat. *Journal of Applied Physiology*, **49**, 1–8.

Sheppard, R.J. (1974). *Men at Work* (Springfield, Illinois: Charles C. Thomas).

Simonson, E. (1971). *Physiology of Work Capacity and Fatigue* (Springfield, Illinois: Charles C. Thomas).

Smith, G.M. and Beecher, H.K. (1959). Amphetamine sulfate and athletic performance. *Journal of the American Medical Association*, **170**, 542.

Smith, J.L. and Jiang, B.C. (1984). A manual materials handling study of bag lifting. *American Industrial Hygiene Association Journal*, **45**, 505–508.

Smith, J.L. and Ramsey, J.D. (1982). Designing physically demanding tasks to minimize levels of worker stress. *Industrial Engineering*, **14**, 44–50.

Smith, L.E. and Edwards, D.K. (1968). Prediction of muscular endurance (work performance) from individual differences in initial post training increments in static strength. *Human Factors*, **10**, 345–350.

Smith, L.E. and Royce, J. (1963). Muscular strength in relation to body composition. *Annals New York Academy of Science*, **110**, 809–813.

Snook, S.H. (1965). *Group Work Capacity: a Technique for Evaluating Physical Tasks in Terms of Fatigue* (Unpublished report, Liberty Mutual Insurance Company, Boston).

Snook, S.H. (1971). The effect of age and physique on continuous work capacity. *Human Factors*, **13**, 467–479.

Snook, S.H. (1978a). The design of manual handling tasks. *Ergonomics*, **21**, 963–985.

Snook, S.H. (1978b). The design of manual handling tasks. Paper presented at *Ergonomics Society Lecture, Cranfield Institute of Technology, Bedfordshire, England*.

Snook, S.H., Campanelli, R.A. and Hart, J.W. (1978). Three preventative approaches to low back injury. *Journal of Occupational Medicine*, **20**, 478–481.

Snook, S.H. and Ciriello, V.M. (1974a). Maximum weights and work loads acceptable to female workers. *Journal of Occupational Medicine*, **16**, 527–534.

Snook, S.H. and Ciriello, V.M. (1974b). The effects of heat stress on manual handling tasks. *American Industrial Hygiene Association Journal*, **35**, 681–685.

Snook, S.H. and Irvine, C.H. (1966). The evaluation of physical tasks in industry. *American Industrial Hygiene Association Journal*, **27**, 228–233.

Snook, S.H. and Irvine, C.H. (1967). Maximum acceptable weight of lift. *American Industrial Hygiene Association Journal*, **28**, 322–329.

Snook, S.H. and Irvine, C.H. (1968). Maximum frequency of lift acceptable to male industrial workers. *American Industrial Hygiene Association Journal*, **29**, 531–536.

Snook, S.H. and Irvine, C.H. (1969). Psychophysical studies of physiological fatigue criteria. *Human Factors*, **11**, 291–299.

Snook, S.H., Irvine, C.H. and Bass, S.F. (1970). Maximum weights and work loads acceptable to male industrial workers. *American Industrial Hygiene Association Journal*, **31**, 579–586.

Snook, S.H. and White, A.H. (1984). Education and training. In *Occupational Low Back Pain*, edited by M.H. Pope, J.W. Frymoyer and G.B.J. Andersson (New York: Praeger), pp. 233–244.

Soule, R.G. and Goldman, R.F. (1969). Energy cost of loads carried on the head, hands, or feet. *Journal of Applied Physiology*, **27**, 687–690.

Soule, R.G., Pandolf, K.B. and Goldman, R.F (1978). Energy expenditure of heavy load carriage. *Ergonomics*, **21**, 373–381.

Spitzer, H. (1952). *Physiologische Grundlagen fur den Erholungszuschlag bei Schwerarbeit* (Darmstadt: REFA-Nachrichten, Heft 2).

Statistical Analysis System (Raleigh, North Carolina: SAS Institute Inc.)

Stevens, S.S. (1960). The psychophysics of sensory function. *American Scientist*, **48**, 226–253.

Stewart, S.F. (1947). Pre-employment examination of the back. *Journal of Bone and Joint Surgery*, **29**, 214–221.

Stobbe, T.J., Bobick, T.G. and Plummer, R.W. (1986). Musculoskeletal injuries in underground mining. *Annals of the American Conference of Governmental Industrial Hygienists*, **14**, 71–76.

Strindberg, L. and Peterson, N.F. (1972). Measurement of force perception in pushing trolleys. *Ergonomics*, **15**, 435–438.

Stubbs, D.A. (1985). Human constraints on manual working capacity: effects of age on intratruncal pressure. *Ergonomics*, **28**, 107–114.

Suggs, C.W. and Splinter, W.E. (1961). Some physiological responses of man to workload and environment. *Journal of Applied Physiology*, **16**, 413–420.

Suzuki, Y. (1980). Human physical performance and cardiocirculatory responses to hot environments during sub-maximal upright cycling. *Ergonomics*, **23**, 527–542.

Switzer, S.A. (1962). *Weight Lifting Capacities of a Selected Sample of Human Males* (Report No. AD-284054) (Wright Patterson Air Force Base, Ohio: Aerospace Medical Research Laboratory).

Syrovy, J. and Gutmann, E. (1970). Changes in speed of contraction and ATPase activities in striated muscle during old age. *Experimental Gerontology*, **5**, 31–35.

Tanz, S.S. (1953). Motion of the lumbar spine: a roentgenographic study. *American Journal of Roentgenography*, **69**, 399–412.

Tauber, J. (1970). An unorthodox look at backaches. *Journal of Occupational Medicine*, **12**, 128–130.

Taylor, J.F. and Towmey, L.T. (1980). Sagittal and horizontal plane movement of the human lumbar vertebral column in cadavers and in the living. *Rheumatology and Rehabilitation*, **19**, 223–232.

Thorstensson, A., Grimby, G. and Karlsson, J. (1976). Force velocity relations and fiber composition in human knee extensor muscles. *Journal of Applied Physiology*, **40**, 1–16.

Thorstensson, A., Larsson, L. Tesch, P. and Karlsson, J. (1972). Muscle strength and fiber composition in athletes and sedentary men. *Medicine and Science in Sports*, **9**, 26–30.

Tichauer, E.R. (1970). *Biomechanics of Lifting* (Report No. RD-3130-MPO-69) (Department of Health, Education and Welfare).

Tichauer, E.R. (1971). A pilot study of the biomechanics of lifting in simulated industrial work situations. *Journal of Safety Research*, **3**, 98–115.

Tichauer, E.R. (1973). Ergonomic aspects of biomechanics in the industrial environment: Its evaluation and control. (NIOSH, Superintendent of Documents, Washington, D.C.)

Tichauer, E.R. (1978). *The Biomechanical Basis of Ergonomics* (New York: John Wiley).

Tichauer, E.R., Miller, M. and Nathan, I.M. (1973). Lordosimetry. A new technique for the measurement of postural response to materials handling. *American Industrial Hygiene Association Journal*, **34**, 1–12.

Troup, J.D.G. and Chapman, A.E. (1969a). The static strength of the lumbar erectores spinae. *Journal of Anatomy*, **105**, 186.

Troup, J.D.G. and Chapman, A.E. (1969b). The strength of the flexor and extensor muscles of the trunk. *Journal of Biomechanics*, **2**, 49–62.

Troup, J.D.G. and Edwards, F.C. (1985). *Manual Handling and Lifting: An Information and Literature Review with Special Reference to the Back* (London: Her Majesty's Stationery Office).

Troup, J.D.G., Leskinen, T.P.J., Stalhammer, H.R. and Kuorinka, I.A.A. (1983). A comparison of intra-abdominal pressure increases, hip torque, and lumbar vertebral compression in different lifting techniques. *Human Factors*, **25**, 517–525.

Troup, J.D.G., Martin, J.W. and Lloyd, D.C.E.F. (1981). Back pain in industry: a prospective survey. *Spine*, **6**, 61–69.

Twomey, L.T. (1979). The effect of age on the ranges of motion of the lumbar region. *Australia Journal of Physiotherapy*, **25**, 257–263.

Tzankoff, S.P., Robinson, S., Pyke, F.S. and Brown, C.A. (1972). Physiological adjustments to work in older men as affected by physical training. *Journal of Applied Physiology*, **33**, 346–350.

Wald, A. and Harrison, L.B. (1975). Stress or effects of static work. *Journal of Occupational Medicine*, **17**, 515–578.

Warwick, D., Novak, G., Schultz, A. and Berkson, M. (1980). Maximum voluntary strengths of male adults in some lifting, pushing, and pulling activities. *Ergonomics*, **23**, 49–54.

Wasserman, K. and Whipp, B.J. (1975). Exercise physiology in health and disease. *American Review of Respiratory Disease*, **112**, 219–249.

Wasserman, K., Whipp, B.J., Koyal, S.N. and Beaver, W.L. (1973). Anaerobic threshold and respiratory gas exchange during exercise. *Journal of Applied Physiology*, **35**, 326–343.

Watson, A.W.S. (1977). The relationship of muscular strength to body size and somatype in post-puberal males. *Irish Journal of Medical Science*, **146**, 307–308.

Whitney, R.J. (1958). The strength of the lifting action in man. *Ergonomics*, **1**, 101–128.

White, A.H. (1983). *Back schools and other conservative approaches to low back pain* (St. Louis: C.V. Mosby).

Williamson, D.K., Chrenko, F.A. and Hamley, E.J. (1984). A study of exposure to cold in cold stores. *Applied Ergonomics*, **15**, 25–30.

Winter, D. (1979). *Biomechanics of Human Movement* (New York: John Wiley).

Wyndham, C.H., Strydom, N.B., Morrison, J.F., Williams, C.G., Bredell, G., Peter, J., Cooke, H.M. and Joffe, A. (1963). The influence of gross weight on oxygen consumption and on physical working capacity of manual laborers. *Ergonomics*, **6**, 275–286.

Yates, J.W., Kamon, E., Rodgers, S.H. and Champney, P.C. (1980). Static lifting strength and maximal isometric voluntary contractions of back, arm and shoulder muscles. *Ergonomics*, **23**, 37–47.

Yoshimura, T., Kiyozumi, R., Nishino, K. and Soeda, T. (1982). Prediction of air pollutant concentrations by revised GMDH algorithms in Tokushima prefecture. *IEEE Transactions on Systems, Man and Cybernetics*, **SMC-12**, 50–55.

Yousef, M.K., Dill, D.B. and Freeland, C.V. (1972). Energy cost of grade walking in man and burro, Equus asinus: desert and mountain. *Journal of Applied Physiology*, **33**, 337–340.

Zadeh, L.A. (1965). Fuzzy sets. *Information and Control*, **8**, 338–353.

Zajonc, R.B. (1965). Social facilitation. *Science*, **149**, 269–274.

Index

Absolute Endurance Index 18
Acceptable weights of lift
  and relationship between strengths 260–2
Action limit 209–10
Aerobic capacity 107–8
Aerobic lifting capacity 108–9
Age 13–15
  and back injuries 1, 2
Altitude 56
Amphetamine sulphate 24
Anaerobic threshold 109–10
Anthropometry 17–18, 62
Asfour's model 113
  limitations of 114
Asymmetrical lifting/carrying 54

Back injury statistics
  compensation costs 4
  in construction and mine workers 5, 6–7
  cost to industry 1, 2–3
  frequency 1
  lost working days due to 2, 5
  in nurses 4–5, 6
  sex and age differences in 1, 2
  types of injury 2, 3
Back lift 51
Back schools 270–1
Back X-rays 193–4
Bink's logarithmic formula 104
Biodynamics, definition of 62
Bioinstrumentation 62
Biomechanical equivalent (BLE) 95–6
Biomechanical stress
  factors affecting 88–90
  design criterion 92–5
  models using 95
Biomechanics, definition of 60–2
Biostatics, definition of 62
Blood lactate level 99, 102–3
Blood pressure 99, 102

Body build 17–18
Body composition 17–18
Body-weight 16–17
Bone mineral content 90
Box-size dimension
  effect on maximum acceptable weight of lift 159
  and interaction effects of frequency 57

Capacity
  and comparison between strength 121–2
  one-handed 184–5
  two-handed 180–4
Compressive strength, sex differences in 89–90
Couplings 41–5
CYBEX II Isokinetic Dynamometer 239–40, 261

Design criterion 111–14
  criterion limit 111
  factors affecting 111
  models using 111
  psychophysical 116–20
Disc cartilage, compressive strength of 89
Dynamic activity 97–9
Dynamic analysis 72–8
Dynamic back extension strength 235
Dynamic elbow flexion strength 235
Dynamic endurance 18–19
Dynamic lift strength 235

El-Bassoussi equations 95
Electromyography (EMG) 88
Endurance
  static and dynamic 18–19
  time 104–7
Energy costs of tasks 100–2
Environment component 54–6
Experience 21–3
External forces on the body 64

Fatigue, physiological 32, 33, 103–7
Force application 46
Frederick's model 111–12
  limitations of 114
Frequency of task, effect on maximum acceptable weight of lift 163

Garg's model 112–13
  limitations of 114
Gravitational forces on the body 64
Ground reaction forces 64
Group method of data handing (GMDH technique) 245, 251, 252
GULHEMP method 196

Hanman method 196–7
Heart rate 99, 102
Heavy manual materials handling
  decreasing 190–1
  eliminating 190
Height level
  effect on maximum acceptable weight of lift 163
  of lift 48–9
Holding tasks, data base for 186–9
Humidity 54–6

Illumination 56
Intaranont's model 113–14
Intra-abdominal pressure (IAP) 52, 85–7
  effects of different weights on 92
Isoinertial strength 19, 226, 227–8
  assessment of 241–2
  prediction of 255–6
Isokinetic strength 19, 226, 227
  assessment of 234–40
  prediction of 251–5
Isometric activity 97, 99
Isometric strength 19–20, 226, 229–33
  arm measurement 230
  assessment of 229–33
  back extension measurement 231
  back measurement 231
  composite measurement 230
  leg lifting measurement 230
  prediction of 243–51
  tests 194
Isotonic strength 19, 226, 227
  assessment of 233–4

Job design/redesign
  NIOSH guidelines for 209–10
  principles of 189–97

Job severity index (JSI) 198–208
  comparison with NIOSH Guidelines for job design/redesign 210–11
  job design/redesign in conjunction with 208–9
  as a job design/redesign device 198–206
  as a job screening/placement device 206–8
  validation 208
Job simulators 196

Kinematics 62
Kinesiology 62
Kinetic lifting technique 267–8
Kinetics 62

Leg lift 51
Lift strength, dynamic 235
Lift strength ratio 211–14
Lifting
  one-handed 148–59, 166
  two-handed 122–48, 159–66
Lifting optimization model 224–5
Lifting task, definition 198
Liquid carrying capacity 184
Lowering
  one-handed 148–59, 166
  two-handed 122–48, 159–66
Lumbar loading 80–8
  asymmetric 84
  symmetric 81–4
Lumbosacral spine, stresses on 80–8

Maximum acceptable frequency of lift 34
Maximum acceptable weight of carry 41
Maximum acceptable weight of lift 40
  box size effect on 159
  comprehensive 163
  effect of height level on 163
  frequency effect on 163
Maximum acceptable weight of liquid 41
Maximum acceptable weight of load (MAWL) 24, 25–6, 118–20
Maximum dynamic strength 262
Maximum lifting capability 260, 261
Maximum permissible limit 209–10
Maximum voluntary contraction (MVC) 23, 99, 226
Maximum voluntary exertion 226
Mean square error (MSE) 246, 247, 251
Medical examinations 194–6
Metabolic energy expenditure rate
  models 276–7
  limitations of 277–8

Mital's model   214–23
MMH system
  age   13–15
  inter-relationship of component of   11–12
  negative-feedback in   9, 10
  response levels in   9–12
  sex   15–16
  worker component   12–13
Muscle forces on the body   64
Musculoskeletal system, stresses on   63–71
  one-segment link   64–6
  internal forces   69–71
  multiple link   68–9
  two-segment link   66–8

NIOSH guidelines   214
  comparison with job severity index for job design/redesign   210–11
  for job design/redesign   209–11
  for personnel screening   197–8
  to training within industrial organizations   269, 270
Noise   56
Non-lifting task, definition   198

Obesity   17
Object shape   40–1
Object size   37–40
Object weight   46
One-handed carrying
  capacity data for   185
  strength data for   184–5
One-handed lifting and lowering
  capacity data for   152–9
  strength data for   148–51
  performance ceiling for   163–6
One-handed pushing and pulling activities
  capacity data for   180
  strength data for   179–80

Personnel screening   197–8
Pervitin   23
1-Phenyl-2-methylaminopropane   23
Physical fitness   23, 107–10
Physical work capacity (PWC)   24, 25–6
  reduction in   14
Physiological fatigue   32, 33, 103–7
Physiological stress, definition   116
Posture/technique   50–4
Poulsen's formula   121
Pregnancy   16
Psychological factors   23–6
Psychological tests   196

Psychophysical approach   32, 33, 116–20
Psychophysically acceptable weights   263
Pulling
  one-handed   179–80
  two-handed   167–79
Pushing
  one-handed   179–80
  two-handed   167–79

Rating methods   196–7
Rating of perceived exertion (RPE) values   42
Repetitive dynamic strengths   262–4
Rest allowances
  computer program for   286–9
  metabolic energy model for determining   278–81
  methods of determining   274–7
  model development for   281–6
    assumption   281
    concept   281
    example   285–6
    structure   282–4
    validation   284–5
  need for   274
Right-hand rule   80

Safe lifting   265–8
  methods of   267
Semi-dynamic activity   97, 99
Sex differences   15–16
  in back injuries   1, 2
  in compressive strength   89–90
Simulated job dynamic strengths (SJDS)   20, 260, 261
Skid resistance testing   43
Slope and traction   50
Space restriction   49–50
Spinal loading
  position of   92
  role of intra-abdominal pressure in   85–7
  size and shape   90–1
  use of EMG in   88
  weight of   92
Static analysis   64
Static endurance   18–19
Static strength, see isometric strength
Statistical Analysis System   245, 246
Strength   19–21
  additivity   256–60
  classification and definition of   226–8
  and comparison between capacity   121–2
  measurement of   228–42
  prediction of   242–56

Strength (cont'd)
  reason for measuring   226
  testing   194
Stress/strain concept   60
Stresses
  biomechanical approach to   60–96
  classification of activity   97–9
  and lifting   99–103
  on the lumbosacral spine   80–8
  object shape   40–1
  object size   37–40
  physiological approach   97–116
Stressful body movements, minimizing   191–2
Super-2 Mini Gym   236–9, 251, 256
Symmetric lumbar loading   81–4
'System'
  'closed' or 'feedback'   9
  definition of   9
System response measures   59

Task component   26–54
  asymmetrical lifting/carrying   54
  couplings   41–5
  distance travelled   50
  exertion   46
  frequency   31–4
  load distribution and stability   46–8
  object shape   40–1
  object size   37–40
  posture/technique   50–4
  slope and traction   50
  task duration   35–6
  vertical lift height   48–9
  workplace geometry   49–50
Task duration   35–6
Temperature   54–6
Three-dimensional modelling   78–80
Training programmes   21–3
  effectiveness of   272–3
  outside industrial organizations   270–2
  within industrial organizations   268–70
Triple scalar product   80
Two-handed carrying
  capacity data for   183–4
  strength data for   180–3
Two-handed lifting and lowering
  capacity data for   130–48
  performance ceiling for   163–6
  strength data for   122–30
  work-rate recommendations for   159–63
Two-handed pushing and pulling
  capacity data for   174–9
  strength data for   167–73

Vertebra, compressive strength of   89–90
Vertical lift height   48–9
Vibration   56

Worker characteristics   12, 13
Workplace geometry   49–50

X-rays, back   193–4